280.75

Engineering Thin Films and Nanostructures with Ion Beams

OPTICAL ENGINEERING

Founding Editor
Brian J. Thompson
University of Rochester
Rochester, New York

Engineering Thin Films and Nanostructures with Ion Beams

edited by
Émile Knystautas

Taylor & Francis
Taylor & Francis Group

Boca Raton   London   New York   Singapore

A CRC title, part of the Taylor & Francis imprint, a member of the
Taylor & Francis Group, the academic division of T&F Informa plc.

Published in 2005 by
CRC Press
Taylor & Francis Group
6000 Broken Sound Parkway NW, Suite 300
Boca Raton, FL 33487-2742

© 2005 by Taylor & Francis Group, LLC
CRC Press is an imprint of Taylor & Francis Group

No claim to original U.S. Government works
Printed in the United States of America on acid-free paper
10 9 8 7 6 5 4 3 2 1

International Standard Book Number-10: 0-8247-2447-X (Hardcover)
International Standard Book Number-13: 978-0-8247-2447-4 (Hardcover)
Library of Congress Card Number 2004058212

Library of Congress Cataloging-in-Publication Data

Engineering thin films and nonostructures with ion beams / [edited by] Emile Knystautas.
    p. cm. -- (Optical engineering ; 92)
    Includes bibliographical references and index.
    ISBN 0-8247-2447-X (alk. paper)
    1. Thin films. 2. Nanostructures. 3. Ion bombardment--Industrial applications. I. Knystautas,
Emile J. II. Optical engineering (Marcel Dekker, Inc.) ; v. 92.

TA418.9.T45E52 2004
621.3815'2--dc22
                                    2004058212

Taylor & Francis Group
is the Academic Division of T&F Informa plc.

Visit the Taylor & Francis Web site at
http://www.taylorandfrancis.com

and the CRC Press Web site at
http://www.crcpress.com

Preface

In the last two decades, many books have been published on ion implantation and ion-beam processing. Why this one now? After all, the advantages of using an energetic ion beam to modify surfaces with a view to enhancing their tribological, electrochemical, optical and magnetic properties have been known for some time.

The aim of this volume is to review the basics of previous work on ion-beam modification of materials and to include enough new material on novel applications to bring newcomers "up to speed" in this exciting area. The authors are all recognized researchers in their respective areas, and the reader will surely benefit from exposure to their expertise. We present a mix of fundamental aspects in addition to very practical topics as they relate to industrial uses of these techniques.

While it used to be that ion-beam-based processes related mainly to simply doping of the "near surface," more recent research centers on the customized (hence the word "engineering" in the title) creation of structures on a fine, i.e., nanometer, scale. Ion beams are now used to aggregate metals and semiconductors into nanoclusters with nonlinear optical properties, to make nanopores of varying dimensions in polymer film alloys and superconductors and to fabricate nanopillars, "nanoflowers" and interconnected nanochannels in three dimensions by the use of sophisticated atomic shadowing techniques, to name just a few.

A Glossary is included at the end of the volume for the benefit of those who may be new to this area and unfamiliar with some of the terms and acronyms used herein. Included in a CD accompanying this volume are video clips taken in an electron microscope that provide striking visual evidence of crater formation and annealing by ion beams.

It is a pleasure to thank all authors for their efforts and professionalism in presenting their contributions.

Émile Knystautas
Québec, Québec
March 2004

Editor

Émile Knystautas was born in Kempten, Bavaria. He received his B.Sc. degree in physics from the Université de Montréal, and M.S. and Ph.D. degrees in physics from the University of Connecticut. Professeur titulaire at Université Laval during the preparation of this book, and a firm believer that one should change jobs every 35 years, he has recently taken up a new position as scientific director of a new nanotechnology center, CIVEN (Coordinamento Interuniversitario Veneto per le Nanotecnologie) in Venice, Italy. From 1978 to 1979, he was a guest worker and consultant at the Atomic and Plasma Radiation Division of the National Bureau of Standards (now N.I.S.T.). Although most of his career has been devoted to fundamental atomic physics studies, especially concerning rather exotic excited states in highly charged ions, his more recent activities also deal with the creation, modification, and characterization of novel materials using ion irradiation. Recent projects have included producing quasicrystals and shape-memory alloys in thin films, making multilayer mirrors for soft x-rays, high-voltage poling of silica for second-harmonic generation, the production of nonlinear optical effects in chalcogenide and other glasses using ion-beam methods, and the use of liquid-crystal-filled nanopores in polymer films in potential photonics applications.

Contributors

John E.E. Baglin
IBM Almaden Research
  Center
San Jose, California

Robert C. Birtcher
Materials Science Division
Argonne National
  Laboratory
Argonne, Illinois

J.D. Demaree
U.S. Army Research Lab
Aberdeen, Maryland

E. Cattaruzza
INFM Department of Physical
Chemistry
University of Venice
Venice, Italy

S.E. Donnelly
Joule Physics Laboratory
Faculty of Science,
  Engineering and
  Environment
University of Salford
Manchester, U.K.

David B. Fenner
Epion Corporation
Billerica, Massachusetts
and
Research Professor
Department of Electrical and
  Computer Engineering
Boston University
Boston, Massachusetts

Robert L. Fleischer
Department of Geology
Union College
Schenectady, New York

Daniel Gall
Department of Material
  Sciences and Engineering
Rensselaer Polytechnic
  Institute
Troy, New York

F. Gonella
INFM Department of Physical
Chemistry
University of Venice
Venice, Italy

James K. Hirvonen
U.S. Army Research Lab
Aberdeen, Maryland

Charles H. Koch
Institute of Materials
  Science
University of Connecticut
Storrs, Connecticut

G. Mattei
INFM Department of Physics
University of Padova
Padova, Italy

Koji Matsuda
Nissin Ion Equipment
  Company
Kyoto, Japan

C. Maurizio
INFM Department of Physics
University of Padova
Padova, Italy

Paolo Mazzoldi
INFM Department of Physics
University of Padova
Padova, Italy

A. Misra
Materials Science and
  Technology Division
Los Alamos National
  Laboratory
Los Alamos, New Mexico

Michael Nastasi
Materials Science and
  Technology Division
Los Alamos National
  Laboratory
Los Alamos, New Mexico

K. Nordlund
Accelerator Laboratory
University of Helsinki
Helsinki, Finland

P.J.T. Nunn
School of Engineering and
  Information Technology
University of Sussex
Brighton, U.K.

Masayasu Tanjyo
Nissin Ion Equipment Company
Kyoto, Japan

P.D. Townsend
School of Engineering and
  Information Technology
University of Sussex
Brighton, U.K.

Contents

Chapter 1

Introduction

Thin films can be produced in many forms and have properties that can differ significantly from their corresponding bulk form. They can be prepared by a host of techniques such as sputtering (single or multiple) layers on a substrate, creating buried waveguides by ion implantation in an optical material, or making complex nanostructures by ion irradiation during vapor deposition.

While ion-beam techniques have been a staple of the semiconductor industry for several decades, their application to other areas, for example metal surface treatment, have not been nearly as successful, generally because of cost considerations. Now, however, with the advent of devices of ever-smaller dimensions, the use of a directed-energy ion beam appears bound to find many novel industrial applications in the custom tailoring of new materials and devices. Such potential applications are too numerous to list here, and any attempt to make predictions at this point about which will pan out and which will not will likely turn out to have completely missed the mark a few years hence.

This book will hopefully provide newcomers to this exciting field with an introduction to its potential and also bring them up to speed on some of the current research in this area.

The first chapter deals with fundamental aspects and examines in detail the effects of a single ion impinging on a thin film. Using a unique "crossed beam" apparatus at Argonne National Laboratory consisting of a powerful transmission electron microscope that views a surface that is bombarded by hundreds of keV heavy ions, Steve Donnelly and his colleagues have studied the dynamics of crater and hole formation on metallic surfaces when *individual* ions impinge on a surface. Comparison with molecular dynamics simulations have given a satisfyingly complete picture of some of the basic mechanisms involved in the formation of craters and their (occasional) annealing by subsequent ion impacts. On the other hand, there are still other matters, such as the emission of nanoclusters, which require further study. Visual evidence of the effects of single-ion impacts is provided in the compact disc accompanying this volume, which contains some stunning video clips of the phenomena discussed. It is suggested that the reader watch these while reading the corresponding text. Rarely can one see sequential phenomena presented so vividly on a microscopic scale.

Magnetic recording is the topic of the following chapter by IBM Almaden's John Baglin, who discusses the ever-increasing demand for higher and higher disk-drive densities and how ion-beam techniques can help to achieve them. After briefly discussing some fundamental results that show the relative roles of ionization and collision processes for various ion beams and energies, he shows how ion-beam mixing (as opposed to ion implantation) can be used in some applications even in an industrial environment, where one might normally expect such a technique to be prohibitively expensive. He points out that spatial resolution issues can also be resolved in the application of ion-beam processing to magnetic storage technology.

For many years one of the standard reference books on ion implantation was the treatise by Jim Hirvonen ["Ion Implantation," J.K. Hirvonen, Ed., vol. 18 of "Treatise on Materials Science and Technology," Academic Press, N.Y., 1980]. In the present volume, with two co-authors, he presents an updated review of ion implantation, ion-beam mixing and

IBAD (ion-beam-assisted deposition), pointing out the strengths and weaknesses of each, as well as a realistic assessment of their applicability to a variety of research and manufacturing applications. In addition, the authors introduce a relatively new technique, GCIB (gas-cluster ion-beam technology) in whose development they played a major role. This powerful new tool has many similarities to the older techniques but also some characteristics that could not have been guessed by straightforward extension from the older ones. Many recent applications of GCIB technology are discussed, especially in the context of an industrial environment.

Peter Townsend's monograph [P.D. Townsend, P.J. Chandler and L. Zhang, "Optical Effects of Ion Implantation," Cambridge University Press, 1994] on the optical applications of ion implantation is now 10 years old, and he contributes herein (along with co-author P.J.T. Nunn) a chapter reviewing these. In addition to discussing the most recent developments in the field, as well as their relevance to industrial applications, he shares the results of many of his own innovative experiments on several aspects of this wide area.

One of the topics mentioned in Townsend and Nunn's review, that of the non-linear properties of metallic nanoclusters in glasses, is further expanded by the Padova group led by Paolo Mazzoldi. Together with their Venetian colleagues (for centuries Venice has been known for its expertise in glass), they trace the history of the optical properties of metallic nanoclusters in glasses back to Faraday, who spoke of metallic inclusions as being responsible for the coloration of glasses. The most recent approach, as described in their chapter, shows how the use of binary alloy nanoclusters allows one to tune the optical properties of glasses by varying the relative composition of such alloys.

The next chapter, by Misra and Nastasi of Los Alamos National Laboratory, discusses an important aspect of thin-film preparation by ion bombardment that is all too often ignored in the literature: the stresses, both tensile and compressive, that can be generated by ion-beam methods, and the problems to which these can give rise (delamination for instance). They discuss the origins of such stresses at the atomic defect level and describe how varying ion-beam energy and dose can modify these to achieve the desired results.

While ion-beam techniques are now standard practice in the semiconductor industry, the demand for micro-devices of ever-smaller dimensions will require considerable refinement. An overview of current problems and their practical solution is provided in the chapter by Koji Matsuda and Masayasu Tanjyo, both with the Nissin Ion Equipment Co. Ltd. in Kyoto. Their discussion centers on the demands of production-line equipment in an industrial, rather than a pure R&D setting.

Daniel Gall's chapter focuses on applying ion-beam techniques combined with physical vapor deposition to the eventual creation of complex nanostructures in transition metal nitrides. He provides examples of how nanopipes can be tailored and how atomic shadowing can create separated columns. Current and future work with deposition at shallow angles to the surface opens up the prospect of made-to-measure nanopillars, zigzag-shaped columns and helices, "nanoflowers," and interconnected nanochannel arrays, to name just a few. Applications are anticipated in magnetic storage devices, photonics, opto-electronics and molecular transport, among others.

For many years, Bob Fleischer and his colleagues at GE–Schenectady have exploited a technique for producing nanometer-dimensioned pores in polymer films. Ion-beam irradiation is first used to loosen or break bonds in the polymer along the ion trajectory, then chemical etching preferentially removes atomic-scale material that is found along the ion tracks in the film. In Chapter 10, he recalls this work and updates it, discussing the mechanisms by which tracks are formed, and hence how the dimensions of the ensuing pores can be controlled. He also discusses track formation in other materials such as intermetallic compounds and oxide superconductors. Aside from the typical applications of these nanopores as filters, there are many others presented, ranging from the study of voltage pulses generated by viruses and sea-urchin sperm to improving the properties of superconductors by creating obstacles to the movement of magnetic flux lines.

The last chapter, by Jim Koch of the University of Connecticut, is a good example of the possibilities of innovation in this field. His work shows how fingerprints can be made permanent and hence more reliable as forensic evidence by recoil-mixing them into the substrate using ion beams. Not only does the record thus become permanent but the fingerprint (even if only a partial one) can then be subjected to very sensitive surface-analytical techniques that can identify not only its shape but also its chemical composition.

A Glossary is included at the end of the book for some terms that may not be familiar to all, given that the intended audience for this volume consists of those who are not already working in this particular field. The definitions provided are "practical" in nature and not intended to be rigorous, aiming rather to facilitate a fluid reading of the book without interruptions to consult references.

Finally, a compact disc that contains several video files to supplement the chapter on single-ion impacts (by Donnelly et al.) is included at the end of the book.

Chapter 2

Single Ion Induced Spike Effects on Thin Metal Films: Observation and Simulation

S.E. DONNELLY, R.C. BIRTCHER, AND
K. NORDLUND

CONTENTS

ABSTRACT

The combination of *in situ* electron microscopy with molecular dynamics (MD) simulations gives important insights into the processes occurring during ion-beam engineering of thin films. This chapter compares and contrasts experimental observations and MD simulations of individual heavy-ion impacts on metal films. These impacts result in the formation of craters and other surface features on metals and the ejection of nano-particles. Images in the manuscript and video sequences on the accompanying CD-ROM illustrate the processes. The simulations of ion impacts match the experiment and give remarkable insight into the processes that give rise to the observed surface structures. Liquid flow and micro-explosions have been unequivocally identified in the MD work and provide an atomic-level understanding of the processes giving rise to cratering. An incomplete understanding exists of the emission of nanoclusters by ion impacts where the experimental size distribution of the emitted particles exhibits a power-law relationship, suggesting that this could be a shock-wave phenomenon. Although this is not, as yet, supported by the MD work, further simulations giving rise to improved statistics on nanocluster emission should enable a better comparison between experiment and simulation and thus serve to test this interpretation.

2.1 INTRODUCTION

Up to a certain energy density, the interaction of an energetic ion with a solid can be successfully described as a series of binary collisions involving the impinging ion and recoiling substrate atoms in what is generally described as a collision cascade. Monte Carlo simulation programs have been extremely successful in using this binary collision approach to estimate statistical parameters such as the distributions of implanted ions and of radiation damage (but neglecting any annealing processes that may take place). Under certain conditions of high energy-deposition density, this approach, however, is inappropriate. As first suggested by Brinkman [1,2], when the mean free path between displacing collisions approaches the interatomic spacing of the substrate, the interaction can no longer be regarded as one involving independent binary collisions and this description breaks down. In such cases, a small highly disturbed region is formed, in which the mean kinetic energy of the atoms may be up to several electronvolts per atom; this is known as an energy or displacement spike. At some time after the initial energy deposition (of order tens of picoseconds), the kinetic energy in the spike may be shared in a relatively continuous distribution by all the atoms within the spike region. Under some conditions this may give rise to an effective temperature within the spike zone significantly above that required for melting — this phase is generally referred to as a thermal spike or a heat spike.

These concepts of displacement and thermal spikes resulting from single ion impacts were first discussed in the scientific literature more than half a century ago; experimentally, however, until much more recently it has been difficult to obtain information about individual spike effects. This is because spikes are both small (typically a few nanometers in diameter) and of short duration (typically around 10 psec). To obtain information on spikes resulting from individual ions thus requires techniques with a high spatial resolution. As far as the time scale is concerned, no technique with adequate spatial resolution has a temporal resolution within orders of magnitude of predicted spike lifetimes. Any measurement is thus always of the effects of the displacement spike, the thermal spike and any ensuing defect annealing processes (both thermal and ion beam assisted) that may take place.

Over the last two decades, as the sizes (and thus volumes) of material resolvable by electron microscopes have become systematically smaller, advances in the speed and capacity of computers have enabled the accurate modeling of larger and larger assemblies of atoms using MD simulations. With this convergence, it is now possible both to image individual spike effects in the transmission electron microscope (TEM) and to model the same events by molecular dynamics. For some years now it has been possible to perform MD simulations of spike effects on "crystallites" of reasonable size and this size increases with every generation of computers. Currently, maximum crystallite sizes possible in MD simulations correspond to primary recoil energies in the range 100–200 keV. This overlaps with the energy range in which experiments are conducted and thus MD simulations can now give significant insights into spike processes — typically up to times of tens of picoseconds or so after the simulated impact. Atomic configurations resulting from an MD simulation can then be exported into TEM "multislice" image simulation software to yield simulated images that can be directly compared with experimental images.

In 1981 in a review of high-density cascade effects, Thompson posed two important questions on the nature of spike processes and these have remained substantially unresolved until the last decade [3]. The questions were: (i) "is it legitimate to use the concept of a vibrational temperature when the number of atoms in the spike (typically on the order of 10^4) may not be sufficient to be described by Maxwell–Boltzmann statistics?" and (ii) "is the duration of the spike (typically on the order of 10^{-11} seconds) sufficient for any major mass transport to occur?"

In this chapter we review recent work, primarily involving TEM, that has enabled these questions to be answered. We also look at MD simulations that have enabled us to develop a more complete atomistic picture of the processes that occur in these energetic single ion impact events. We believe that this understanding may lead to important advances in the engineering of thin films with ion beams. The size of a spike region is typically a few nanometers — just the right size for materials modification for the purposes of nanotechnology. Future uses of ion beams may thus increasingly employ single ion impact effects. For instance, a recent paper reports on the use of a focused ion beam system that is gated to allow the passage of individual ions to a specimen. This system (used to control the position of dopants in MOSFET devices) enables single ions to impact on a specimen with a spatial accuracy of about 60 nm [4]. With a slightly increased spatial precision, it may one day be possible to engineer materials using spike processes from individual heavy ions delivered to precisely defined locations.

2.2 CRATER AND HOLE FORMATION

2.2.1 *Ex Situ* Studies of Crater Formation

In 1981, Merkle and Jäger used TEM to examine Au surfaces that had been irradiated with Bi and Au ions with energies in the range 10–500 keV and discovered that, for energies above 50 keV, nanometer size craters were formed on the irradiated surfaces [5]. Figure 2.1a shows two micrographs from this work in which craters can be clearly seen. The authors also observed features that they identified, using stereoscopic techniques, as "lids" protruding from the surface at steep angles. Crater formation was observed to be a relatively rare event, occurring in only about 1% of the ion impacts, although that percentage was observed to increase when molecular ions (Bi_2^+) were employed. Crater sizes were typically about 5 nm and some were faceted as seen in the right-hand micrograph of Figure 2.1a. Crater formation was attributed to spike processes but the authors concluded that craters were formed due to evaporation of Au atoms. For this to occur, the relatively high Au sublimation energy must be given to each atom that is removed to form the crater, implying that only statistically unlikely high energy-density cascades at or very near to the surface could be responsible. In this model, all the atoms removed from the crater were sublimed from the surface and thus contributed to the overall sputtering yield. Merkle and Jäger assumed that spikes also occurred under the surface at a sufficient distance such that some of the material between the spike and the surface was not melted. The resulting local pressure rise due to the confined thermal spike was then sufficient to shear the material around the edges of an approximately disc-shaped region, resulting in the formation of the observed crater lids.

A study involving much lower energy ions at low temperatures was carried out in 1983 by Pramanik and Seidman [6]. They used field-ion microscopy (FIM), rather than TEM, to image a W tip irradiated with Ag and W atomic and molecular ions, and observed void-like near-surface contrast, which they attributed to "surface voids" (a description that would appear to be synonymous with "crater") formed as a result of the nucleation and growth of vacancies within a displacement cascade. The size of these features was in the range 2–3 nm.

In 1987 English and Jenkins, in a TEM study that was aimed primarily at studying the collapse of the core of displacement cascades to vacancy loops, found craters on Mo

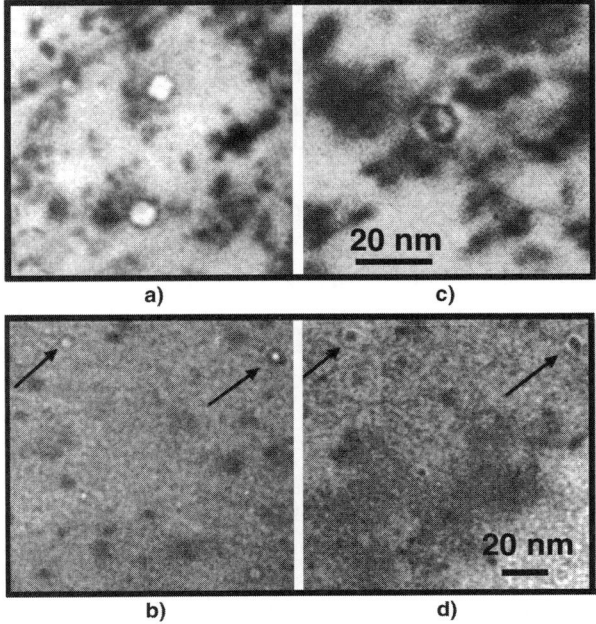

Figure 2.1 Examples of craters seen on ion-irradiated metal surfaces. (a) Au irradiated with Bi⁺ ions: Au (100) surface imaged in underfocus in the left-hand image and Au (111) surface in overfocus in the right-hand image. Faceting can be seen in both cases. (From K.L. Merkle and W. Jäger, *Phil. Mag.*, A 44 (1981) 741. With permission.) (b) Mo irradiated with Sb_3^+ ions with two craters arrowed and visible in underfocus (left-hand image) and overfocus (right-hand image). (From C.A. English and M.L. Jenkins, *Mater. Sci. Forum* 15–18 (1987) 1003. With permission.)

surfaces irradiated with Mo, W and Sb ions [7]. For irradiation with monoatomic ions, craters were only observed at high temperature (670 K); however, craters were observed to form at room temperature when the Mo was irradiated with Sb_2^+ and Sb_3^+. An example of these craters is shown in Figure 2.1b in both underfocus and overfocus. The number of craters created per ion varied from 0.4% for the monoatomic ions to as much as 3% for Sb_3^+. The origin of these features was attributed to the collapse of individual cascades or subcascades at the surface.

At this stage, in the late 1980s, the consensus of scientific opinion was that spike effects *did* occur in heavy ion collisions and that craters could be formed in some circumstances due either to sublimation of atoms from the surface during thermal spikes or to collapse of a cascade core at the surface.

2.2.2 *In Situ* Studies of Crater Formation

2.2.2.1 Gold

In 1995, two of the present authors were involved in a study of changes induced in populations of helium bubbles in gold and aluminum by 400 keV Ar ion irradiation [8]. This study was carried out in a high-voltage electron microscope in which *in situ* ion irradiation could be performed. In this study, discrete bubble jumps were observed to result from individual Ar ion impacts but such movement was not observed in similar experiments on Al. An important conclusion of this study was that the discrete bubble jumps resulted from the interaction of bubbles with the melt zone formed during the thermal spike phase of a contiguous cascade. Such melt zones were not expected to occur in aluminum.

In a follow-on study, effects of heavy ions on a variety of metals were observed using a similar *in situ* facility with higher resolution [9–11]. The microscope used was a Hitachi H-9000 (TEM) operating at 300 keV in which the ion beam is oriented 30° from the microscope axis [12]. Figure 2.2 shows a picture of this facility. In these experiments the specimen was tilted 15° toward the ion beam so that both ions and electrons were incident on the specimen at 15° to the foil normal. Specimens were irradiated with Xe⁺ ions at energies in the range 50–400 keV at dose rates between 10^{10} and 10^{12} ions/cm² per second. The Au films were of thickness 62 ± 2 nm, made by thermal evaporation onto heated NaCl. As in the work of Merkle and Jäger, craters were made visible in TEM in the same way as voids and bubbles by means of their phase-contrast under controlled amounts of objective lens defocusing. Images were obtained in bright field, on a region of the (some-

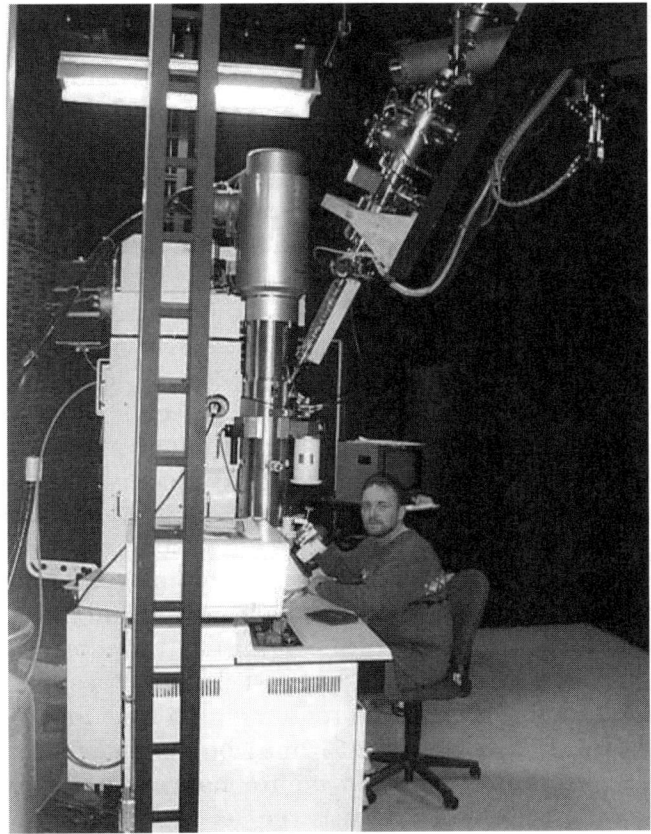

Figure 2.2 One of the authors (S.E. Donnelly) using the IVEM/accelerator facility in the Materials Science Division at Argonne National Laboratory. The ion beam line from the accelerators on the floor above makes an angle of 30° with the axis of the Hitachi H-9000 transmission electron microscope.

what bent films) in which no Bragg reflection was strongly excited. Under these conditions, underfocusing the objective lens by typically 1000 nm yields reasonably sharp images in which the crater is lighter than the background and is delineated by a dark Fresnel fringe. Similarly, a small mound or particle on the surface appears darker than the background with a light fringe around it. A similar degree of *over*focus gives rise to images in which this contrast is reversed. i.e., small craters appear darker than the background and small particles appear lighter than the background. In addition to normal photographic recording, images from a Gatan 622 video camera and image-intensification system were viewed with total magnifications of approximately 2 million, and recorded on video tape with a time resolution of 33 ms (1/30 sec — a single video frame). However, in most of the images shown in this section, to reduce video noise in the images an average has been made of eight successive video frames following (or preceding) an event of interest.

Although images of craters and other surface features resulting from individual ion impacts have been published in a number of papers, it is difficult to appreciate the nature of the observed processes from static images. Unfortunately, it is generally not possible to provide video sequences in journal articles; however, accompanying this chapter is a CD-ROM containing short clips of video sequences from some of our experiments as well as from some MD simulations (to be discussed later). We would suggest that before proceeding further with this chapter, the reader watch the video sequence entitled craters.avi. Note that it may be necessary to copy the file to your computer's hard drive to ensure smooth playback. The clip shows extracts from three separate experiments, in which gold films were irradiated at room temperature with 50, 200 and 400 keV Xe ions, respectively. Note that the microscope was operated at a magnification of 100,000× with the video system providing an additional magnification (typically about 10–20×) dependent on the display size. The area visible on the video clip corresponds to an actual area of 110 × 85 nm.

At the dose rates used in these experiments (2.5×10^{11} ions/cm^2/sec), approximately 20 ions impact on the area viewed every second and, with creation rates of between 0.02 and 0.05 craters per ion on Au, this results on average in a new crater appearing in the area viewed every 1–2 sec. However, as will be discussed later, craters are unstable under irradiation and are rapidly filled in by material transported from other impact sites. The video recording thus gives the impression of a surface exhibiting almost fluid-like properties on which a crater (sometimes along with expelled material) suddenly appears and then disappears over several seconds, during which time new craters appear. A frame-by-frame analysis or experiments at lower dose rates, however, reveal that both crater creation and the flow that causes crater annihilation are discrete processes resulting from single ion impact effects.

Figure 2.3 shows three images, digitized from videotape, of craters resulting from impacts of individual (a) 50 keV, (b) 200 keV and (c) 400 keV Xe ions on Au, respectively. In each case, the crater appeared between successive video frames and thus resulted from a single ion impact. The use of stereo-scopic techniques reveals that at 400 keV, craters appeared on both the entrance and exit surfaces of the film. This is consistent with the results of simulations to determine ion-induced damage in gold, using the Monte Carlo code TRIM [13], which indicates that the damage distributions at this energy are such that significant energy would be expected to be deposited at the back surface. The craters in Figure 2.3 have a fairly simple symmetrical form; however, Figure 2.4 shows a selection of craters with far more complex morphologies. Inspection of these images reveals clearly that there has been significant mass transport of material from the

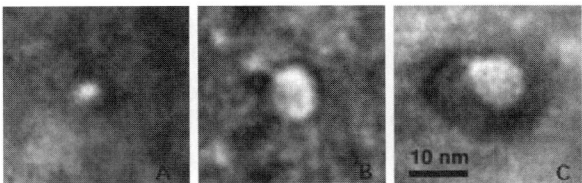

Figure 2.3 Images (digitized from video recordings) of craters resulting from single ion impacts of Xe on Au at energies of (A) 50 keV, (B) 200 keV and (C) 400 keV. (From S.E. Donnelly and R.C. Birtcher, *Phys. Rev.* B 56 (21) (1997) 13599. With permission.)

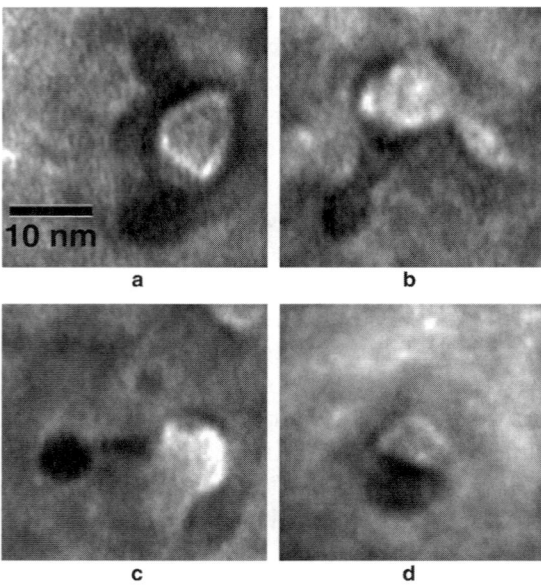

Figure 2.4   Examples of large craters on Au resulting from single ion impacts of Xe ions at energies of (a) 200 keV, and (b, c) 400 keV; (d) small craters occasionally seem to be accompanied by a plug of material having the approximate form of the crater. The images have been digitized from video recordings. (From S.E. Donnelly and R.C. Birtcher, *Phil. Mag.* A 79(1) (1999) 133. With permission.)

impact site, answering Thompson's question (ii) in the Introduction in the affirmative. In addition, the expelled material in each of the images shown does not have the same shape as its crater, implying that it has not been expelled as a solid plug. In Figures 2.4a–c the form of the expelled material indicates that molten material has been expelled from the impact site and that surface tension forces have acted — producing, for instance, an apparently quenched droplet in Figure 2.4b and a separated and seemingly spherical particle in Figure 2.4c. Regardless of whether the spatial and temporal dimensions of the spike are sufficient to permit the use of Maxwell–Boltzmann statistics, the images indicate that macroscopic concepts such as melting and flow in response to surface tension forces, and quenching, provide a satisfactory description.

Although the above description seems appropriate in the majority of craters where expelled material could be seen, in a number of instances small craters occasionally appeared to be accompanied by a solid plug of material having the approximate form of the crater. Such a crater is illustrated in Figure 2.4d. This is similar to the "lid" images recorded by Merkle and Jäger discussed in the Introduction. For such craters, we follow the interpretation of Merkle and Jäger that these result from spikes a sufficient distance below the surface, such that a solid disc of material is punched out by the large pressure increase that accompanies the thermal spike; an event that can be visualized as a "micro-explosion" below the surface.

MD simulations of 10 keV self-ion impacts on gold by Averback and Ghaly [14] indeed indicate that large pressures may occur in a spike. Figure 2.5 shows both the temperature and pressure variation reported in this paper. The figure indicates that pressures in the range 5–8 GPa exist at the core of a very small spike from 0.65 to 4.9 psec following ion impact. The pressure spike would be expected to result in the punching of a dislocation loop (approximately) when:

$$P_{spike} > \mu b/R \qquad (2.1)$$

where μ is the shear modulus of the gold, b is the magnitude of the burgers vector of the loop and R its radius (assumed to be the radius of the spike) [15]. On the basis of this simple description, a spike 5 nm in diameter will punch out loops for pressures in excess of 3 GPa; a significantly lower value than predicted by simply considering the theoretical shear modulus as proposed by Merkle and Jäger [5]. When close to the surface, this process may result in the translation of material to the surface via a sequence of loops punched out toward the surface or by the explosive ejection of a solid disc of material (where all the bonds at the disc edge are essentially broken simultaneously).

Finally, evidence of faceting of both craters and extruded material was occasionally observed. For example, faceting can be seen on the crater in Pb shown in Figure 2.6c. Although our assumption was initially that faceting would take place as a result of diffusion processes either

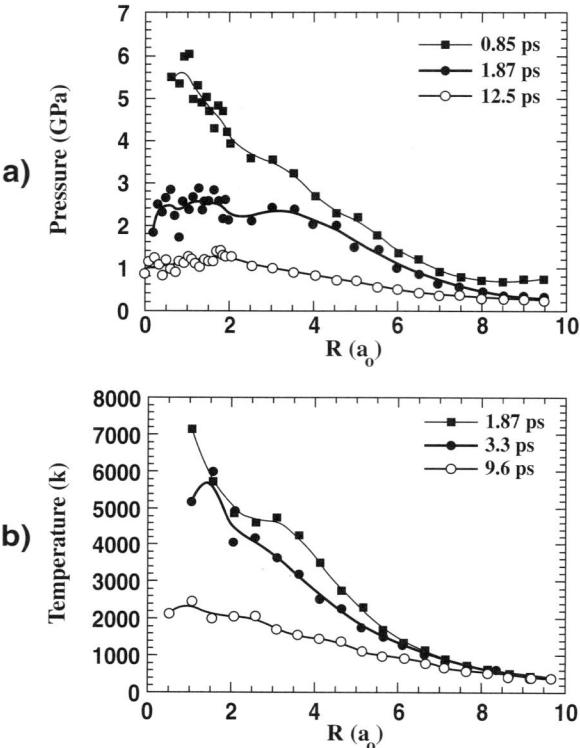

Figure 2.5 Plots of (a) pressure and (b) temperature profiles in the vicinity of a cascade due to a 10 keV Au ion in Au. Distance is plotted in lattice constants radially outward from the center of energy. For Au the lattice constant is 4.08 Å. (From R.S. Averback and M. Ghaly, *J. Appl. Phys.* 76 (1994) 3908. With permission.)

during the decay of the thermal spike or subsequently at room temperature, recent MD simulations have shown that faceting will occasionally occur directly as the crater is formed and the gold atoms are expelled. This will be discussed further in Section 2.4.

Figure 2.6 shows craters observed on the surfaces of (a) indium, (b) silver and (c) lead following Xe ion irradiation. Note that, as with gold, craters are made visible using phase

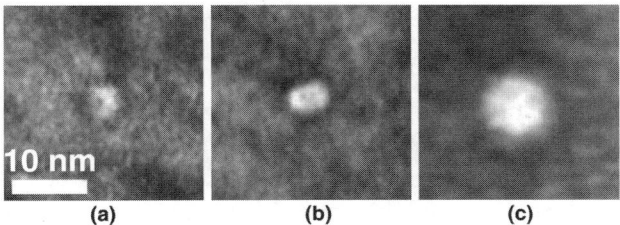

Figure 2.6 Examples of craters resulting from (a) 400 keV Xe ion irradiation of In at 17 K, (b) 100 keV Xe ion irradiation of Ag at room temperature, and (c) 200 keV Xe ion irradiation of Pb at room temperature. (From S.E. Donnelly and R.C. Birtcher, *Phil. Mag.* A 79(1) (1999) 133. With permission.)

contrast by controlled defocusing of the objective lens. The specific details of cratering in these three materials will be briefly discussed below.

2.2.2.2 Silver

Crater formation on Ag surfaces was observed to be qualitatively similar to that for gold but with a significantly lower creation efficiency of 0.6%. Large craters of up to 10 nm in diameter and having an irregular shape were occasionally observed.

2.2.2.3 Lead

Images of craters in Pb were less clear than in the case of Au, largely due to the higher degree of defocus that had to be used for imaging. The defocus used was typically 9 μm under focus compared with 1 μm for Au. The necessity for a higher degree of defocus almost certainly indicates that the craters were shallower for Pb than for Au. Unlike gold, there was little evidence of material ejected from the craters visible as particles on the surface. Approximately 0.7% of ion impacts resulted in craters in lead. As with gold and silver, the observed craters are thermally stable at room temperature when the ion irradiation is halted, but under conditions of continual irradiation are annihilated discretely by subsequent ion impacts.

A final important point with regard to Pb is that, until the specimen had been bombarded with a dose of approximately 4×10^{14} ions/cm^2, no cratering was observed. At lower doses, inspection of the diffraction pattern from the specimen indicated that an amorphous (oxide) layer was present on the surfaces of the specimen. Beyond this dose, as cratering began to occur, the diffraction pattern revealed that this layer had been removed by sputtering. The clear implication of this observation is that cratering is suppressed if an oxide layer is present on the specimen. This is consistent with the finding by Merkle and Jäger [5] that a thin amorphous layer of carbon deposited on gold films reduces the incidence of cratering on the gold surface.

2.2.2.4 Indium

Experiments were initially carried out on indium at room temperature and these failed to reveal any cratering. Some oxide appeared to be initially present on the indium surface as in the case of lead but even after examination of the diffraction pattern indicated that this had been removed, cratering still did not occur. The absence of cratering at room temperature, however, does not necessarily imply that craters did not form, as the existence of a crater for a sufficiently long period to be observed may involve four distinct processes, namely: crater creation by spike-induced local melting and flow, annealing during the quenching phase of the spike, thermal annealing at the ambient temperature and full or partial annihilation by subsequent ion impacts.

For a crater to be recorded on videotape it must persist for the order of one video frame (33 msec). At the dose rates used, the rate of annihilation by subsequent ion impacts should always have been sufficiently low to permit craters to be observed; however, as far as room temperature thermal annealing was concerned, it is possible that the high homologous temperature ($0.69\,T_m$) may have resulted in a very rapid annealing of craters by surface and bulk diffusion processes

— although craters were being created, they may not have survived sufficiently long to be recorded. To test this possibility, indium specimens were irradiated with 400 keV Xe⁺ ions at 17 K using a cryogenic specimen holder in the TEM. A frame-by-frame analysis of the video recording of this experiment revealed the very occasional formation of craters at this temperature. Specifically, two craters were observed during 30 min of ion irradiation implying a creation efficiency (very approximately) of 5×10^{-5} craters/ion. These craters were observed to be discretely annihilated by subsequent ion impacts as was the case for the other metals. Crater annihilation will be discussed in the next section.

2.2.3 Crater Annihilation

In general the static images give an incomplete account of crater behavior. The video-recorded sequences show, particularly for the higher energies, a rapidly changing crater population with approximately seven craters in the field of view (110×85 nm) at any time for the 400 keV Xe irradiation of Au. The "life cycle" of a crater at room temperature is illustrated in Figure 2.7. A discrete change occurring between video frames is clearly the result of a single ion impact. (To improve picture clarity by increasing the signal to noise ratio, each image consists of an average of four video frames taken before and after a discontinuous change). The figure shows a crater approximately 7.5 nm in diameter formed as a result of a single 400 keV Xe ion impact. The crater appears to be partially faceted and remains essentially unchanged for 50 video frames (1.6 sec) after which time a discrete event causes flow of material into the crater, resulting in its partial obliteration. A few seconds later (not shown), the remaining contrast disappears in a second discrete event. Note that in this particular case, no new crater was observed to form within the field of view (approximately six times the area shown) at the moment that the crater was filled in. However, in other cases, material expelled during the creation of a crater is seen to fill-in a second nearby crater. It is important to note that at room temperature we do not see any gradual infilling of

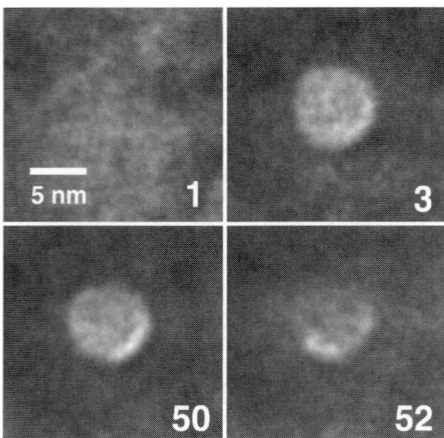

Figure 2.7   The creation and subsequent annihilation of a crater as a result of impacts of individual 400 keV Xe ions. The numbers are video frame numbers (i.e., time steps in units of 1/30 s). (From S.E. Donnelly and R.C. Birtcher, *Phys. Rev.* B 56 (21) (1997) 13599. With permission.)

craters due to, for instance, thermally activated surface diffusion processes. Rather, small craters typically disappear in one discrete event and larger craters in two or more.

Under steady-state irradiation at room temperature, at a dose rate of 2.4×10^{11} ions/cm^2/sec, craters have a lifetime in the range 1–12 s for all three ion energies studied. Lifetime measurements on 14 small craters with a mean diameter of 4.4 nm yielded a mean lifetime of 4.9 sec for craters annihilated in a single step. With the assumption that all ion impacts are capable of annihilating existing craters, the Xe ion has an annihilation cross section for small craters of approximately 85 nm^2, i.e., an ion impact within a radius of approximately 5 nm of the center of a small crater will annihilate the crater.

Although examples here have been limited to Au, the same qualitative behavior was observed for craters on Ag and Pb.

Although thermally activated crater annihilation processes appear not to be important on Au at room temperature,

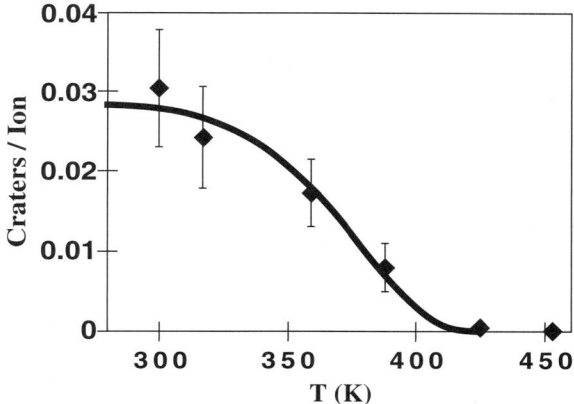

Figure 2.8 Temperature dependence of cratering on Au by 200 keV Xe ions. (From R.C. Birtcher and S.E. Donnelly, *Mater. Chem. Phys.* 54 (1998) 111. With permission.)

at higher temperatures such processes begin to play a role. Elevated temperature *in situ* irradiation experiments were performed on Au and Pb using a heating stage in the TEM. The crater production efficiency as a function of temperature is shown in Figure 2.8 for 200 keV Xe ions [16]. Craters were not observed (or more accurately crater persistence was less than 33 ms) above some critical temperature (approximately 420 K). At elevated temperatures, it appears that the observed cratering is a result of competition between the creation process, the direct ion-induced annihilation process discussed above and thermal recovery processes that occur on the time scale of a video frame. The craters observed are those that last at least one video frame or 1/30 sec. As the temperature is increased there is an increasing probability that a newly produced crater will shrink below a detectable size or not survive long enough to be recorded on video.

An interpretation of these observations will be presented in Section 2.2.5.

2.2.4   *In Situ* Studies of Hole Formation

Although most of our *in situ* studies have been concerned with crater formation and annihilation, our most striking experiments were carried out on single ion impacts on a thin Au foil (< 50 nm in its thinnest areas although the precise thickness was not known) [9]. In these experiments, many single ion impacts resulted in the creation of holes all the way through the foil. The conditions for these experiments were essentially the same as those for the cratering experiments; the main difference being that the TEM specimens were somewhat thinner and were prepared by jet-polishing of 99.999 at.% pure Au with grain size greater than 10 μm having a (110) texture (cf. evaporated films for the cratering experiments). The films were irradiated with 200 keV Xe ions with dose rates in the range 1 to 25×10^{10} ions/cm per second.

We would suggest that before proceeding further with this chapter, the reader watch the video sequence named holes.avi on the CD-ROM included with this book. In this video sequence, the large light area at the bottom right of the image is the hole that resulted from the electrochemical thinning process around which is the thinnest area of the Au. Patches of similar gray level that appear and disappear are thus also holes. Other patches of light contrast (but darker than the holes) are craters, similar to those discussed previously and seen on the craters.avi video clip. Contrast is not optimized for craters in this sequence as the specimen is only very slightly underfocused. Finally, darker areas may arise from defects within the foil, from expelled material that locally increases the foil thickness or as a result of diffraction effects due to bending of the foil.

Typically, the holes have diameters between 5 and 10 nm, two examples of which are shown in Figure 2.9. Approximately 0.5% of the Xe ions produced holes. Measurements in the area shown in Figure 2.9, made with a 50 nm diameter electron beam, indicate that no holes were created in regions with thickness greater than about 50 nm. This thickness is consistent with Monte Carlo calculations, using TRIM-95 [13], of the ability of 200 keV Xe ions to produce damage though

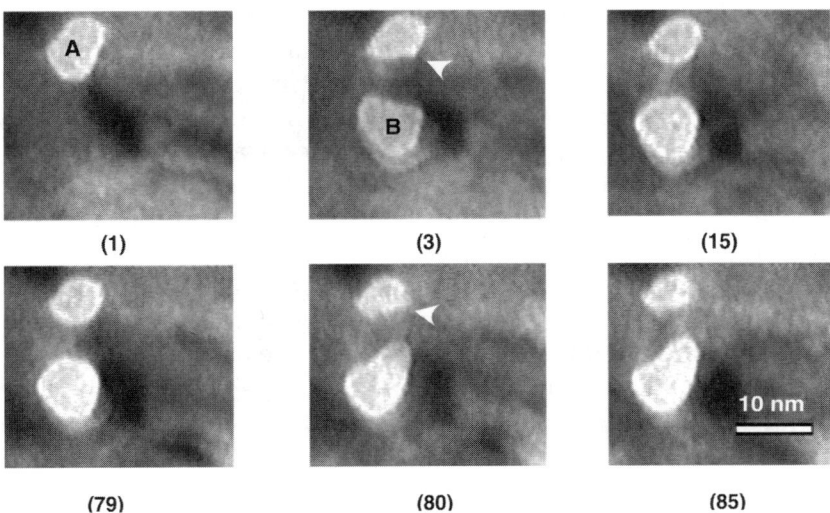

Figure 2.9 Microstructural evolution in Au during 200 keV Xe irradiation at room temperature. Parts (1) and (3) illustrate the creation of a hole (B) by the impact of a single 200 keV Xe$^+$ ion. Parts (3) and (15) illustrate the "rounding" of holes A and B as a result of plastic flow processes caused by single ion impacts. Parts (79), (80) and (85) illustrate a single ion impact modifying both holes. The numbers under each part of the figure refer to the video-frame from which the image was taken, with the first frame of the figure being numbered (1). Frame rate is 30 fps. (From R.C. Birtcher and S.E. Donnelly, *Phys. Rev. Lett.* 77 (1996) 4374. With permission.)

the entire depth of an Au foil. On the order of 1% of all ions stop in the last 2 nm of a 50 nm thick specimen. This suggests that it is only when ion damage extends through the entire specimen thickness that a hole may be formed.

Figure 2.9-(3) shows a hole (B) appearing next to a previously created hole (A). (The number under each section of the figure refers to the video frame from which the image was taken, with the first frame of the figure being numbered 1. The contrast within the holes is due to noise in the imaging system.) The image of an ion-induced hole is inevitably recorded, because of the nature of the image recording process, long after the cascade that produced it has ended and as such, includes any annealing that occurs during the cascade quenching. An ion strikes the entire area shown in Figure 2.9 on average every 0.3 sec (10 frames), which is very much longer than the cascade lifetime of a few tens of picoseconds. As in the case of craters, the holes appear between successive video frames (within a time period of 1/30th sec) and have been made by single ion impacts. Assuming a foil thickness between 20 and 50 nm, between 20,000 and 50,000 gold atoms were removed to create hole (B). This would imply an enormously high sputtering yield if the atoms were ejected from the gold surface. However, the change in image contrast suggests that these atoms have been moved to the specimen surface. Although expelled material is likely to be closely associated with a hole, small particles also occasionally appear far from any hole. This was also observed in the cratering experiments. The MD simulations indicate (Section 2.4) that there are three different mechanisms that can produce such isolated particles on the surface.

A second, perhaps equally striking, observation is the change in the shape and size of holes during continued irradiation, as also shown in Figure 2.9. Although holes have been annealed by the quench phase of their own cascade, additional ion impacts may produce further annealing. These changes occur in both large steps, such as when hole B is formed, and small steps as illustrated between Figures 2.9-(3) and 2.9-(15). The partial filling of hole A resulted from plastic flow of material away from the site of hole B, as indicated by the arrow on Figure 2.9-(3). No further major changes are observed until frame 80 when an ion impact causes an enlargement of hole B, and material from the impact partially fills hole A in the region indicated by the arrow on Figure 2.9-(80). In general, filling of holes does not depend on a second hole being nearby, and frequently the impact event responsible for a change does not itself produce a visible feature.

Another clear example of plastic flow is displayed in some still frames from the video clip in Figure 2.10. This figure shows the details of the filling of a hole during irradiation. An ion strikes the area shown in Figure 2.10 on average every

10 nm

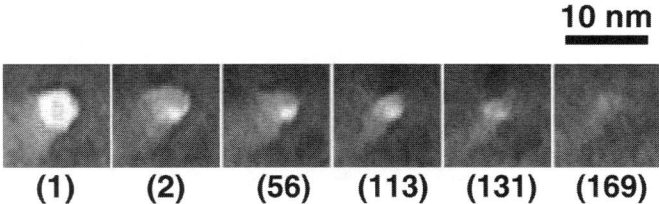

(1)   (2)   (56)   (113)   (131)   (169)

Figure 2.10 Filling of a hole in Au during 200 keV Xe irradiation at room temperature. Only frames showing changes are displayed. The image remains unchanged on intermediate frames such as from 3 to 55. Changes to the hole occur in steps as a result of individual ion impacts. The numbers under each part of the figure refer to the video frame from which the image was taken, with the first frame of the figure being numbered (1). Frame rate is 30 fps. (From R.C. Birtcher and S.E. Donnelly, Phys. Rev. Lett. 77 (1996) 4374. With permission.)

57 frames. On the video clip at this dose rate of 2.5×10^{11} Xe/cm²/sec, the motion of gold into the hole appears liquid-like; however, frame-by-frame examination reveals that, just as in the case of the craters, the process occurs in discrete steps associated with individual ion impacts. It should be noted that when the ion irradiation is stopped, flow of material ceases and there are no further changes to the morphology of the gold foil.

Finally, Figure 2.11 shows the effect of a single ion impact near the edge of the jet-thinned perforation in the foil. A finger of material is extruded into the hole (arrow on Figure 2.11b) but without another hole being formed. The defocus contrast suggests that the protrusion is inclined with respect to the surface. Although holes may be enlarged or shrunk by ion impacts, the net trend is hole growth and such protrusions disappear.

Many ion impact events change the shape of existing holes. Rounding of holes (e.g., in Figure 2.9) is a persistent process that takes place in discrete events. Changes in hole shape are generally due to cascade events that do not produce new holes but generate plastic flow near to existing ones.

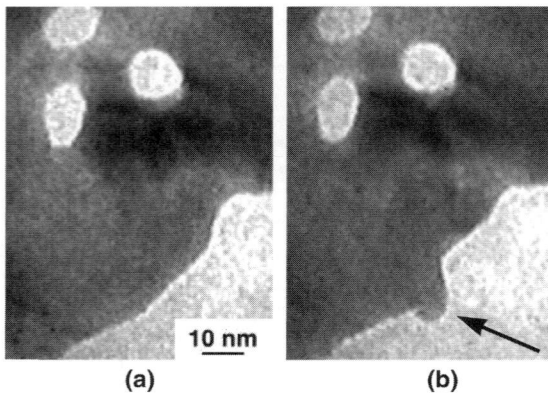

Figure 2.11   A protrusion formed at the edge of the Au foil as a result of an impact with a single 200 keV Xe ion at room temperature. The images in (a) and (b) are two successive video frames from the video recording shown in holes.avi. Frame rate is 30 fps.

Although an energetic ion impact may initially give rise to an explosive outflow of material, during the quenching phase of the molten zone to the solid state, surface tension forces will act on any free surfaces involved with the melt zone. This gives rise to changes in the shape of edges (including protrusions of the type discussed above) and the tendency of holes to become more circular. Similar effects have recently been seen in MD simulations and will be discussed in Section 2.4.

Although most experiments were carried out at room temperature or above, some were also carried out at low temperature. Figure 2.12 shows examples of single 200 keV Xe ion impacts on Au at 50 K. Each of the events shown occurred in the time interval between the capture of two successive video frames. Figure 2.12a shows material near the edge of the initial perforation in the specimen before irradiation. After the start of the irradiation a hole appeared (Figure 2.12b). This hole remained intact until another nearby ion impact produced a new hole and injected material, arrowed in Figure 2.12c, into the first hole. Another ion impact, Figure 2.12d, ejected a finger of material from the edge of the specimen into the initial perforation in the spec-

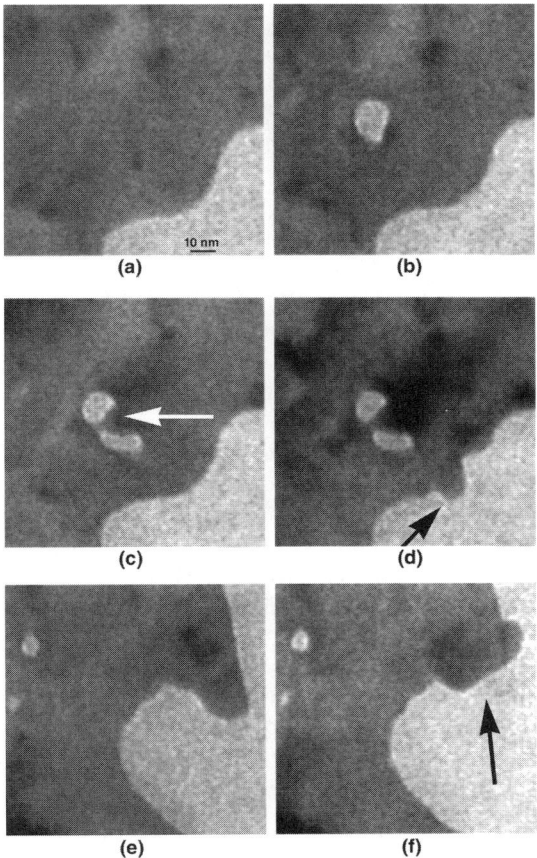

Figure 2.12 Pairs of successive video frames, digitized from an experiment carried out at 50 K (see text), showing: (a, b) hole formation, (c, d) creation of a protrusion and (e, f) significant modification of film morphology. Frame rate is 30 fps.

imen. The effect of a single ion impact on a protrusion into the initial perforation in the specimen, located in a second area slightly separated from that shown in Figure 2.12a, is shown in Figures 2.12e and f. Within the time of a single video frame, the protrusion appears to have melted at its center and quenched into a bent shape.

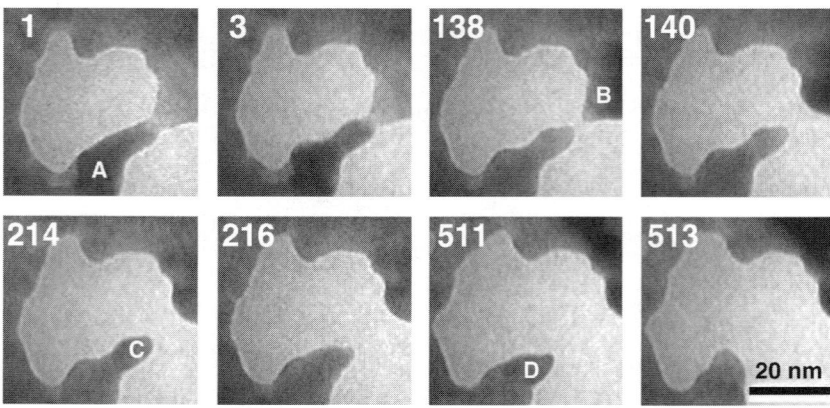

Figure 2.13 Discrete structural changes to an Au foil during 200 keV Xe irradiation at room temperature. The numbers in the upper left corner of each figure refer to the video frame from which the image was taken, with the first frame of the figure being numbered (1). Frame rate is 30 fps. (From R.C. Birtcher and S.E. Donnelly, *Phys. Rev. Lett.* 77 (1996) 4374. With permission.)

Even when the Au foil is at a temperature of 50 K, the processes that occur under heavy ion impacts on gold give the impression of a material at elevated temperature.

It has been a long-standing observation in our laboratories that thin areas of irradiated gold specimens disappear. This is a consequence of plastic flow during heavy-ion irradiation even if holes do not form. The dynamics of this process are shown in Figure 2.13 and can be seen clearly in the part of the video clip holes.avi from which these images were taken. An ion strikes the area shown in Figure 2.13 on average every four video frames. Our *in situ* observations reveal that the thickening occurs in pulses as a result of ion beam-induced plastic flow. When observed at a high dose rate this process appears to be similar to changes that occur when thin gold foils are heated to close to the bulk melting temperature and surface tension forces cause the material to flow. Frame-by-frame analysis such as that shown in Figure 2.13, however, shows this process to be a pulsed localized flow identical to

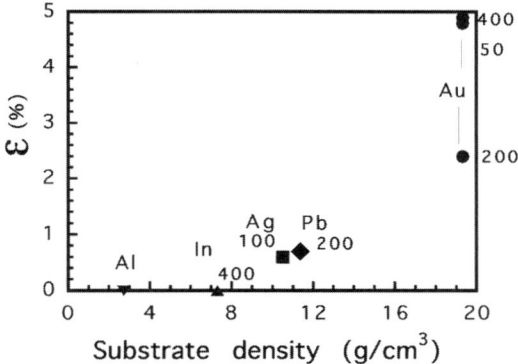

Figure 2.14 Measured cratering efficiencies under Xe ion bombardment, , as a function of density of the irradiated metal. (From S.E. Donnelly and R.C. Birtcher, *Phil. Mag.* A 79(1) (1999) 133. With permission.).

that responsible for the rounding of holes. Between frames 1 and 3 of Figure 2.13, the foil morphology changes due to an ion impact probably in the region indicated by the letter "A." The structure remains relatively constant until an impact in region "B" on frame 138 causes another significant modification. Similar discrete events occur on frames 214 and 511 in the regions marked with the letters "C" and "D." As in the case of hole production, only a small fraction of ion impacts result in major changes.

2.2.5 Craters and Holes — Discussion

Values of crater creation efficiency, ε, scale monotonically with density rather than with atomic number or atomic weight as illustrated in Figure 2.14, which summarizes our findings for the four metals studied. The numbers beside each of the points indicate the respective Xe ion energies. The values for ε (craters per incident Xe ion expressed as a percentage) have been obtained by determining the number of craters produced in a given time interval; with existence for one video frame being sufficient for a crater to be recorded. The figure indicates a threshold material density for crater production (by Xe ions) on the order of 7 g/cm^3. The data are also tabulated in Table 2.1.

The correlation with density (rather than atomic mass) is perhaps not altogether surprising as the energy density within cascades depends on this parameter. However, the distance of an individual spike from the surface is clearly also important.

In an attempt to model cascade-induced cratering [11], we follow Walker and Thompson [17] in defining the core volume of an individual cascade, V_c, as an ellipsoid with radii equal to the longitudinal straggling σ_x and the transverse straggling σ_y and σ_z, i.e.:

TABLE 2.1 Comparison of measured and calculated values of cratering efficiency, ε, for Xe ions on the four materials of interest. Other parameters tabulated are atomic number, Z; atomic weight, M; density, ρ; Xe ion energy, E_o; irradiation temperature, T; and energy per atom necessary for melting, θ_M.

	Z	M	(g/cc)	E_o (keV)	T (K)	θ_M (eV/atom)	ε (%) measured	ε (%) x_c = 5 nm	ε (%) x_c = 10 nm
Ag	47	18	10.50	100	293	0.20	0.6	0.5	18
In	49	15	7.31	400	17	0.07	0.00005	0	0.04
Au	79	17	19.32	50	293		4.8	28.8	82
				200	293	0.23	2.4	1.9	15
				400	293		4.9	6.7[a]	17
Pb	82	27	11.35	200	293	0.08	0.7	0.3	2.7

[a]Note that rear-surface cratering makes a significant contribution to this value (6.3%).
Source: S.E. Donnelly and R.C. Birtcher, *Phil. Mag.* A 79(1) (1999) 133.

$$V_c = (4/3) \, \pi \, \sigma_x \, \sigma_y \, \sigma_z \qquad (2.2)$$

The mean energy density θ (eV/atom) within an individual cascade is then given by:

$$\theta = 0.2 \, \nu(E)/V_c \, N \qquad (2.3)$$

where $\nu(E)$ is the amount of energy in the cascade deposited in collision events, and N is the atomic density of the metal. The 0.2 represents the fraction of total energy in the cascade that is deposited within the ellipsoid defined by the straggling.

Using a version of the Monte Carlo code TRIM, modified to yield individual cascade dimensions [18,19], we have calculated the mean longitudinal and lateral straggling for 10,000 ion trajectories, for each of the ion/substrate/energy combinations of interest. In addition, we have obtained an approximate value for $\nu(E)$ by assuming that it is equivalent, in each cascade, to the number of atoms displaced within the cascade multiplied by the displacement energy used in the calculation. We have then calculated values for V_c and θ, using the above equations. Note that no attempt has been made to examine subcascades; the model assumes a single cascade per incident ion and fits an ellipsoid to the damage distribution using the calculated straggling values. For any trajectories where distinct subcascades may occur, the calculation would be expected to yield a fairly inaccurate result for θ.

We have then assumed that only cascades with energy densities sufficient to cause melting will be able to cause cratering; the critical values of energy density at melting, θ_M, being obtained as:

$$\theta_M \text{ (eV/atom)} = E_M + k (T_M - T_o) \qquad (2.4)$$

where E_M is the latent heat of melting and T_M the melting temperature for each of the materials, T_o is the temperature at which the experiment was conducted, and k is Boltzmann's constant. We have then simply determined the percentage of total cascades with centers lying within a critical distance x_c of the surface, for which $\theta \geq \theta_M$. Note that in the case of gold, with an initial film thickness of 62 nm, both surfaces were included and reduction in film thickness arising from sputtering of the films was also taken into account.

Table 2.1 shows both the measured and calculated values of ε, for x_c equal to 5 nm and 10 nm. The agreement with the measured values is significantly better using the smaller distance. Although the experimental value of ε is greater at 50 keV than at 200 keV, it is nonetheless much smaller (4.8%)

than the calculated value of almost 30%. However, the experimental crater count at this energy is undoubtedly an underestimate, due to the considerable difficulty in resolving, and obtaining sufficient contrast to record very small craters. (In the calculations no "cut-off" is carried out to eliminate small craters.) Note also that the calculated cratering efficiency on the irradiated surface *decreases* with increasing energy as the ion penetration increases, but in the case of gold, cratering on the *rear* surface results in a larger value of ε at 400 keV than at 200 keV.

The precise values obtained for θ in these calculations are, of course, dependent on the definition of the core volume of the spike so that too much significance should perhaps not be attached to the particular value of x_c at which reasonable agreement is obtained. However, the theoretical analysis does yield the important result that the data cannot be modeled by assuming that only high energy-density cascades (e.g., $\theta_c > 1$ eV/atom) give rise to cratering. Instead, it is clear that cascades with energy densities less than 1 eV, in close proximity to the surface are largely responsible. For example, consideration of the distribution of cascade centers with depth indicates that to obtain agreement with experimental data for 200 keV Xe on Au, almost all near-surface cascades must yield craters. Finally, although no attempt has been made to correlate crater size with energy density, it is likely that the relatively rare large craters (e.g., ≥ 10 nm) do result from high energy-density cascades. The necessary and sufficient condition for cratering to occur thus appears to be simply that atoms within a cascade close to the surface must be given sufficient energy to cause melting.

In a recent publication on molecular dynamics modeling of ion-induced spike effects in solids, Averback and Ghaly [14] developed a simple model for spike-induced viscous flow effects and concluded that the parameter $(F_D/e_m)^2$, where F_D is the energy deposited per unit length of ion track (assumed linear) and e_m is the average energy of an atom at the melting point (somewhat higher than θ_M), could be used to determine whether cratering due to surface melting would occur for specific systems. Application of this criterion to a range of materials led to the conclusion that cratering due to melting and viscous flow would be expected to occur in a number of relatively low-density materials including Si. Although not specifically addressed in their paper, application of the Averback-Ghaly criterion to Al also predicts cratering in this material. Application of our analysis to low-density materials such as these, however, indicates that cratering would not be expected to occur due to the low value of θ occurring in these materials. Indeed, experiments in our laboratories on Al have consistently yielded no cratering whatsoever.

Although density is incorporated into the parameter F_D used by Averback and Ghaly, their assumption of linear ion track almost certainly leads to an underestimate of the importance of substrate density in determining whether melting and flow will occur. Surface features resulting from single ion impacts are seen using STM on a number of low-density materials, but we believe that these are not due to melting and flow.

Estimates of the recoil cascade size using TRIM-95 [13] yield, for 200 keV Xe ions incident on 50 nm thick gold, cascades with an approximately cylindrical volume (through the foil thickness) on the order of 5–10 nm in diameter into which almost all of the incident ion energy is deposited. After approximately 10^{-11} sec, atoms within this volume have a mean energy of about 2 eV. Calculations using the thermal spike model of Kelly [20], in which the integrated thermal sputtering from the surface is calculated over the spike lifetime, indicate that during the short spike lifetime, only a small number of atoms at the surface could evaporate. This is not sufficient to produce a crater or a hole. However, the thermal spike may cause melting and yielding. The sudden melting of this volume of material, with its concomitant volume change of approximately 10%, may give rise to an explosive flow of gold atoms to either (or both) surface(s). We believe that events such as that illustrated in Figures 2.9-(1) and 2.9-(3) are the first observed occurrences of such an ion-induced, pulsed, localized flow process. However, earlier MD simulations by Ghaly and Averback of cascades close to the surface due to a 20 keV Au ion incident on Au [21] had already suggested the existence of such a plastic flow process. They found a cascade region on the order of 6 nm in diameter by 6 nm in depth in which atoms each had a mean energy of approximately 2 eV. At 7.0 psec after ion impact a crater remained, which they described as resulting from plastic flow of the melt zone to the surface. The MD simulations discussed later in this chapter provide additional insight into the cratering process, now at energies directly comparable with the experiments.

In addition to the rapid, short-range processes identified in the MD calculations, longer-range annealing effects are experimentally observed. These may include enhanced surface diffusion as well as direct material transfer by a flow process.

Ion-induced, pulsed localized plastic flow is a fundamental process that may change the surface morphology of any irradiated metal in which displacement cascades are strongly localized such that molten zones form. For TEM observations of thin foils, holes result from only a small percentage of ion impacts with others causing cratering on one surface or the other (visible under defocus contrast). Simultaneously, many ion impacts result in plastic-flow annealing of previously created craters and holes (to be discussed in the next section). Any post-irradiation examination of surfaces, such as that carried out by Merkle and Jäger [5], in which craters on a gold surface were assumed to result from single ion impact sputtering, significantly underestimates the actual number of crater-creation events that occurs while misidentifying the origin of the craters. Pulsed localized flow is also important during gas bubble evolution and provides an explanation for ion-induced bubble motion observed during Xe irradiation of Au containing He bubbles [8]. We believe that plastic flow may also be responsible for other processes occurring in gold under ion irradiation such as grain-boundary movement. Since a single ion impact may affect tens of thousands of gold atoms, plastic flow may be the single most important mechanism in determining the behavior of gold under ion irradiation. The degree to which this phenomenon depends on material properties and affects other materials is not yet known but is under investigation in our laboratories.

2.2.4.1 Crater and Hole Annihilation — Discussion

The rapid annealing of craters and holes at elevated temperatures (under circumstances where plastic flow of material from the creation of a nearby crater is not seen) is most likely to result from surface diffusion processes. Let us assume there is a population of freely migrating surface defects (adatoms) at the elevated temperature (we will discuss below what their source is). Taking the crater to be a hemisphere of radius r that anneals by the arrival of surface diffusing defects, the number of missing atoms in a crater, n, is described by:

$$dn/dt = - P_n \, P_i \, D_o \, e^{Q/KT} \qquad (2.5)$$

where the number of lattice sites around the rim of the crater is given by:

$$P_n = 2 \, \pi r/a = (12\pi^2)^{1/3} \, n^{1/3} \qquad (2.6)$$

where a^3 is atomic volume and recovery by surface diffusion described by a radiation-induced surface adatom concentration P_i that is proportional to the ion flux, a diffusion rate D_o and an activation energy Q. The solution to equation 2.5 is:

$$n^{2/3} - n_o^{2/3} = -\{2/3 \, (12\pi^2)^{1/3} \, P_i \, D_o \, e^{Q/KT}\} \, t \qquad (2.7)$$

The lifetime of a crater of initial size r_o is given by the time interval between creation and the moment when n goes to zero:

$$t = 1/2 \, (r_o/a)^2 \, e^{-Q/KT}/P_i \, D_o \qquad (2.8)$$

Crater lifetime depends on the initial crater size, irradiation flux and the temperature.

In the experimental observations of crater production, all cratering events are counted regardless of their duration providing that they survive long enough to be recorded on a single video frame, t > 1/30 sec. Only craters with sizes larger than a minimum size, given by the relationship in Equation

(2.8), have the required lifetime. Given the minimum size required for survival at a given temperature, the total number of craters that will be recorded can be calculated from the size distributions at any temperature where they are thermally stable. This approach has been used to determine the activation energy Q, by fitting the measurements in Figure 2.8, and the result is shown as the solid line in that figure. The activation energy determined from this fit is 0.76 eV for Au (assuming a jump-attempt frequency of 10^{13}/sec).

We now discuss the possible sources of the migrating adatoms. It is well known that collision cascades produce interstitials in the bulk. These are mobile even at 5 K in Au, so any interstitials produced would be very likely to quickly migrate to the surface and become adatoms. However, MD simulation studies by Nordlund et al. indicate that in connection with cratering, the number of interstitials produced is only a small fraction of the number of surface "vacancies" that comprise the crater holes [22]. Hence interstitials cannot account for most of the crater infilling. Moreover, the interstitial explanation has the drawback that the activation energy of 0.76 eV should correspond to the adatom migration energies. We are not aware of a direct measurement of an adatom migration energy in Au, but in the other metals where it has been measured, it tends to be less than 0.5 eV [23]. Instead, we believe the activation energy of 0.76 eV corresponds to the energy of *emission* of adatoms from the crater rims. An adatom at the outer edge of an atom layer on the rim needs to break only two or three bonds to become a freely migrating adatom. In a simple bond-counting picture, taking into account that atoms in the bulk in Au have 12 nearest neighbors and that the cohesive energy of Au is 3.93 eV, the activation energy can be estimated to be 0.65–1.0 eV, well in line with the measured value. In this picture, the emission of adatoms is the rate-limiting step, so the exact value of the (lower) adatom migration energy is not relevant.

An overall interpretation of the observed annihilation behavior is thus as follows. At room temperature and below (on Au), impacts that result in the formation of craters may annihilate other nearby craters by plastic flow. However, impacts also give rise to a large number of adatoms on the crater rims, which can emit free adatoms by thermal activation and thus annihilate or part-annihilate the hole of the same crater, or that of another one in the vicinity. At a critical temperature, the crater lifetime (determined by competition between creation and annihilation) becomes smaller than one video frame (1/30 sec) and crater formation is no longer observed.

2.3 NANOCLUSTER EMISSION

2.3.1 Craters and Nanoparticles

Production of a crater on the surface of an irradiated material indicates the occurrence of near-surface energy deposition leading to a displacement cascade. As has been discussed above, a small fraction of these events result in the generation of a nanoparticle either connected with a crater as an apparent lid (Figure 2.4d) or unconnected and lying isolated on the surface [5,10,11]. Very occasionally more than one particle is seen to be attached to a crater; however, the vast majority of craters do not have an attached particle. As previously discussed, cratering events may result from either plastic flow of material away from the site of energy deposition or from a subsurface micro-explosion leading to the ballistic ejection of a solid plug. Craters with attached lid-like particles (as opposed to droplet-type structures) fall into the second category. Given that most craters do not have an attached particle, it is not clear whether such particle formation is rare or whether it is relatively common, but the majority of particles are completely ejected. In the latter case, such nanoparticle ejection may make a (nonlinear) contribution to the overall sputtering yield. Nonlinear sputtering effects from dense displacement cascades have been recognized in high sputter yields [24], the ejection of small atomic clusters during sputtering of Au [26] and increased yields for molecular ions over single ions [27,28].

2.3.2   Nanoparticle Collection

To investigate the significance of the ballistic ejection of nanoparticles, *in situ* TEM sputtering experiments were performed on gold and platinum [29]. The arrangement used to study crater formation was modified to include a thin carbon collector foil either above or below the metal target. The electron and ion beams passed through both the Au thin film and the carbon collector foil. In both experimental geometries the carbon collector foil and the target film were approximately 40 μm apart. By adjusting the specimen height in the TEM it was possible to view either the carbon foil or the target film with the other contributing an out-of-focus background to the image. In this way, the sputtered material accumulated on the collector foils could be observed with a resolution of approximately 1 nm. TEM images were recorded both, during irradiation, with a Gatan 622 video camera and image-intensification system and, during interruption of the irradiation, on photographic film. The video images were recorded with a time resolution of 1/30 sec (a single video frame). Images recorded on video have a total field of view into which approximately 20 ions arrived every second (30 video frames). Experimental details, including the total ion dose, were recorded on each video frame. Post-irradiation, high-resolution TEM observations of the Au nanoparticles on the carbon collector were made with a JEOL-4000 EX operating at 400 keV and a JEOL 3010 operating at 300 keV.

In situ viewing and video recording during irradiation reveals that particles were ejected from both surfaces of the metal target. Arrival of nanoparticles on the collector is illustrated in Figure 2.15, showing three images extracted from single frames of a video recording over a time during which 116 ions impacted the specimen in an area the size shown in the figure. Dark spots in the figure are Au nanoparticles. Frame-by-frame examination of the recordings made during irradiation reveal that particles appear on the carbon collector within the 1/30 sec required for the recording of a single video frame. Nanoparticle arrivals at the collector foil are well separated in time and space because ions impact the viewed area

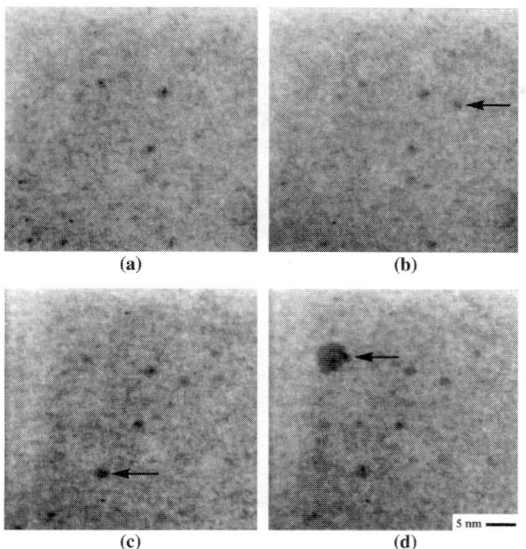

Figure 2.15 Nanoparticle collection during 400 keV Xe irradiation of Au. Frames (a and b) and (c and d) are separated in time by a video frame. (From R.C. Birtcher, S.E. Donnelly and S. Schlutig, *Phys. Rev. Lett.* 85 (2000) 4968. With permission.)

at a rate that is less than once every video frame and the particle production rate is about 1%. The total area of view in the video is 9.35×10^5 nm^2. During the time required to record an image, the mean number of ions arriving in this area is approximately 0.14, significantly less than one, leading to the conclusion that nanoparticle ejection is caused by single ion impacts. This sequence is typical of all sputtering events observed. Note that the appearance of a new nanoparticle does not disturb existing nanoparticles. Even the arrival of the large nanoparticle in the last frame leaves unchanged the two smaller nanoparticles to its left and above. No nanoparticles appeared after the ion beam was turned off.

Bright-field TEM micrographs of nanoparticles collected on the carbon-coated grids following separate 400-keV Ne, Kr and Xe irradiations to fluences of approximately 10^{18} ions/m^2 are shown in Figure 2.16. As can be readily seen in the

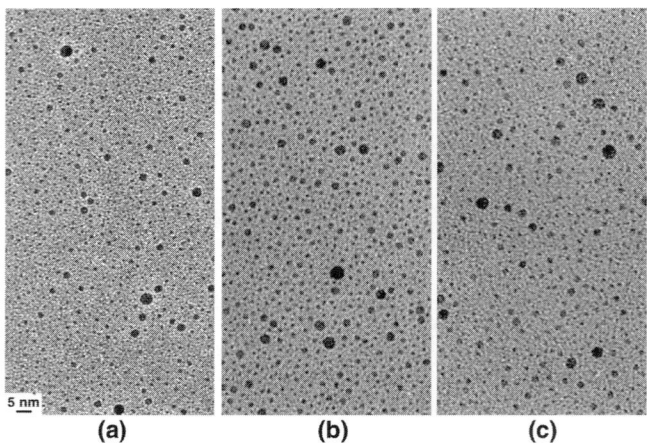

Figure 2.16 Bright-field TEM micrographs of nanoparticles collected on the carbon-coated grids following separate 400-keV Ne, Kr and Xe irradiations to fluences of approximately 10^{18} ions/m². (From L.E. Rehn, R.C. Birtcher, S.E. Donnelly, P.M. Baldo and L. Funk, *Phys. Rev. Lett.* 87, (2001) 207601-1. With permission.)

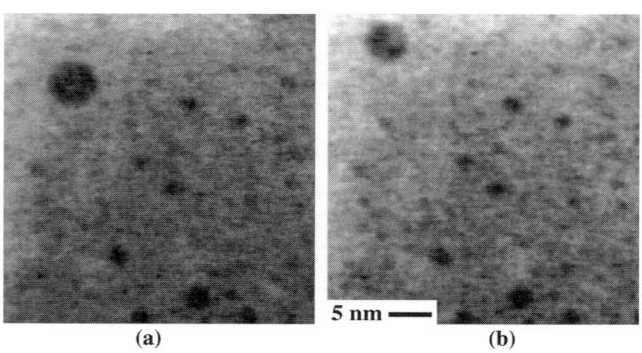

Figure 2.17 Change in position of a nanoparticle caused by a single 400 keV Xe ion impact. Frames (a) and (b) are separated in time by a video frame. (From R.C. Birtcher, S.E. Donnelly and S. Schlutig, *Phys. Rev. Lett.* 85 (2000) 4968. With permission.)

micrographs, the sizes of the collected particles are broadly distributed from just barely discernible images somewhat less than a nm in lateral dimension up to particles as large as approximately 8 nm in extent. There are more small particles than large ones. The average particle size (i.e., particles with sizes above the resolution limit of 1 nm) is on the order of 3 nm. This is consistent with the average crater size of 3.3 nm. It is uncertain whether all particles stick to the collector foil; however, no particles were observed to disappear once deposited on the foil. The TEM imaging conditions were adjusted so that the contrast in Figure 2.16 arises primarily from differences in total thickness. Hence, the fact that the larger particles are noticeably darker in the micrograph demonstrates that the particles are three-dimensional entities. The three-dimensional nature of the collected particles was subsequently confirmed from high-resolution TEM (HRTEM) images. At high doses, the new particles were subsequently incorporated into Au deposited on the carbon film and became indistinguishable. The incorporation at high doses of particles into existing Au islands makes any observation difficult and *in situ* observations a requirement.

2.3.3   Radiation Effects on Nanoparticles

With a few exceptions for the largest particles, the nanoparticles on the carbon collector foil do not exhibit other signs of themselves being struck by ions during continued irradiation. Resputtering by the keV Xe ions that had passed through the Au film was also negligible — unlike sputtering caused by ionization during MeV ion irradiation of Au islands evaporated on carbon foils [28]. When the nanoparticles are large, it is possible for a Xe ion impact to produce a high-energy recoil event in the particle that causes it to jump a fraction of its diameter. Such an event is illustrated in Figure 2.17. Shown in this figure are successive video frames of a 5 nm diameter particle displaying movement that occurred during the time required to record a video frame. Although the image of the nanoparticle indicates that its diameter decreased, its volume cannot be accurately estimated because of possible change of its thickness. During the time required to record the jump, statistically fewer than 3×10^{-3} ions impacted an area four times the area of the nanoparticle, leading to the conclusion that its motion was caused by a single ion impact. No nanoparticle motion was detected after the ion irradiation was stopped. During *ex situ*, high-resolution TEM observations with 300 and 400 keV electrons, Au nanoparticles on the collector foils did not move although their internal structure changed as has been reported previously [30]. During the ion irradiations, the microscope resolution was insufficient to determine whether such structural changes occurred. The motion events recorded during ion irradiation are due to ion impacts and not the electron beam used to observe the nanoparticles. Simple calculations show that nanoparticle motion cannot be the result of simple momentum transfer between a Xe ion and the Au nanoparticle.

As would be expected, such motion events are rare even for such large particles, and only five were noted during the many irradiations. Although the experiments reported here stopped at low doses, other experiments have revealed that continued nanoparticle deposition to higher density on the collector foil eventually results in particle coalescence and incorporation of newly deposited nanoparticles into existing material. This results in the growth of a continuous Au thin film on the collector. As the Au coverage and thickness increases, the effects of ion impacts are seen on the deposited material. This is associated with the ability of a particle or film to retain the energy deposited in atomic recoils in a displacement cascade caused by a single ion impact. The implication is that the nanoparticle motion was the result of the surface melting or micro-explosion caused by a displacement cascade retained by the particle.

2.3.4 Nanoparticle Ejection Rates

The low nanoparticle ejection rate allows accurate determination of the ion dose at which each particle is first observed. The rates of nanoparticle ejection depended on ion type and energy [31]. Examples are shown in Figure 2.18 for nanopar-

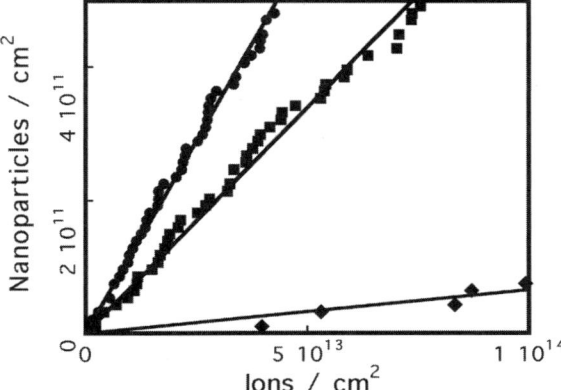

Figure 2.18 Accumulation of nanoparticles ejected from the exit surface of a 62 nm thick Au film during irradiation with 400 keV Ne (diamonds), Kr (squares) and Xe (circles) ions. (From R.C. Birtcher, S.E. Donnelly and S. Schlutig, *Nucl. Instrum. Methods* B, 215 (2004) 69. With permission.)

ticles ejected from the back side of the 62 nm thick Au foil. The detection limit for nanoparticles is approximately 1 nm, so these rates do not include either small clusters or individual Au atoms. For all irradiations, nanoparticle deposition commences from the start of the irradiation, and over the dose range studied, the accumulation appears to be linear. From least-square fits to the data, the rates of nanoparticle ejection were determined. These rates are shown in Figure 2.19a as a function of ion energy for different ions. For the energy range studied, the rate of nanoparticle ejection increases both with increasing ion energy and with increasing ion mass. The lines in Figure 2.19a are fits with a function that varies as the square of the ion energy with a multiplicative factor that varies with ion mass. It must be kept in mind these rates are for ions that have traversed the 62 nm thick Au foil.

The dependence of nanoparticle ejection on ion mass is extracted from the multiplicative constants determined from fitting the curves in Figure 2.19a. Dependence of the multiplicative fitting parameters on ion mass is shown in Figure

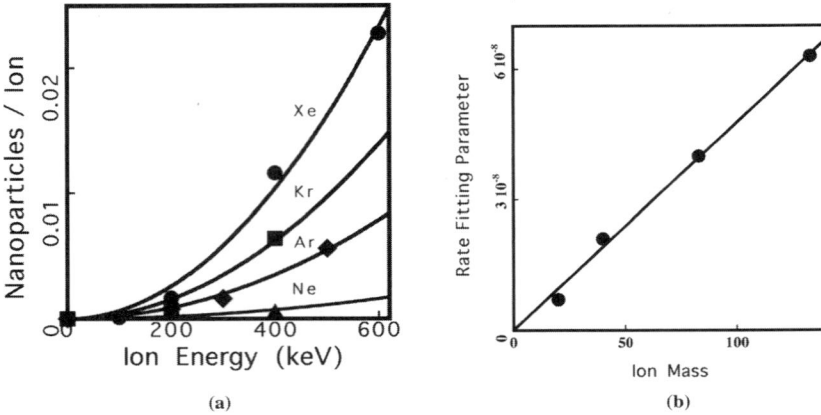

(a)   (b)

Figure 2.19   (a) Rate of nanoparticle ejection from the exit surface of a 62 nm thick Au film as a function of ion energy for Ne (triangle), Ar (circles), Kr (squares), and Xe (diamonds). The lines in Figure 2.3a are fits with a function that varies as the square of the ion energy with a multiplicitive constant. (b) Variation of the fitting parameter in (a) as a function of ion mass. (From R.C. Birtcher, S.E. Donnelly and S. Schlutig, *Nucl. Instrum. Methods* B, 215 (2004) 69. With permission.)

2.19b. The fitting parameter increases linearly with ion mass. The nanoparticle ejection rate from the back side of a 62 nm thick Au film at room temperature is described by:

$$R = 0.00048 \, M \, E^2 \qquad (2.9)$$

where E is the incident ion energy in MeV and M is the atomic mass of the ion. Since ejection occurs from the back side of the foil, this description involves a convolution of ion energy loss through the foil with the cross-section for the ion to create an energetic displacement cascade near the exit surface. Because of this detail, this description has limited applicability.

2.3.5   Relationship of Nanoparticle Ejection to Cratering and Cascade Events

The observed rate of nanoparticle ejection can be related to the probability for an ion impact to produce a crater on Au.

As discussed earlier, TEM observations revealed a few nano-particles attached to crater rims as well as on the surface not associated with a crater. An estimated 3.3% of 400 keV Xe ions produce a crater on the surface of Au. In the nanocluster experiments, this can be most closely compared to an irradiation with 600 keV Xe ions that had on average approximately 330 keV near the back surface of the Au specimen and ejected 0.023 nanoparticles/ion. Using Equation (2.9) to extrapolate to impacts caused by 400 keV Xe ions at the exit surface increases this estimate to 0.033 nanoparticles/ion. This suggests that essentially all craters observed without an associated nanoparticle on the irradiated surface have ejected a nanoparticle.

To directly compare nanoparticle ejection with the atomic recoil process, the probability for recoil events with energies greater than 50 keV centered between 5 and 10 nm from the specimen surface was determined by Monte Carlo calculations using the TRIM computer code [13]. The recoil energy and depth of recoils were chosen to represent the near-surface, high-energy displacement cascades of the type that cause melting of the cascade volume and are expected to be responsible for cratering and nanoparticle ejection [11]. Calculations were made for all irradiations used, and the rates of nanoparticle ejection by all ions and energies are compared to the probability for such high-energy recoil events in Figure 2.20. The straight line is a linear fit to the data that yields 0.42 nanoparticles per near-surface cascade. Thus in this model, one half of the near-surface displacement cascades produced by a Au recoil with 50 keV or higher energy ejected a nanoparticle that was subsequently observed on the collector foil. This is a lower limit whose accuracy is limited by a lack of a reliable estimate of the fraction of ejected nanoparticles that are not observed because they fail to strike or fail to stick to the collector, disintegrate in flight, or are smaller than the TEM detection limit. The number of cascades estimated for an irradiation depends on the displacement energy used in the TRIM calculations. If this parameter is taken to be 60 keV, the efficiency is one, but such an exercise has limited meaning. The important point is that, as with cratering, the

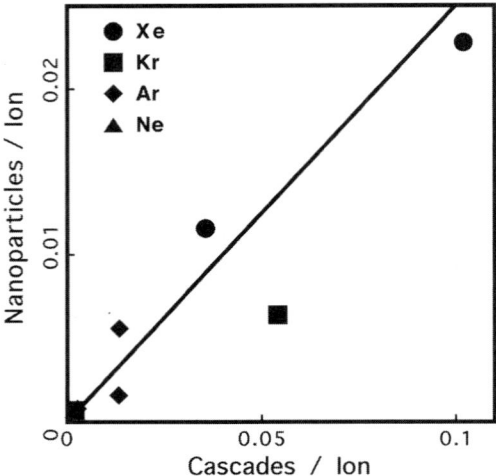

Figure 2.20 Rate of nanoparticle ejection from Au by Ne, Ar, Kr and Xe as a function of the number of near-surface displacement cascades calculated using TRIM with a 50 keV displacement energy. The line is a linear fit that yields 0.42 nanoparticles per near-surface cascade. (From R.C. Birtcher, S.E. Donnelly and S. Schlutig, *Nucl. Instrum. Methods* B, 215 (2004) 69. With permission.)

rate of nanoparticle ejection scales with the probability for high energy recoil events in the near-surface region.

2.3.6 Nanoparticle Ejection Mechanisms

The emission of clusters of atoms during ion sputtering was first reported more than 40 years ago [32]. This observation of stable, intact clusters elicited considerable surprise because the energies involved in the ion-target collisions that generate sputtered atoms are typically much larger, ranging up to many keV, than the 1–2 eV binding energies holding the clusters together. Hence it is difficult to understand how stable clusters can form during ion sputtering. Ever since their discovery, the unsolved fundamental question of the cluster emission mechanism has attracted considerable experimental and theoretical attention. Recent relevant reviews can be found in the literature [22,33,34].

Mass spectrometry has been used extensively in prior experimental studies of cluster emission during sputtering primarily because this technique allows the size distributions of clusters, at least for those comprising up to about 40 atoms, to be determined both accurately and conveniently [22,33–35]. Since there were serious concerns regarding possible fragmentation during measurement, particularly for larger clusters, and since larger clusters cannot be monitored as efficiently with mass spectrometry, it appears that only populations of smaller clusters have been investigated previously. It has been shown [35, 36] that the size distributions, $Y(n)$, of the emitted clusters obey an inverse power law, i.e.,

$$Y(n) \propto n^{-\delta} \qquad (2.10)$$

where n is the number of atoms in a given cluster. The total sputtering yield is defined as the total number of atoms emitted from the surface per incoming ion, regardless of whether they are emitted individually or in clusters. The exponent, δ, was found to correlate [34–36] with the total sputtering yield, such that higher sputtering yields result in smaller values of δ. The values of δ that have been determined experimentally using mass spectrometry techniques (for clusters containing up to approximately 40 atoms) range between 4 and 8.

No simple theoretical model has been able to reproduce both the power-law size distributions and the high values of δ found experimentally. Statistical models, which assume that the observed intact clusters arise from the re-aggregation of atoms after they have been ejected individually from the surface, and which would appear most appropriate for predicting the distribution of the smaller clusters, yield an exponentially decaying number of clusters with increasing cluster size [33,34]. A power-law distribution is predicted by the shock-wave model of Bitensky and Parilis [37], and an approximate power-law distribution by the thermodynamic-equilibrium model of Urbassek [38]. However, these models yield values of 2 and 7/3, respectively, for the δ in Equation (2.9), and they have thus been rejected based upon the previously reported experimental results lying between 4 and 8.

2.3.7 Ejected Nanoparticle Size Distribution

The particle size distributions, Y(n), i.e., the number of clusters containing n atoms, were determined after removing the Au target foil [40]. First, an image similar to the ones shown in Figure 2.16 was digitized at a final image resolution of 10 pixels/nm. Sizes of individual nanoparticles were obtained from the scanned images. The measured lateral dimensions were converted into the number of atoms for a given cluster by assuming that the particles were hemispheres with the density of bulk Au. All visible particles in an area of ~10^5 nm^2 were sized, i.e., approximately 650 and 200 particles, respectively, for the Au and the Ne irradiations. The results were grouped into bins having 0.5 nm steps; the procedure was repeated following each irradiation. The total nanoparticle density determined by counting individual nanoparticles in the scanned images had a one-to-one correlation with the integrated arrival rates determined from the *in situ* video recordings. Subsequent irradiation of the particles without a target foil did not produce particle growth. Taken together, these findings demonstrate that size changes following deposition were insignificant.

The size distributions that were obtained are displayed in Figure 2.21. Size distributions were determined for the particles collected from the 400-keV Kr irradiation, and for 400-keV Ne and Au irradiations to fluences of 6.2×10^{18} ions/m^2 and 3.8×10^{17} ions/m^2, respectively, and for a 500-keV Ar irradiation to a fluence of 2×10^{18} ions/m^2. Two important findings emerge from Figure 2.21. First, the displayed power-law fits to the data from the four individual irradiations all have the same slope despite large differences in the total sputtering yield (e.g., ~5 for the Ne and >100 for the Au). Second, the magnitude of this slope is −2 within remarkably tight error limits; the average standard deviation determined for the four runs was < 0.05.

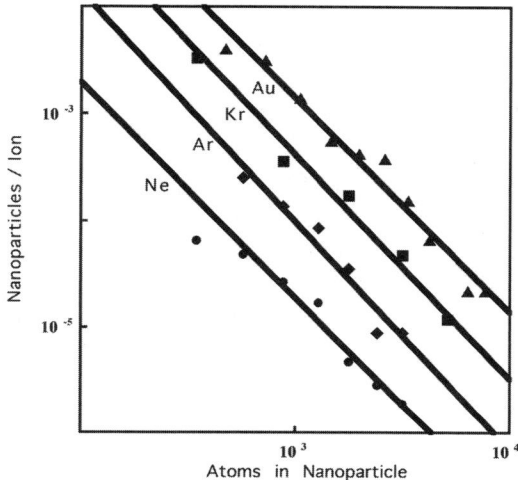

Figure 2.21 Number of collected nanoparticles as a function of size (n) for 400-keV Kr irradiation shown in Figure 2.16, and for 400-keV Ne and Au irradiations to fluences of 6.2 × 1018 ions/m2 and 3.8 × 1017 ions/m2, respectively, and for a 500-keV Ar irradiation to a fluence of 2 × 1018 ions/m2. (From L.E. Rehn, R.C. Birtcher, S.E. Donnelly, P.M. Baldo and L. Funk, *Phys. Rev. Lett.* 87, (2001) 207601-1. With permission.)

2.3.8 Shock Wave Model

Early after the discovery of nanoparticle ejection, it was recognized that a mechanism involving the cooperative motion of atoms was probably responsible for the emission of intact clusters during ion bombardment. Expanding on this concept, and noting that clusters were formed with the greatest probability under energetic heavy-ion bombardment, Bitensky and Parilis [37] formulated an analytical model for cluster emission based upon the formation and emergence of shock waves that originate from energetic subsurface collision cascades. Essentially, the arrival and reflection of a shock wave at a sample surface places the latter under tension. If this tension exceeds a critical value, known as the fracture strength, the surface fractures and fragments are ejected. The Bitensky and Parilis model predictions are in excellent agreement with the results displayed in Figure 2.21. Their shock-wave model predicts a power-law decrease for the cluster size distribution with an exponent of −2. The inverse-square dependence arises because the critical parameter in the fracture process is the energy required to create new surface area, which for a three-dimensional particle increases as its characteristic dimension squared.

The experimental verification that relatively large atom clusters, i.e., those up to approximately 10,000 atoms, are generated during high-energy ion bombardment also offers a possible explanation for why many previously reported investigations yielded power-law size distributions with considerably higher exponents. Very simply, a higher exponent means that smaller clusters are present with disproportionately higher probabilities in the mass spectrometry studies. Such an excess of smaller clusters is of course what would occur if all the larger clusters do not remain completely intact. The very fact that the inter-atomic bond energy is so small in comparison with the energies involved in the displacement events implies that some clusters can be expected to fragment after being ablated.

We also note from the data displayed in Figure 2.21 that the size of the largest nanoparticle that is ablated from the target surface increases systematically, and substantially, with increasing ion mass. Although the smallest observable particle size will depend strongly on TEM imaging conditions, and therefore can vary considerably from sample to sample, the largest particles are the easiest to observe, and the easiest to characterize. With this fact in mind we see that some particles containing in excess of 8000 atoms were generated by the Au irradiation, while the largest for the Kr irradiation contained approximately 5000 atoms, and it decreased even further, to only approximately 3000 atoms, for the Ar and Ne bombardments. This increase in the largest observed nanoparticle size can be readily understood in terms of the maximum energy available in the displacement-cascade-induced shock wave, which also increases with increasing ion mass.

2.3.9 Relationship of Nanoparticle Ejection to Sputtering

Ejected nanoparticles make an important contribution to ion sputtering yields from gold. The measured total yield of transmission sputtering by 400 keV Xe on 62 nm thick Au is 33 [25]. The number of atoms sputtered in nanoparticles can be determined by a summation over the atoms in all particles represented in the nanoparticle size distribution. At a rate of 0.011 nanoparticles/ion for 400 keV Xe ions, the nanoparticle contribution to the sputtering yield is 9. The difference of 24 is essentially the same as the sputtering yield of 22 calculated for linear recoil processes [39] by a Xe ion with the average exit energy. Given the uncertainty in these measurements, nanoparticle ejection accounts for the missing component in theoretical estimates of sputtering. The importance of this effect for a given ion/substrate combination can be determined using calculations based on a Monte Carlo approach using TRIM [13] as describe above and, although the work here has been confined to gold substrates, such calculations should be applicable to any combination of ion and substrate.

It might appear surprising at first that an element as ductile as Au could actually fracture in the manner described by the shock-wave model of Bitensky and Parilis. To understand this, it is important to note the exceedingly fast time scale on which the shock wave develops and impacts the surface. The pressure pulse generated by a highly energetic displacement cascade is fully developed within a time less than approximately one picosecond. Traveling at the speed of sound, a criterion that can actually be exceeded by a strong shock wave, the shock wave would reach the surface from a depth of approximately 100 Å also in a time of approximately 1 psec. This extremely short time scale of less than a few picoseconds is not sufficient for dislocations to either develop or move. Hence even normally highly ductile materials will become brittle at such high strain rates.

2.3.10 Synthesis

This latter observation implies that the ion ablation technique described above should be useful for synthesizing nanoparticles of a wide variety of alloy compositions and phases, i.e., normally brittle materials are not required. To examine this concept [41], we prepared a layered Co/Pt target surface. We deposited a 30 nm thick layer of Pt on NaCl by e-beam deposition followed by a 1.5 nm layer of Co. These materials were selected because we felt that with their large differences in atomic number (27 vs. 78) and crystal structure (hcp vs. fcc), our *in situ* TEM technique could be used to image the CoPt layered particles that should be produced with this target. Furthermore, the selection of Pt, a dense material, for the substrate is consistent with the idea that highly energetic displacement cascades are needed for development of the shock waves.

An HRTEM micrograph of one of the larger Co/Pt particles that was synthesized is displayed in Figure 2.22. The cross-fringed lattice image confirms that the collected particles are indeed three-dimensional and crystalline. Unfortunately, our attempts to separate out the different layers of Pt and Co using TEM techniques have not yet proven successful; we are therefore turning to atomic force microscopy in order to obtain more conclusive evidence for the existence of a layered structure. However, *in situ* experiments did reveal a significant change in the selected area diffraction pattern following a 20-min anneal at 200°C, consistent with the expectation that the particles had not experienced any large thermal excursion during the shock-wave induced ablation process. No change in diffraction pattern was observed during a similar anneal at 100°C.

2.3.11 Summary of Nanoparticle Experiments

TEM measurements have shown that single ion impacts result in craters and nanoparticle ejection. The larger clusters (n ≥ 500) that are sputtered from the surface by high-energy ion impacts have an inverse power-law size-distribution with an exponent of –2 that is independent of the total sputtering

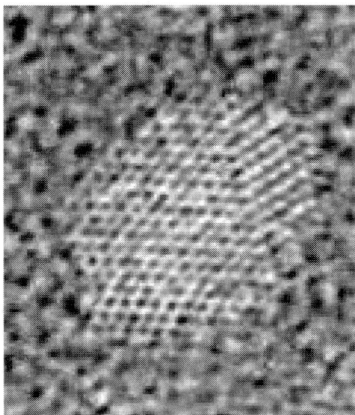

Figure 2.22 An HRTEM micrograph of a Co/Pt particle ejected by 400 keV Xe from a bilayer consisting of a 30 nm thick layer of Pt and a 1.5 nm layer of Co. The cross-fringed lattice image confirms that the collected particles are indeed three-dimensional and crystalline. (From L.E. Rehn, R.C. Birtcher, S.E. Donnelly, P.M. Baldo and L. Funk, *Mater. Res. Soc. Symp. Proc.* 704 (2002) 181. With permission.)

yield and thus, the ion mass and its energy. This inverse-square dependence is consistent with the idea that the larger clusters are formed when shock waves, generated by subsurface displacement cascades, impact and ablate the surface, as predicted in the model of Bitensky and Parilis [37]. An important corollary from these results is that the nanoparticles that are generated consist of simple fragments of the original surface, i.e., those that have not undergone any large thermal excursions. The ion ablation technique discussed here should therefore be useful for synthesizing nanoparticles of a wide variety of alloy compositions and phases. The process generates a broad distribution of nanoparticle sizes, up to approximately 8 nm in dimension. The efficiency for the process can quickly be estimated from TRIM [11] calculations.

2.4. MD MOLECULAR DYNAMICS
SIMULATIONS OF CRATER PRODUCTION

2.4.1 Monte Carlo Simulations versus Molecular Dynamics

As evident from earlier discussion in this chapter, the Monte Carlo computer code TRIM [13] is widely used to study ion irradiation events. In using TRIM, it is very important to understand and keep in mind its range of applicability. As an introduction to the reasons for also using molecular dynamics simulations (MD), we will briefly remind the reader of these limitations.

TRIM simulates binary collisions between atoms in a random medium (where the only input parameters for the material are its mass, density and the repulsive potential between atoms). This approximation works reasonably well in dealing with the collisional phase of a collision cascade, when all atoms have kinetic energies clearly exceeding the threshold displacement energy (about 35 eV in Au). When the atoms have energies below this, however, the collisions become inherently many-body in character and TRIM is no longer valid. However, because atoms with kinetic energies of some 10 eV can no longer move far from their initial site on picosecond time scales, TRIM can be used effectively to estimate the spatial extent and energy of a collision cascade (as done previously in this chapter). This argument also, of course, assumes that channeling is not important; a reasonably safe assumption unless the initial beam direction is directly in a channeling direction.

As discussed previously, the production of craters is a thermal spike process. Thermal spikes develop on picosecond time scales, long after all the energetic recoils have thermalized [42], and hence TRIM cannot predict the actual evolution of a spikes. To model the actual thermal spike, we thus use molecular dynamics simulations. Contrary to TRIM and other binary collision models, they treat all interactions that the atoms experience and can, with a good interatomic potential, in principle treat properly all thermodynamic effects in a material. In particular, for common simple metals it is known that MD can describe well both the liquid and solid phase, melting and high-pressure effects. By adding a realistic repulsive potential and electronic stopping model (which can be exactly the same as in TRIM), MD can thus treat both the energetic collision phase, the thermal spike phase and cooling phase of collision cascades in metals.

Although MD in principle can thus do everything TRIM can and more, it is very much slower than TRIM, and hence cannot completely replace it. But with present-day parallel computers and efficient algorithms, it is perfectly possible to simulate the full development of a ~ 200 keV bombardment event in dense metals. Therefore MD can be used to study cratering under exactly the same bombarding conditions as in current experiments.

2.4.2   Channeling Effects

As indicated above, if a significant number of ions are channeled during a given irradiation, the use of TRIM is questionable. Channeling can also complicate the comparison of MD simulations with experiments, since even though MD can deal with the phenomenon, a one-to-one comparison with experiments can only be made when the implantation angles are either exactly known or are known to be in a "good" non-channeling or so-called random direction.

For the particular case of the (001) channel in fcc metals, and energies of the order of 100 keV, the channel width is surprisingly large, requiring careful selection of both the tilt (θ) and twist (φ) angles to obtain a good non-channeling direction. In Table 2.2 the mean ion ranges and straggling values for 50 keV Xe bombardment of (001) Au are given. The values have been calculated using the MDRANGE code [43] for a temperature of 0 K but using the zero-point atom vibration amplitude as given by the MARLOWE code. As is evident from Table 2.2, for tilt (θ) angles below 15°, channeling has a major effect on the ion ranges. Note also that even the choice of the twist (φ) angle can affect the range strongly even when the tilt angle is fixed. Since cratering depends on the cascade

TABLE 2.2 Mean ranges (R) and straggling values (σ) for 50 keV Xe irradiation of the Au (001) surface. The values are calculated using the MDRANGE code, which is an ion range calculation code that accounts for the crystal structure [39]. The values are calculated at 0 K but using the zero-point atom vibration amplitude as given by the MARLOWE code.

θ	φ	E	R	σ
0	0	50	1174 ± 27	736
5	20–30	50	509 ± 12	431
10	20–30	50	114 ± 2	118
15	20–30	50	81 ± 3	91
20	20–30	50	74 ± 2	65
25	20–30	50	71 ± 3	79
30	20–30	50	79 ± 3	85
25	0–10	50	169 ± 7	227
25	10–20	50	77 ± 3	87
25	20–30	50	71 ± 3	79
25	30–40	50	75 ± 3	75

density, these effects can affect cratering probabilities and crater sizes as well.

2.4.3 MD Simulation Method

In molecular dynamics computer simulations the equations of motion of an N-body system are solved numerically over a short time step [44]. By iterating this procedure, one can obtain an arbitrarily accurate description of the time evolution of the system of atoms within the interaction model used. For most atomic systems, the only significant source of uncertainty in the simulations is the interatomic potential — all other uncertainties in the simulations are numerical in character and can be made as small as desired.

For the most commonly studied metals, such as the noble ones, the embedded-atom method (EAM) has been found to give a quite good description of a wide range of materials properties, including those relevant in thermal spikes and for defects [45]. The EAM potential and ordinary MD does not correctly describe high-energy collisions between atoms, electronic stopping or electron-phonon coupling, but during the last 10 years much experience has been gained with methods to augment EAM and MD to treat these correctly [46,47].

In surface simulations, we create a slab of atoms (typically but not necessarily cubic in shape) in the correct crystal structure. We use periodic boundaries in the x and y directions, and a fixed bottom layer in the z direction. Thus the top layer in the z direction becomes a free surface facing a perfect vacuum.

In simulating ion irradiation, we place an incident ion in a random position a few angstroms above the surface, and give it a kinetic energy of 0–200 keV toward the sample. The incident angle is chosen in an off-channeling direction close to the surface normal. Most of our simulations have been carried out for a (001) surface, but we have also simulated a few irradiation events for (111) surfaces, and found behavior very similar to that for the (001) surfaces. In most cases, we prefer to run our simulations at 0 K to be certain that observed effects are due to the collision cascade rather than to ambient temperature thermal annealing occurring simultaneously with the cascade cooling. However, we have also simulated a few cases at 300 K, and found very similar behavior and crater sizes to those at 0 K.

During the cascade runs, the atom motion in the middle of the cell is allowed to evolve freely under direct solution of the Newtonian equations of motion (the NVE thermodynamic ensemble), apart from the electronic stopping that is applied as a frictional force affecting all atoms with a kinetic energy higher than 5 eV. At the periodic x and y boundaries and above the fixed bottom z region, we cool down the system to the ambient temperature in a few atomic layers. The simulation cells had at least 16 atoms per eV of incident ion energy, which was enough to prevent cell heating beyond the melting point or pressure wave reflection from the borders strong enough to affect the cascade outcome. Moreover, the temperature scaling at the boundaries absorbed most of the pressure wave produced in cascades.

We always carry out several simulations for each ion energy-projectile-target combination to obtain representative statistics. Due to the strongly coupled nature of the atomic systems, the cascades are highly chaotic, and varying the initial position of the impactor even slightly can cause the resulting cascades to behave very differently.

2.4.4   Formation of Ordinary Craters

We first qualitatively discuss how MD provides important insights into the formation of the most common, approximately spherical form of crater. This provides a basis for understanding the formation mechanisms of the more exotic features sometimes also observed in experiments, which will be discussed in Section 2.4.5.

2.4.4.1 Surface Damage Mechanisms

Based on the MD simulations of cratering that we have carried out to date in Co, Cu, Ag, Au, Mo, W and a few of their alloys, we divide the surface damage mechanisms in dense metals as follows (this division is originally due to Averback et al. [22]). The mechanisms are illustrated schematically in Figure 2.23. The first effect (Figure 2.23a), sputtering of single recoil atoms by ballistic collisions, is well understood from binary collision theory and simulations. For light materials and ions penetrating deep into the sample, ballistic sputtering often is the only surface effect of a cascade. The other three surface effects all require the formation of a true thermal spike, i.e., a region with liquid-like character where the liquid thermalizes before it recrystallizes. The second effect, plastic flow of hot liquid onto the surface, can result when a cascade is centered inside the sample, but is bounded by the surface so that liquid atoms can flow onto the surface (Figure 2.23b) [21]. The third effect, "micro-explosions," occurs when the liquid zone is so close to a surface that the pressure wave from the cascade essentially ruptures the surface (Figure 2.23c). In this case, pockets of hot liquid can explode out from the surface as a direct result of the collision cascade [48,49].

In the fourth kind of surface damage effect, coherent displacement (Figure 2.23d) [50], several atom layers are displaced toward a surface in a coherent manner due to slip of atom planes induced by the high pressure in the liquid-like

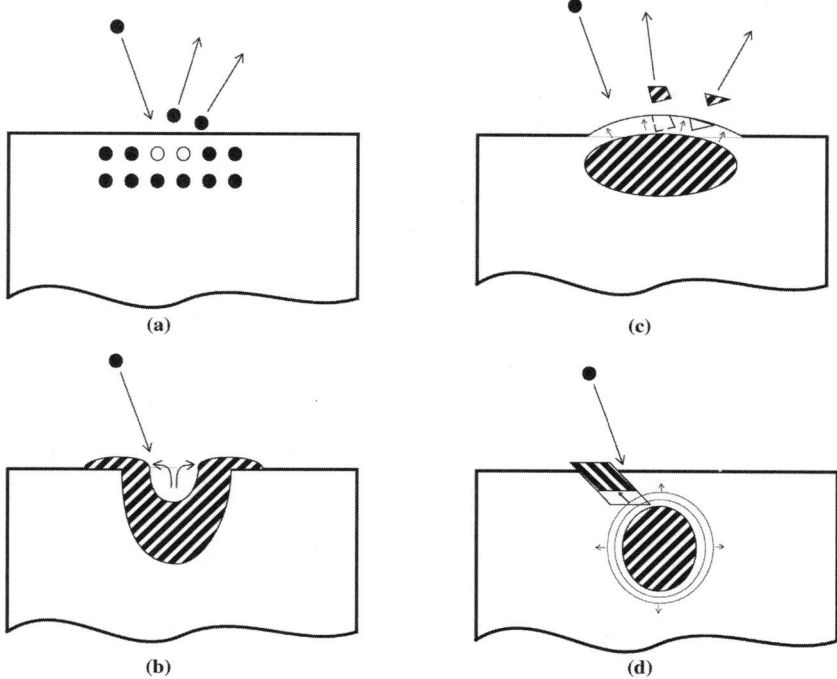

Figure 2.23 Schematic illustration of surface damage mechanisms (see text).

heat spike. This effect can occur when the heat spike is completely below the surface, deep enough not to lead to a micro-explosion, but still close enough to allow for the slip of the atom planes.

2.4.4.2 Basic Crater Formation Mechanism

Craters are produced by effects 2 and 3 described above, i.e., liquid flow and micro-explosions. Although sometimes a cascade can clearly involve only one of these, in practice most cratering events simultaneously involve both, and it is not easy, or even necessary, to make a clear distinction between the two. Coherent displacement does not produce craters, but we often do observe that coherent displacements have

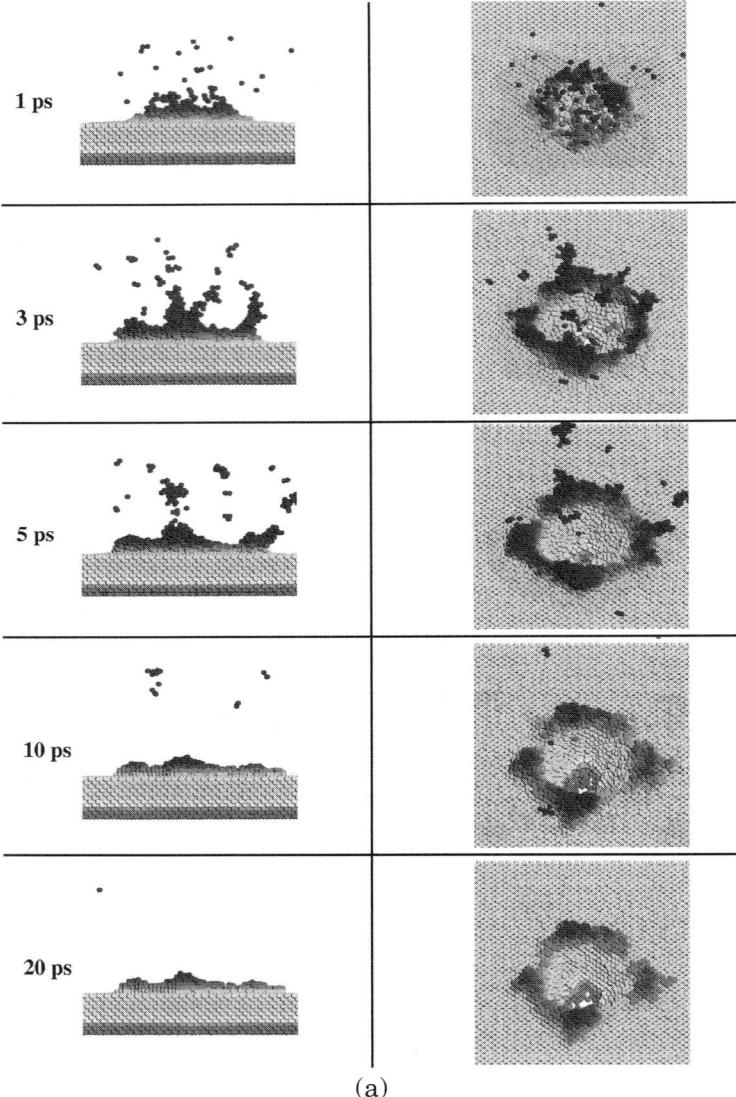

(a)

Figure 2.24 Two series of snapshots of crater formation. In both figures each sphere illustrates an atom position. (a) Crater formation by the impact of a 20 keV 13 atom Pt cluster impacting on Pt. The colors give the height of the atoms, with lighter atoms being the lowest below the surface, and darker atoms the highest above.

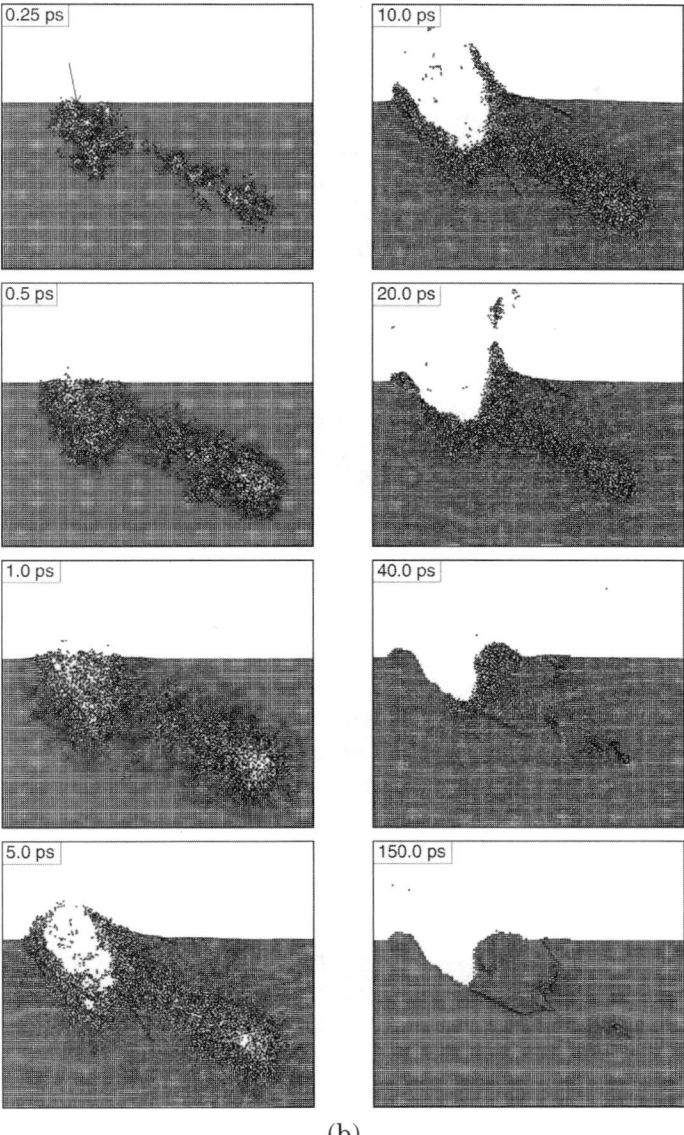

(b)

Figure 2.24 Continued (b) Crater formation by a 100 keV Xe ion impacting on Au, shown as a cross-sectional slice of 8 atom layers in the (−2 1 0) plane cut through the central part of the simulation cell. The arrow indicates the original motion direction of the incoming projectile, projected on the (−2 1 0) plane.

occurred at the outskirts of the cascade, somewhat widening the crater rim.

The formation of craters is easiest to understand from snapshots or animations of a cratering MD simulation. Figure 2.24 offers two different series of snapshots of crater formation. Part (a) shows a side view (left) and tilted top view (right) of cratering induced by a 20 keV 13 atom Pt cluster impacting on Pt. Part (b) illustrates crater formation by a 100 keV Xe ion hitting Au, shown as a cross-sectional slice of eight atom layers in the (−2 1 0) plane in the central part of the simulation cell. On the attached CD-ROM, the animation xeau100.avi is a depiction of the same event, in the same cross section. Also, the animations slice.avi and liq.avi illustrate a cratering event produced by 50 keV Xe on Au from a different viewpoint. The animation slice.avi shows a tilted view from above. The colors are the kinetic energy scale of the atoms. The animation clearly shows that during the ballistic phase of the cascade, not much happens at the surface, but on longer time scales, after 1 psec, thousands of atoms flow to the surface. Finger-like protrusions also form, but they collapse back to the surface toward the end of the event.

The animation liq.avi, which is of exactly the same event as slice.avi (and with the same viewing angle), is highly instructive for understanding the reason for the liquid flow. This animation shows only those atoms that have kinetic energies clearly higher than normal thermal energies. In this animation, one can clearly see the initial fractal-like shapes produced by the ballistic phase of the cascades. After the ballistic phase, a large liquid-like zone forms first inside the material. This zone then expands toward the surface simply because a hot Au liquid has much lower density than a cold Au crystal, leading to the curious features seen in the first animation slice.avi. In most cases, this leads to the production of a roughly circular crater, such as the one illustrated in Figure 2.25.

To ensure that these craters indeed correspond to those seen in the TEM experiments, we have carried out TEM image simulation on a few representative craters. This was done using the JEMS simulation package [51], which correctly

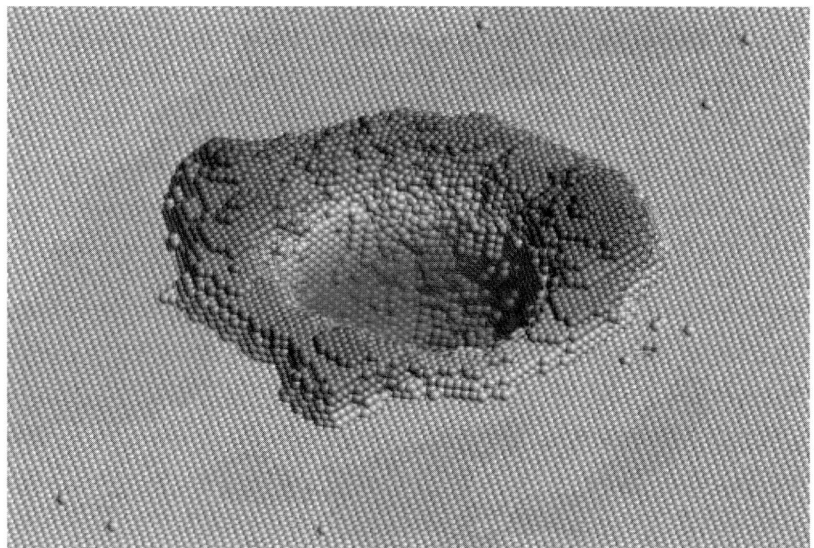

Figure 2.25 Typical roughly spherical shape of a crater. The colors illustrate the height scale, with lighter atoms being the lowest and darker atoms the highest.

takes into account dynamical scattering effects by employing the multislice technique. The image simulations showed that under imaging conditions similar to the ones used in the experiments (underfocused objective lens), the adatom regions above the surface are indeed darker, and the crater wells lighter, than the surrounding region. This is illustrated in Figure 2.26. Moreover, it shows that the size of the crater seen after the image simulation is well in line with that which can be determined visually from the atomic information simply by looking at where the outer edge of the rim is located. Hence it is not necessary to carry out the computationally expensive TEM image simulations on all craters for quantitative comparison with experiments.

We have simulated and analyzed a large number of craters. In Figure 2.27 we show a direct comparison between simulated and experimental crater radii. For the highest crater sizes, both the simulations and experiments show a lev-

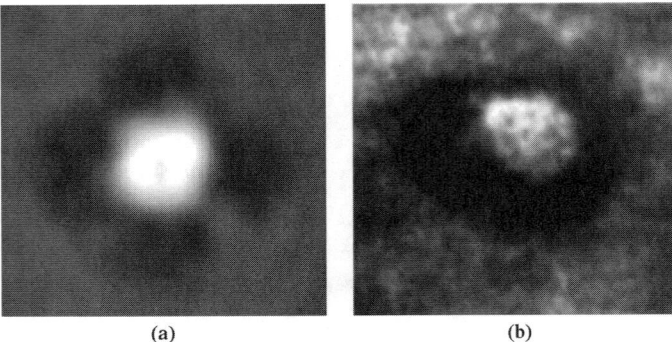

(a)   (b)

Figure 2.26 Left: Simulated TEM image of a constructed on the basis of the outcome of an atomistic simulation. Right: Experimental TEM image.

eling off in the crater size, and even the absolute values agree well. This gives us great confidence that the crater formation mechanism we observe in the simulations is indeed correct.

The reason why the crater size saturates is that at this energy, the cascades start to split into separate subcascades (as seen in Figure 2.24). There is a maximum size that a cascade can attain before it splits into subcascades, which in Au is reached at roughly 100 keV. Hence the average cascade size does not grow any more above this energy.

2.4.5 Formation of Exotic Crater Structures

In addition to the simple spherical crater features, both the simulations and experiments show the formation of many other more exotic structures. We will now discuss the formation of these structures based on the simulations.

We first describe the formation of finger-like features over the craters. As stated above, the crater formation starts by expansion of the very hot liquid-like zone formed in the collision cascade. This expansion has previously been reported on some occasions to lead to the emission of large chunks of matter from the cascade. It also leads to the flow of liquid that forms the rims of ordinary craters. As an intermediate

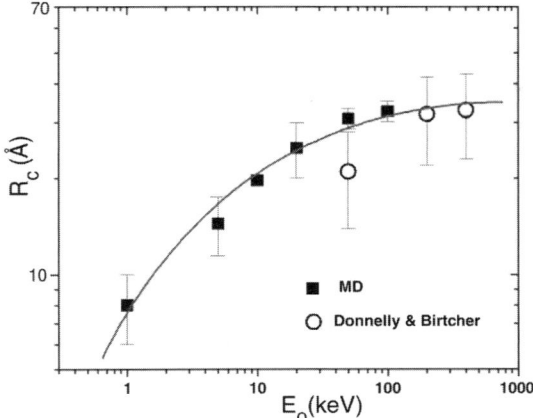

Figure 2.27 Quantitative comparison between simulated and experimental crater radii for Xe ion irradiation of Au. The solid squares are MD simulations by Nordlund et al., the open circles experimental results by Donnelly and Birtcher.

situation to these two extremes, we have frequently observed that large, extended atom regions are formed high (up to some 50 nm) over the crater, and that these are not sputtered due to the cohesive energy between the atoms. After the thermal spike phase of the cascade, the fingers contract to minimize the surface energy, but in some cases a clearly distinguishable protrusion remains stable over relatively long time scales. Simulated and experimental images of such protrusions are illustrated in Figure 2.28.

An interesting variation may result when the finger is long enough and has outward momentum. In this case, it collapses far from the crater and forms a clearly distinguishable adatom ridge ending in a blob of matter. The end result of such a sequence of effects is illustrated in Figure 2.29, which also shows an experimental image of a similar structure.

Finally, we have also observed a slingshot-like effect. If the finger is slightly thinner in the middle, it may break because the surface energy can be reduced in this way. The

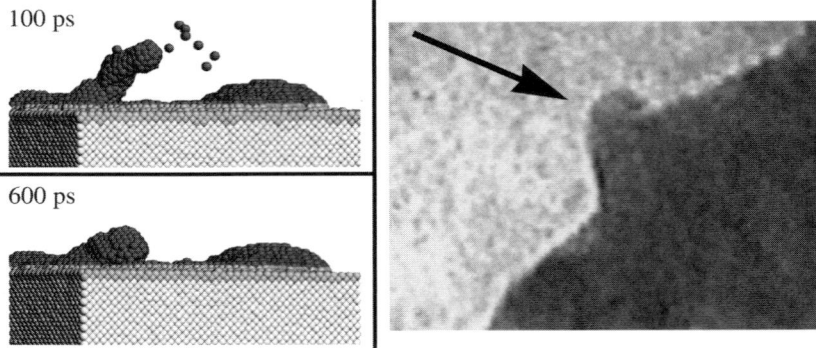

Figure 2.28 Left: Side view of the formation of a protrusion in a cascade. At 100 psec a finger-shaped protrusion has formed in the cascade. By 200 psec the finger had emitted a few sputtered clusters, and contracted to a shape that remained stable for at least 400 psec at room temperature, forming the protrusion shown in the bottom left figure. Right: Experimental image of a similar protrusion observed during heavy ion irradiation of gold. The atoms in the simulation images are colored in a gray scale such that atoms above the surface are darker and below the surface lighter.

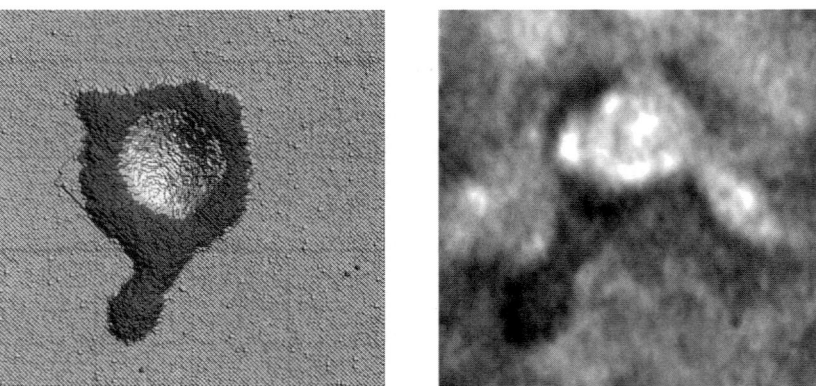

Figure 2.29 An adatom ridge formed during a simulated cratering event. The right part of the figure shows an experimental TEM image of a similar adatom ridge. The gray scale is as in Figure 2.28.

end of the finger then becomes a free atom cluster. If this emission occurs when the cohesion has already started to move the end of the finger downwards, the cluster will not sputter but will be redeposited on the surface far from the initial crater. This sequence of events is illustrated in Figure 2.30.

Possible evidence for such an effect is seen in experiments as isolated adatom islands, one of which is illustrated in Figure 2.30. We have also observed two other possible mechanisms in the simulations by which these islands could form. One is the coherent displacement effect discussed above. The other is liquid flow associated with a recoil that goes deeper in the sample and does not produce a crater. We do not currently have reliable enough statistics to judge the relative probability of each variety.

2.4.6 Analysis Based on MD

Based on the MD and TRIM simulations, we have analyzed the cratering probability and the energy density at which cratering occurs [52]. The main conclusions from this work can be summarized as follows.

On the basis of the experimental work alone, we stated that a single energy density derived from TRIM can sometimes be used to predict when cratering occurs. However, the simulations indicate that this is true only for the large craters formed in truly massive thermal spikes. The simulations show that craters form also at much lower energies, down to as low as 1 keV (see Figure 2.27). At the lower ion energies an energy density of about $0.5 \, n \, U_0$ is needed for cratering to occur, where n is the atomic density and U_0 the cohesive energy. At higher energies, much lower densities, $\sim 0.2 \, n \, U_0$, suffice, in agreement with the experiments.

Comparing the experimental and theoretical cratering probabilities is complicated by the uncertainties associated with channeling. Predicting subcascade formation probabilities is also very complicated. However, in Ref. [52] a combined use of MD and TRIM permitted the parameterization an analytical model, and the prediction of a cratering probability of

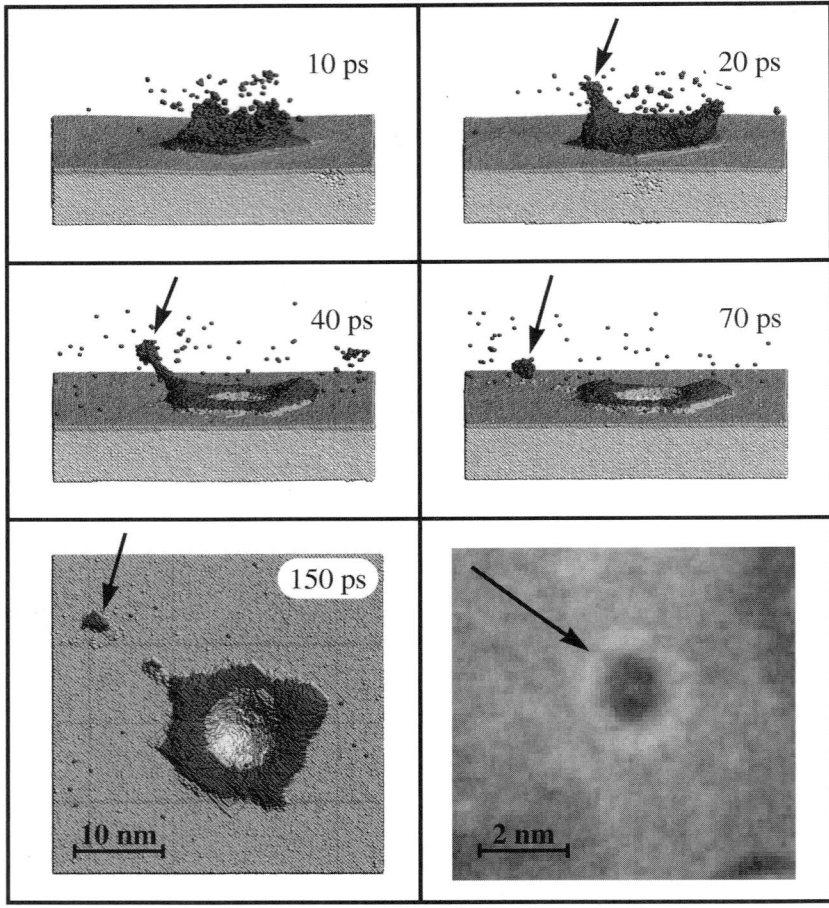

Figure 2.30   An atomic "slingshot" effect, illustrating how a finger formed in a cascade can break up and lead to the ejection of an atom cluster downward, resulting in the production of an isolated adatom island on the surface (bottom left). The bottom right part of the figure shows an experimental TEM image of an isolated adatom island; note that the simulated and experimental islands are about the same size. The gray scale is as in Figure 2.28.

0.07 ± 0.05 for 400 keV Xe irradiation of Au. This agrees well with the experimentally observed probability of 0.03.

In molecular dynamics simulations it is also possible to examine the way in which a single material parameter affects cratering, by creating artificial modifications of an interatomic potential, where only one central property is altered. We did this in two ways. First, we simply modified the cohesive energy U_0 of Au by directly multiplying the interatomic potential with a modifying factor f. We found that in this modification, the crater size scales as $1/U_0^2$ [52]. The same behavior has also been previously reported by Aderjan and Urbassek [53]. We then proceeded with a more advanced modification, where we altered the melting point of the material, T_{melt}, without affecting the cohesive energy [54]. This can be done by changing only the repulsive part of the potential at interatomic separations shorter than the equilibrium one. In this modification we found that the crater size is inversely proportional to the T_{melt}. Taken together, the two studies indicate that the crater size scaling is best given as $1/(U_0 \times T_{melt})$. Since the melting point scales with the cohesive energy in the simple multiplicative modification, this result is perfectly consistent with the previous observation of $1/U_0^2$ scaling. However, the explicit dependence on the melting point found in the latter study gives direct evidence that cratering really is associated with liquid flow. The value of the melting point determines the size of the liquid region and hence the crater size.

2.4.7 Observations of Nanocluster Ejection

In agreement with the experiments, the MD simulations clearly show that nanoclusters can be ejected during the cratering events. This is evident from Figure 2.24, as well as the animations provided. The ejection of large nanoclusters is associated with the formation of the finger-like features discussed above, and hence is also well explained by the liquid flow/micro-explosion picture of crater formation.

An important aspect of the cluster emission, which cannot easily be observed in the experiments, is the role of cluster fragmentation. MD simulations of cluster emission consistently show that many of the clusters fragment quickly after emission. In a recent systematic study of this effect [55], it was found that for 15 keV Xe impacts on Ag, about 90% of the clusters larger than four atoms fragmented. For 20 keV Xe impacts on Au, the same figure was 50%. The majority of the fragmentation occurred within 100 psec, so the experimentally observed sputtering certainly measures the yield after fragmentation. Since some of the fragments may obtain a velocity toward the surface, this can actually reduce the final sputtering yield from the initial "nascent yield" value. However, in the aforementioned test cases, this effect was less than 10% in both cases. Since sputtering yield measurements usually have uncertainties of the same order, the redeposition effect is not likely to be a significant concern in most cases.

The significant role of cluster fragmentation does raise the question of whether it is justified to analyze the experimentally observed sputtering yield exponents (discussed previously in this chapter) using analytical sputtering models that always describe the nascent sputtering yield only. From the simulation of the Xe impacts on Ag and Au mentioned above, it does seem that the nascent and final sputtering yield power-law exponents are roughly the same. This gives some justification to comparing experiments directly with nascent yield exponents. However, understanding the full relation between nascent and final power-law exponents, as well as why the exponents are different in different cluster size regions, clearly will demand further study.

2.5 CONCLUSIONS

This chapter has presented experimental observations and molecular dynamics simulations of the formation of craters and other surface features on metals as a result of individual heavy ion impacts, and we hope that the video sequences on the accompanying CD-ROM have served to illustrate the ion impact processes in a manner not possible simply using static images. The simulations have been carried out using ions in essentially the same energy range as that used in the experimental work, and have thus given a remarkable insight into the processes that give rise to the observed surface structures. Mechanisms that were tentatively guessed at on the basis of the experimental observations — liquid flow and micro-explosions — have been unequivocally identified in the MD work and have provided a relatively complete understanding of the processes. This is not entirely the case for the emission of nanoclusters, however, where there is arguably incomplete agreement between experiment and simulation. However, additional MD work, giving rise to improved statistics on nanocluster emission, should enable a better comparison between experiment and simulation to be made. It will be interesting to see whether the power-law relationship observed experimentally is also found in the MD simulations and whether the interpretation of this as a shock-wave phenomenon is borne out by simulation.

As far as the engineering of thin films with ion beams is concerned, we believe that the combination of high-resolution microscopy techniques (particularly where *in situ* ion irradiation is possible) with molecular dynamics simulations has given important insights into the processes occurring during heavy ion irradiation of dense metals. The same approach could yield similarly important results for other ion/thin film combinations.

ACKNOWLEDGMENTS

We thank B. Kestel for specimen preparation, R.E. Cook for assistance with analytical measurements, E. Ryan, L. Funk and S. Ockers for assistance with the *in situ* TEM experiments, and R. Csencsits and C.A. Faunce for assistance with HREM observations. This work has been supported by the U.S. Department of Energy, Office of Science, under Contract W-31-109-Eng-38. Finally, one of the authors (S.E.D.) acknowledges the provision of financial support from the Materials Science Division at Argonne National Laboratory, which has enabled regular extended visits to that laboratory.

REFERENCES

1. J.A. Brinkman, *J. Appl. Phys.* 25 (1954) 951.

2. J.A. Brinkman, *Am. J. Phys.* 24 (1956) 246.

3. D.A. Thompson, *Radiat. Eff.* 56 (1981) 105.

4. T. Shinada, H. Koyama, C. Hinoshita, K. Imamura and I. Ohdomari, *Jpn. J. Appl. Phys.* 41 (2002) L287.

5. K.L. Merkle and W. Jäger, *Phil. Mag.*, A 44 (1981) 741.

6. D. Pramanik and D.N. Siedman, *J. Appl. Phys.* 54 (11) (1983) 6352.

7. C.A. English and M.L. Jenkins, *Mater. Sci. Forum* 15–18 (1987) 1003.

8. S.E. Donnelly, R.C. Birtcher, C. Templier and V.M. Vishnyakov, *Phys. Rev.* B 52 (1995) 3970.

9. R.C. Birtcher and S.E. Donnelly, *Phys. Rev. Lett.* 77 (1996) 4374.

10. S.E. Donnelly and R.C. Birtcher, *Phys. Rev.* B 56 (21) (1997) 13599.

11. S.E. Donnelly and R.C. Birtcher, *Phil. Mag.* A 79 (1) (1999) 133.

12. C.W. Allen, L.L. Funk, E.A. Ryan and S.T. Ockers, *Nucl. Instrum. Methods* B 40/41 (1989) 553.

13. J.F. Ziegler, J.P. Biersack and U. Littmark, *The Stopping and Ranges of Ions in Solids*, Pergamon Press, New York, 1985. See http://www.srim.org

14. R.S. Averback and M. Ghaly, *J. Appl. Phys.* 76 (1994) 3908.

15. G.W. Greenwood, A.J.E. Foreman and D.E. Rimmer, *J. Nucl. Mater.* 4 (1959) 305.

16. R.C. Birtcher and S.E. Donnelly, *Mater. Chem. Phys.*, 54 (1998) 111.

17. R.S. Walker and D.A. Thompson, *Radiat. Eff.* 37 (1978) 113.

18. M.W. Bench, I.M. Robertson and M.A. Kirk, *Nucl. Instrum. Methods* B 59–60 (1991) 372.

19. M.W. Bench, 1992, Ph.D. Thesis, University of Illinois.

20. R. Kelly, *Radiat. Eff.* 32 (1977) 91.

21. M. Ghaly and R.S. Averback, *Phys. Rev. Lett.* 72 (3) (1994) 364–367.

22. K. Nordlund, J. Keinonen, M. Ghaly and R.S. Averback, *Nucl. Instrum. Methods* B 148 (1–4) (1999) 74.

23. H. Brune, *Surf. Sci. Rep.* 31 (3–4) (1998) 125–229.

24. H.L. Bay, H.H. Andersen, W.O. Hofer and O. Nielsen, *Nucl. Instrum. Methods* 132 (1976) 301.

25. R.C. Birtcher and S.E. Donnelly, *Mater. Res. Soc. Proc.* 585 (2000) 117.

26. H.H. Andersen and H.L. Bay, *J. Appl. Phys.* 45 (1974) 953.

27. H.H. Andersen, A. Brunelle, S. Della-Negra, J. Depauw, D. Jacquet and Y. Le Beyec, *Phys. Rev. Lett.* 80 (1998) 5433.

28. H.H. Andersen, H. Knudsen and P. Moller Petersen, *J. Appl. Phys.* 49 (1978) 5638.

29. R.C. Birtcher, S.E. Donnelly and S. Schlutig, *Phys. Rev. Lett.* 85 (2000) 4968.

30. S. Iijima and T. Ichihashi, *Phys. Rev. Lett.* 56 (1986) 616.

31. R.C. Birtcher, S.E. Donnelly and S. Schlutig, *Nucl. Instrum. Methods* B, 215 (2004) 69.

32. R.E. Honig, *J. Appl. Phys.* 29 (1958) 549.

33. W.O. Hofer, in R. Behrisch and K. Wittmaack (Eds.), *Sputtering by Particle Bombardment III*, Springer, Berlin, 1991, p. 15.

34. C. Staudt, R. Heinrich and A. Wucher, *Nucl. Instrum. Methods* B 164–165 (2000) 677.

35. S.R. Coon, W.F. Calaway, J.W. Burnett, M.J. Pellin, D.M. Gruen, D.R. Spiegel and J.M. White, *Surf. Sci.* 259 (1991) 275.

36. A. Wucher, M. Wahl and H. Oechsner, *Nucl. Instrum. Methods* B 82 (1993) 337.

37. I.S. Bitensky and E.S. Parilis, *Nucl. Instrum. Methods* B 21 (1987) 26.

38. H.M. Urbassek, *Nucl. Instrum. Methods* B 31 (1988) 541.

39. P. Sigmund, *Phys. Rev.* 184 (1969) 383.

40. L.E. Rehn, R.C. Birtcher, S.E. Donnelly, P.M. Baldo and L. Funk, *Phys. Rev. Lett.* 87 (2001) 207601-1.

41. L.E. Rehn, R.C. Birtcher, S.E. Donnelly, P.M. Baldo and L. Funk, *Mater. Res. Soc. Symp. Proc.* 704 (2002) 181.

42. T. Diaz de la Rubia, R.S. Averback, R. Benedek and W.E. King. *Phys. Rev. Lett.* 59 (1987)1930. See also erratum: *Phys. Rev. Lett.* 60 (1988) 76.

43. K. Nordlund, *Comput. Mater. Sci.* 3 (1995) 448.

44. M.P. Allen and D.J. Tildesley. *Computer Simulation of Liquids*, Oxford University Press, Oxford, UK, 1989.

45. M.S. Daw, S.M. Foiles and M.I. Baskes, *Mater. Sci. Rep.* 9 (1993) 251.

46. K. Nordlund, M. Ghaly, R.S. Averback, M. Caturla, T. Diaz de la Rubia and J. Tarus, *Phys. Rev.* B 57(13) (1998) 7556.

47. K. Nordlund, L. Wei, Y. Zhong and R.S. Averback. *Phys. Rev.* B 57 (1998) 13965.

48. H.M. Urbassek and K.T. Waldeer, *Phys. Rev. Lett.* 67 (1991) 105.

49. M. Ghaly, K. Nordlund and R.S. Averback, *Phil. Mag.* A 79 (1999) 795.

50. K. Nordlund, J. Keinonen, M. Ghaly and R.S. Averback, *Nature* 398 (6722) (1999) 49.

51. P. Stadelmann, JEMS computer code, 2002. http://cimewww.epfl.ch/people/stadelmann/

52. E. Bringa, K. Nordlund and J. Keinonen. *Phys. Rev.* B 64 (2001) 235426.

53. R. Aderjan and H.M. Urbassek. *Nucl. Instrum. Methods Phys. Res.* B 164–165 (2000) 697.

54. K. Nordlund, K.O.E. Henriksson and J. Keinonen, *Appl. Phys. Lett.* 79(22) (2001) 3624.

55. K.O.E. Henriksson, K. Nordlund and J. Keinonen, *Phys. Rev. B*, Accepted for publication, (2004).

Chapter 3

Ion Beam Effects in Magnetic Thin Films

JOHN E.E. BAGLIN

CONTENTS

ABSTRACT

The physical and technological consequences of ion implantation and ion irradiation in semiconductors, structural metals and polymers have been studied in depth over the past 30 years. However, only in relatively recent times have we started to explore the remarkable interactions and opportunities for new technologies resulting from ion beam treatment of magnetic materials. This chapter reviews the basic processes of ion beam interactions in solids, and proceeds to explore the role of these processes in custom-tailoring the magnetic properties (e.g., moment, coercivity, anisotropy, magnetoresistance) of thin film magnetic materials, such as permalloy or CoPtCr, having special relevance to applications in magnetic information storage and the development of advanced disk and head structures for high-density disk drives. The mechanism of ion beam modification of anisotropy and coercivity in multi-layer Co–Pt or Fe–Pt thin film media for perpendicular recording is explored in depth. High spatial resolution (tens of nanometers) obtained with ion irradiation using stencil masks and/or projection ion beam tools enables robust, cost-effective full-disk patterning of storage media at densities exceeding 100 Gb/in^2. The extension of this technique to patterned self-assembled FePt nanoparticle media is demonstrated. Finally, we display the application of ion beam patterning in fabricating GMR read and write heads characterized by uniquely narrow track width performance.

3.1 INTRODUCTION

The question of how ion beam processing might be used to custom-tailor magnetic properties of thin film materials has only recently been studied extensively, spurred by the possibilities for application of magnetic materials in nanoscale devices. This is especially relevant for the fast-moving technology of high-density information storage, where disk drive densities of 100 Gb/in^2 appear to be reachable, and still higher densities are sought. The issue may also be relevant in the future evolution of chip-based memory and logic devices that depend on magnetic thin-film elements for functionality and non-volatile information storage.

In this Chapter, we shall present some of the recent research on radiation effects in magnetic thin films, and discuss some possible applications in the information storage technologies.

3.2 ION BEAM INTERACTION PROCESSES

The fundamental interactions occurring when a fast ion enters a solid material have been extensively documented and modeled, together with their physical consequences for the properties of the test material. Most of the applications have been related to the technologies of semiconductor devices and chip integration; however, their possible application in magnetic thin films had received little attention until recently.

To clarify our subsequent discussion, let us briefly identify some of those significant concepts.

3.2.1 Ion Energy Deposition

Upon penetrating the solid, the ion initiates a "collision cascade," a succession of encounters with host atoms, involving ionization of those atoms and bond-breaking, or nuclear collisions in which the ion is scattered elastically and the struck atom recoils. If the recoiling atom receives sufficient energy, it escapes from its local electronic bonding and migrates, producing its own "cascade," while the original ion continues its own random path. After losing energy to the host via multiple collisions, the ion and the recoil particles eventually come to rest. They may occupy pre-existing lattice vacancies, create their own vacancies, or they may lodge interstitially. Since the final result is the deposition of all the energy of the primary ion into the host, this energy will mainly appear as lattice phonons, or heat, in the host.

The statistical nature of the energy deposition process means that its modeling is appropriate for Monte Carlo simulations, such as SRIM [1], or more detailed molecular dynamics simulations [2–5].

Surface sputtering loss may result when occasional recoil atoms in the cascade return to the free surface with sufficient energy to escape from the host. Heavy incident ions at low energy, whose cascade recoils are prolific, and near to the surface, are most likely to produce significant sputtering loss.

Ion beam mixing results at an interface between two thin films of different materials. The accumulated recoil displacements of interface atoms across the interface result in an intermixed interface, analogous to the result of a thermal interdiffusion process.

Ion implantation is the addition of the incoming ions to the host material, at a depth near the end of the ion's projected range, R_p. The process may be used to adjust the local elemental abundance in the host material, in the selected area irradiated. Incorporation of such implanted species into a crystalline phase of the host (or its segregation into islands) normally requires a subsequent high-temperature annealing process.

3.3 ION IRRADIATION EFFECTS IN PERMALLOY FILMS

An empirical study of the effects of ion irradiation upon 150 nm thick films of $Ni_{80}Fe_{20}$ was reported by Baglin et al. [6].

TABLE 3.1 Irradiation of [$Ni_{80}Fe_{20}$ (150 nm)/glass]

Run/ Sample No.	Ion species	Dose (×10^{16} ion/cm²)	Energy (MeV)	Collisional ΔE_{coll} (arb. units)	Ionization ΔE_{ioniz} (arb. units)	Concentration of implanted species (at.%)	Comments
1	He+	2.0	2.0	0.004	4.0	–	Ionization ΔE
2	He+	4.0		0.008	8.0	–	
3	C+	1.9	Multiple	0.6	2.0	1.4	Active implants
4	N+	2.1		0.9	2.0	1.4	
5	O+	2.4		1.3	2.0	1.7	
6	Si+	1.0		3.1	2.0	1.3	
7	Ar+	1.3	Multiple	3.8	2.0	0.9	Inert implants
8	Ar+	2.6		7.6	4.0	1.8	
9	Xe+	0.9		23.8	2.0	0.6	
10	Xe+	1.8		47.6	4.0	1.2	
11	Xe+	0.5	1.5	10.0	2.0	–	Collisional ΔE
12	Xe+	1.0		20.0	4.0	–	

Note: Irradiation conditions for each sample, including a comparative metric for the calculated total energy deposited in the NiFe film by means of collisional (ΔE_{coll}) or ionization (ΔE_{ioniz}) processes, are shown. The units of ΔE_{coll} are arbitrary, as are those for ΔE_{ioniz}. Samples 3–10 were irradiated successively at multiple energies to obtain uniform depth distribution of the accumulated ΔE.

The films were deposited by DC-magnetron sputtering in the presence of a magnetic field, to establish in-plane magnetic anisotropy in each film. In an effort to identify separately the effects of ion energy deposition by ionization, ballistic damage, and/or chemical modification, identical samples were exposed, at room temperature, to the beam conditions shown in Table 3.1, which also shows the relevant energy-deposition parameters, as derived from SRIM. Magnetic hysteresis loops (for both hard and easy axes) were then compared for the as-deposited and irradiated samples. A set of these data is displayed in Figure 3.1.

For irradiation by the inert ions He+ and Xe+ (samples 1, 2, 11, 12), no significant change in hysteresis loops was produced by He+ runs (purely electronic energy deposition, ΔEe), while for Xe+ with the same ΔE_e, but a large ΔE_{recoil},

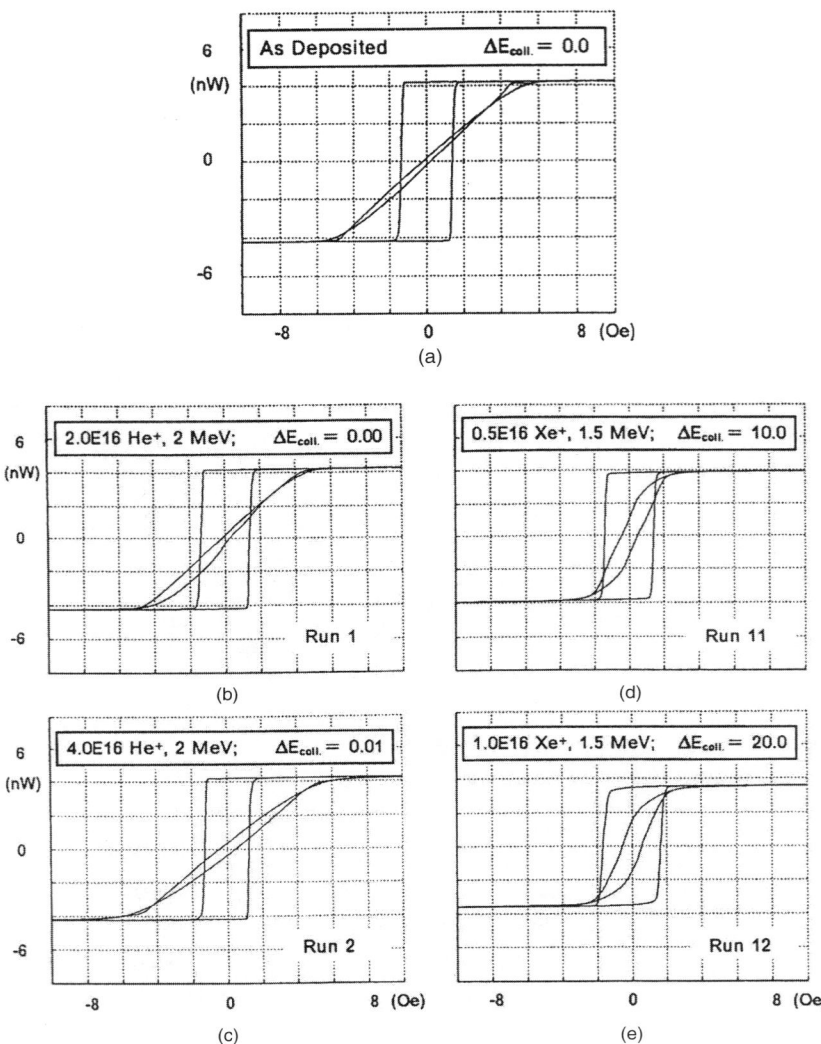

Figure 3.1 Hysteresis loops (easy axis and hard axis) for 150 nm films of NiFe: (a) as deposited; (b–e) after transmission irradiation with He$^+$ or Xe$^+$ ions; (f–j) after homogeneous implantation with Ar$^+$ or Xe$^+$ ions. In each case, the ion species, dose and energy are specified, together with the relative metric ΔE_{coll} describing the total collisional energy delivered within the film by the ion beam. (From J.E.E. Baglin, M.H. Tabacniks, R. Fontana, A.J. Kellock and T.T. Bardin, *Mater. Sci. Forum* 248–249, 87–93, 1997.)

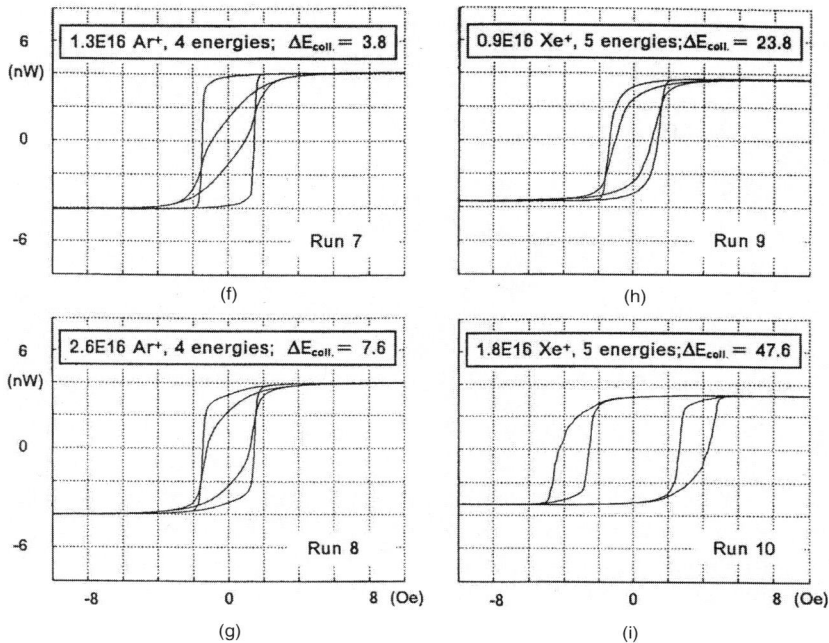

Figure 3.1 (continued)

much of the initial uniaxial anisotropy was lost; coercivity increased somewhat, the squareness of the loop was lost, and a few percent of saturation was found (after allowing for a small sputtering loss of the film thickness). These observations would be consistent with a chemical disordering process in the originally oriented NiFe, as a result of ballistic displacements produced by the Xe^+ ions and their secondary recoil atoms, while leaving the cubic order and (111) orientation of the NiFe lattice unchanged (as confirmed by x-ray diffraction [XRD]). Additionally, some ion-induced segregation [7,8] may occur.

Those samples (nos. 7–10) subjected to high dose implantation (1 or 2 at.%) by Ar^+ or Xe^+ displayed, by contrast, a rapid loss of magnetic anisotropy and, for high-dose Xe^+, a doubling of coercivity; however, XRD revealed no phys-

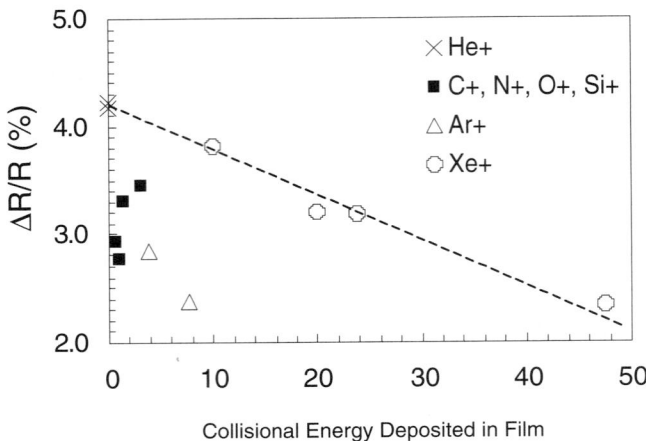

Figure 3.2 Magnetoresistance $\Delta R/R$, as a function of ΔE_{coll} for an ion implanted 150nm NiFe film. (From J.E.E. Baglin, M.H. Tabacniks, R. Fontana, A.J. Kellock and T.T. Bardin, *Mater. Sci. Forum* 248–249, 87–93, 1997.)

ical disordering. Evidently, strain resulting from the implanted argon or xenon contributed to the chemical disordering process.

The magnetoresistance, $\Delta R/R$, of the films proved to be strongly degraded by irradiation, as shown in Figure 3.2. However, the change in $\Delta R/R$ was clearly not determined solely either by the total collisional energy deposited or by the ionization energy loss, or by the atomic percent of the implanted ions. It was strongly dependent on the species of implanted ions, with C, N, O and Si having greater effects than Xe.

This experiment serves to suggest the primary ability of ion irradiation of magnetic alloys to (i) create chemical disorder by recoil displacements, while leaving the basic lattice structure largely unaffected, and (ii) introduce precipitates and/or strain effects when ions are implanted at the level of several atomic percent. We shall find that these themes recur as this review proceeds.

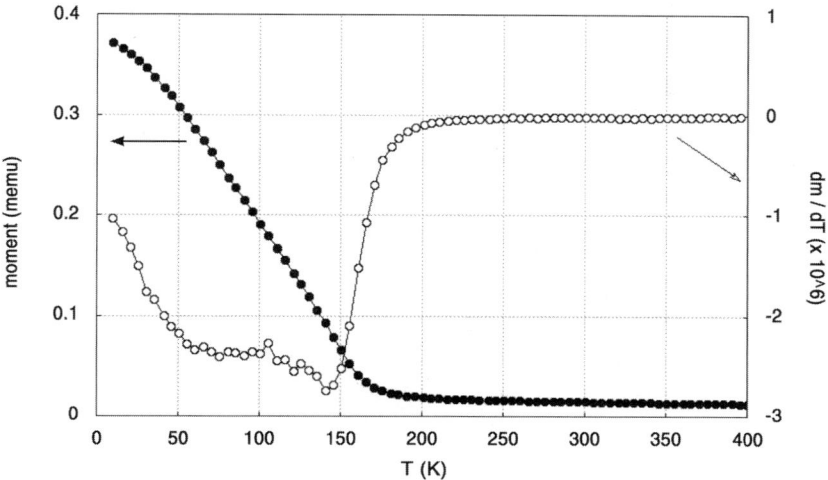

Figure 3.3 Magnetic moment vs. temperature for NiFe (50 nm) films, after implantation with 5×10^{16} Cr$^+$/cm^2 at each of two energies (40 keV and 125 keV). The derivative of the data highlights the magnetic transition at $T_c \sim 170$ K. (From L. Folks, R. Fontana, B. Gurney, J. Childress, S. Maat, J. Katine, J.E.E. Baglin, A. Kellock and P. Saunders, *J. Phys. D: Appl. Phys.* 36, 2601–2604, 2003.)

3.4 ION IMPLANTATION TO MODIFY FILM COMPOSITION

Folks et al. [9] have reported that doping of NiFe alloys with ion implanted Cr$^+$ causes the Curie temperature to be greatly reduced. Implantation of Cr$^+$ leads to the formation of solid solutions of Ni–Fe–Cr, creating films that are paramagnetic at room temperature. Figure 3.3 shows the decline of magnetic moment with temperature for a 50 nm thick NiFe film after the implantation of about 20 at.% Cr (a dose of 5×10^{16} Cr$^+$/cm^2 at 40 keV and at 125 keV). It is found that, at room temperature, the moment and coercivity decrease steadily with the dose of Cr$^+$, reaching zero with the onset of paramagnetism. Ion implantation of Cr$^+$ in selected (patterned) areas of a continuous NiFe film could thus be used to produce locally paramagnetic regions — a technique proposed, for example, in trimming the edge profiles of magnetic write heads, as in

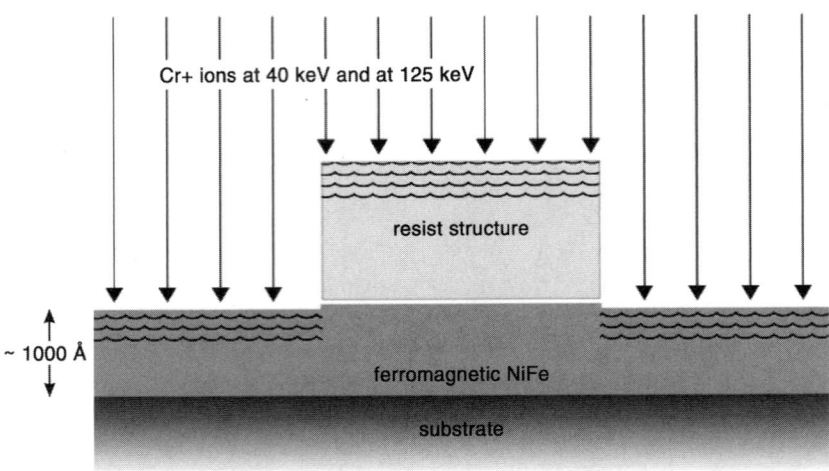

Figure 3.4 Schematic of an ion beam patterning process for a NiFe continuous film, generating locally paramagnetic regions. (From L. Folks, R. Fontana, B. Gurney, J. Childress, S. Maat, J. Katine, J.E.E. Baglin, A. Kellock and P. Saunders, *J. Phys. D: Appl. Phys.* 36, 2601–2604, 2003.)

the notching process used to constrain the stray field from the write gap (Figure 3.4).

As a practical matter, it should be pointed out that implantation at very high doses ($\geq 10^{17}$ ions/cm^2), of boron or chromium or other metals, such as one requires to alter alloy phase composition by several atomic percent, may be informative for research, but it is expensive and time-consuming, and invites sputtering losses in the sample. As a manufacturing strategy it would be very costly, unless confined to patterned applications (small areas) and thin layers. We caution that using ion implantation to make local changes of several percent in alloy compositions is likely to be uneconomic for manufacturing. This stands in contrast to the interface mixing techniques discussed in the next section, where very low doses (e.g., $\leq 10^{14}$ ions/cm^2) may serve a sound and cost-effective technological function for patterned systems.

3.5 MODIFICATION OF ANISOTROPY AND COERCIVITY

3.5.1 Co/Pt Multilayers

Weller et al. [10] have demonstrated the magnetic properties of multi-layered structures deposited as series of ~ 10 layers of Co(3 Å) and Pt(10 Å) at 210–250°C on a SiN_x/Si substrate precoated with 200 Å of Pt. The film displays substantial uniaxial magnetic anisotropy ($4–8 \times 10^7$ erg/cm³, as determined by torque magnetometry), leading to a coercivity H_c of ~ 10 kOe. This might therefore be a suitable medium for perpendicular recording, in which an individual bit is represented by magnetization up or down in a localized region.

The total uniaxial magnetic anisotropy, K_u, for the multilayer may be written:

$$K_u = K_v - 2\pi M_s^2 + 2 K_s/t_{Co} \qquad (3.1)$$

where K_v is a total volume anisotropy contribution (estimated at $1.2–2.6 \times 10^6$ erg/cm³), $2\pi M_s^2$ is the average demagnetization energy per unit Co volume (~ 1×10^7 erg/cm³), t_{Co} is the Co layer thickness, and K_s is the interface contribution arising from the orbital overlap of Co-3d and Pt-5d wave functions (estimated at 0.4–0.8 erg/cm²). K_s/t_{Co} is thus the controlling term that determines K_u.

Ion irradiation of this multilayer system produces profound changes in its magnetic properties, as illustrated in Figure 3.5, which shows typical polar Kerr hysteresis loops before and after the sample was exposed to irradiation by 700 keV N⁺ at various doses. The process leads to a systematic reduction of perpendicular coercivity H_c, while the magnetic moment remains substantially unchanged. The plot of Figure 3.6 displays a steady drop in H_c/H_c^0 with ion dose, while the perpendicular remanence (M_R/M_s) (due to squareness of the loops) remains intact until H_c/H_c^0 is heavily reduced.

Ballistic intermixing of Co and Pt atoms at the layer interfaces, caused by ion irradiation, is expected to be the principal mechanism for this decline of H_c/H_c^0, which is found to track with a decline of K_s. Ultimately, at high doses, this leads to a transition of the easy axis from out-of-plane to the

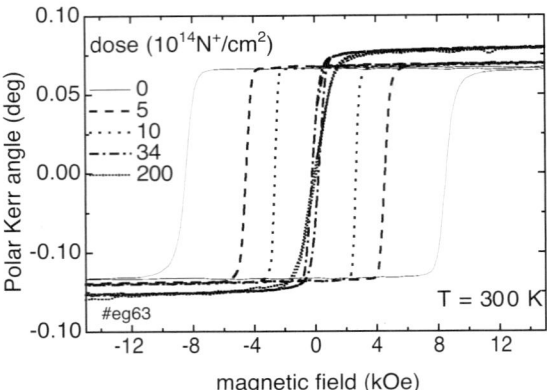

Figure 3.5   Polar Kerr hysteresis loops for a $[\mathrm{Co/Pt}]_n$ multilayer film, following its exposure to 700 keV $\mathrm{N^+}$ irradiation at various doses. (From D. Weller, J.E.E. Baglin, A.J. Kellock, K.A. Hannibal, M.F. Toney, G. Kusinski, S. Lang, L. Folks, M.E. Best and B.D. Terris, *J. Appl. Phys.* 87, 5768, 2000.)

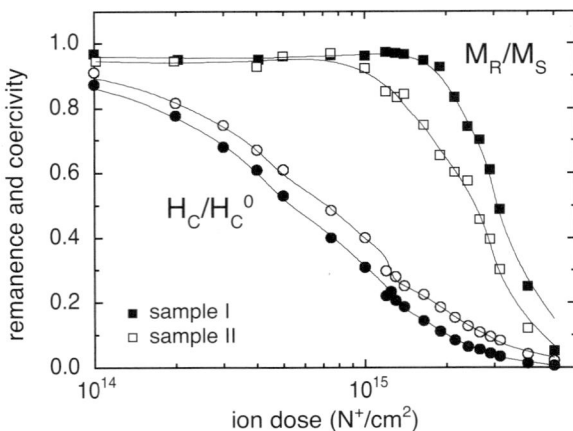

Figure 3.6   Perpendicular remanence (M_R/M_s) and normalized coercivity (H_c/H_c^0), as functions of dose of 700 keV $\mathrm{N^+}$ ions. The two sets of data refer to samples grown at 230°C (sample I) and 210°C (sample II), whose initial anisotropy constants differ by a factor of two. (From D. Weller, J.E.E. Baglin, A.J. Kellock, K.A. Hannibal, M.F. Toney, G. Kusinski, S. Lang, L. Folks, M.E. Best and B.D. Terris, *J. Appl. Phys.* 87, 5768, 2000.)

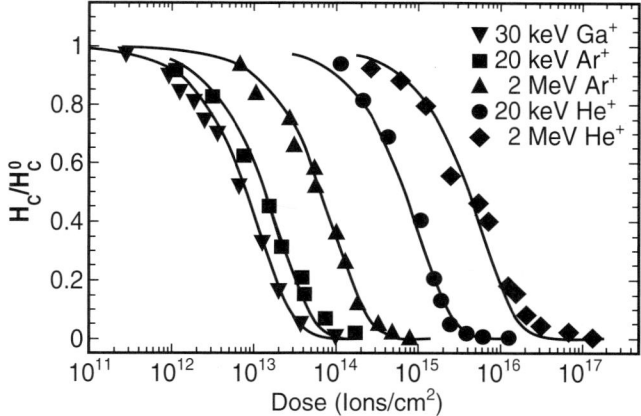

Figure 3.7 Dependence of perpendicular coercivity on ion beam species and dose for [Co/Pt]$_n$ multilayer samples. Coercivities are normalized to the as-deposited value. (From C.T. Rettner, S. Anders, J.E.E. Baglin, T. Thomson and B.D. Terris, *Appl. Phys. Lett.* 80, 279, 2002.)

in-plane orientation. Experimentally, XRD data have also supported the correlation of interface disordering with the observed magnetic changes.

A comprehensive experimental study of the dependence of H_c/H_c^0 for Co/Pt multilayers on the dose of bombarding ions of various energies and species has been made by Rettner et al. [11] as shown in Figure 3.7. The results, for 30 keV Ga$^+$, 20 keV Ar$^+$, 2 MeV Ar$^+$, 20 keV He$^+$ and 2 MeV He$^+$, span three orders of magnitude in the dose required to accomplish the transition. The curves all display the form of a simple exponential function of ion dose.

3.5.2 Chemically Ordered FePt

The alloy FePt, in its chemically ordered fct (L1$_0$) phase, can be visualized as a naturally ordered system of alternating atomic layers of Fe and Pt normal to the *c*-axis. Thus, a suitably oriented film of polycrystalline fct FePt displays magnetic anisotropy and radiation dose sensitivity similar to

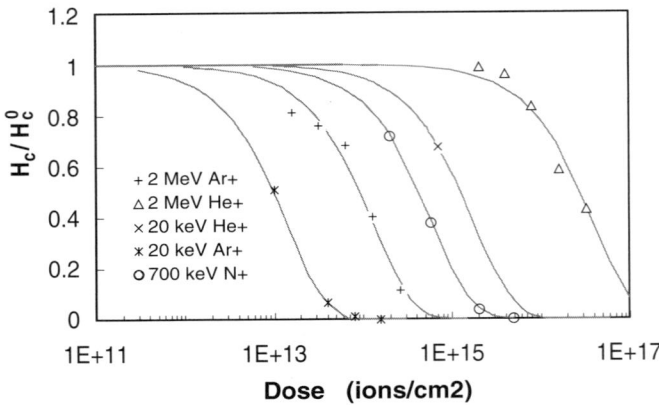

Figure 3.8 Dependence of perpendicular coercivity on ion beam species and dose, for chemically ordered FePt films. (From J.E.E. Baglin, C.T. Rettner, B.D. Terris, D.K. Weller, J.-U. Thiele, A.J. Kellock, S. Anders and T. Thomson, *Proc. SPIE 4468*, 1–7, 2001.)

those of Co/Pt multilayers. (A FePt fcc phase that forms at lower temperatures has almost identical lattice parameters, but lacks "chemical order," i.e., the lattice sites are interchangeably populated with either Pt or Fe atoms. This phase is not magnetic, and it could be regarded as the ultimate result of chemically disordering the magnetic "layered" fct phase.) For the FePt system, the dependence of H_c/H_c^0 (from polar Kerr measurements) on the dose of ions of various energy and species was reported by Baglin et al. [12], and the results are shown in Figure 3.8. As in the case of Co/Pt multilayers, the data in each case are consistent with a simple exponential change with dose.

3.5.3 Ballistic Displacement Model

Baglin et al. [12] have proposed a physical model for ion beam disordering processes responsible for the decline of interface anisotropy K_s (and thus of H_c/H_c^0) in multilayered systems, of which $[Co/Pt]_n$ and FePt are prime examples.

Only ballistic displacements in a system of alternating layers are considered. Each layer j contains a concentration c_j of atoms of the magnetic species. Then, according to the model of MacLaren and Victora [13,14], based on the Néel nearest-neighbor coupling model, the magnetic interface anisotropy K_s for n layers may be expressed as

$$K_S / K_S^0 = 0.5 \sum_{j=1}^{n} (c_j - c_{j+1})^2 \qquad (3.2)$$

where K_s^0 is the anisotropy for perfectly ordered layers.

3.5.3.1 Ion-Induced Disorder

Ion irradiation produces displacements mostly by recoil cascade collisions (as demonstrated by the Monte Carlo simulation program SRIM [1]), and the directions of such displacements are effectively random. We make the assumption that thermodynamic driving forces have negligible effect on the location of displaced atoms, i.e., ion-assisted (or defect-assisted) thermal diffusion or alloy phase formation are negligible.

Under these conditions, we may consider that the ensemble of all randomly displaced atoms will always be completely disordered among themselves. Thus, although an individual Co (or Fe) atom displaced to a new site may make a new pair-coupling contribution to the system anisotropy, the ensemble of randomly displaced Co (or Fe) atoms will make no contribution to anisotropy of the sample. The net anisotropy of the sample is thus attributable only to the population of the remaining *undisturbed* Co (or Fe) atoms. Thus, each ballistic displacement of a Co or Fe atom in layer j may be considered as a reduction of c_j, defining c_j as the surviving concentration of undisturbed magnetic atoms in layer j.

From SRIM simulation, we derive μ, the number of Co or Fe atoms displaced per angstrom of depth, averaged throughout the entire sample, due to each incoming ion. Within any such 1 Å layer in the sample, there are initially N_{mag} atoms/cm^2 of the "magnetic" species (Fe or Co). If we irradiate a sample with a dose of D ions/cm^2, we expect μD displacements of magnetic atoms per cm^2 per angstrom, regardless of whether those atoms are "ordered" or not. We now define the quantity n_{mag} as the number of magnetic atoms/cm^2 per angstrom that are *undisturbed* from their original ordered locations. The number of previously undisturbed magnetic atoms/cm^2 per angstrom displaced during this exposure will then be $(n_{mag}/N_{mag}).\mu D$. Hence, we can write

$$\partial(n_{mag})/\partial D = -\mu.(n_{mag}/N_{mag}) \qquad (3.3)$$

Thus,

$$n_{mag} = n_{mag}{}^0 \cdot \exp(-\mu D/N_{mag}) \qquad (3.4)$$

For each of the structured layers j of Equation (3.1), the concentration c_j scales with n_{mag}. Hence, we conclude that each layer j is depleted in its concentration of undisturbed magnetic species exactly as described by Equation (3.3). Therefore,

$$c_j = c_j{}^0 \cdot \exp(-\mu D/N_{mag}) \qquad (3.5)$$

where $c_j{}^0$ is the initial undisturbed concentration in layer j. Substituting in Equation (3.1), we obtain

$$K_s = K_s{}^0 \cdot \exp(-2\mu D/N_{mag}) \qquad (3.6)$$

It should be pointed out that, assuming the concepts of Ref. [16], the simple relation of Equation (3.6) has general applicability in any structures in which some degree of planar order is responsible for magnetic anisotropy. It should be equally valid, for example, in describing either the $[Co/Pt]_n$ constructed multilayers or the FePt(111) L1$_0$ chemically ordered phase. Only the values of N_{mag} and μ will be case-dependent.

3.5.4 Comparison with Experimental Data

If the total anisotropy is dominated by interface terms for these layered structures, as we shall assume for the moment, then the perpendicular anisotropy will scale with K_s, and if we further assume that K_u scales with coercivity in these cases, then we can write

$$H_c = H_c^0 \cdot \exp\left(-2\mu \, D/N\right) = H_c^0 \cdot \exp\left(-\beta \, D\right) \quad (3.7)$$

where
$$\beta = 2\mu/N \quad (3.8)$$

The solid lines in Figures 3.7 and 3.8 represent empirical best fits for Equation (3.7). The predicted exponential dose dependence is clearly supported by the observations. Table 3.2 compares the fitted value of β with the value deduced for each system and ion species using SRIM with its default displacement energies of 25 eV for Fe, Co and Pt.

TABLE 3.2 Values of β in Units of $[10^{-15} \, \text{cm}^2]$

Ion Beam	[Co/Pt]$_n$			FePt		
	β_{expt}	β_{SRIM}	$\beta_{expt.}/\beta_{SRIM}$	β_{expt}	β_{SRIM}	$\beta_{expt.}/\beta_{SRIM}$
Ga$^+$(30 keV)	84	9.9	8.5	–	–	–
Ar$^+$(20 keV)	51	6.0	8.5	70	4.8	14.6
Ar$^+$(2 MeV)	10.4	1.6	6.6	7.5	0.87	8.6
N$^+$(700 keV)	1.2	0.23	5.3	1.6	0.21	7.6
He$^+$(20 keV)	0.99	0.15	6.7	0.55	0.092	6.0
He$^+$(2 MeV)	0.14	0.0023	61	0.025	0.0026	9.6

Note: The table compares values of β derived from SRIM with those derived from experimental data. Experimental uncertainty in β is estimated to be ± 10%. The Co/Pt data are reproduced from Ref. [11].

Two conclusions are implied:

1. Scaling of the magnetic changes among various ion species and energies is qualitatively consistent with the association of ballistic displacements with the observed change of H_c. For example, ions having a higher rate of displacement damage require smaller doses to change H_c.

2. The observed ion doses required to create magnetic changes are in most cases a factor of about 5–10 lower than would be expected based on simple SRIM-mod-

eled ballistic displacement, and the layer-disordering model of Ref. [13].

3.5.5   Enhancement Mechanisms

The data point for 2 MeV He+ on Co/Pt appears to be anomalous. However, with the exception of that point, both experimental data and the model deliver very similar sets of results for the two systems Co/Pt and FePt (L1$_0$), which suggests that their coercivity changes under irradiation have a common physical mechanism. The observed variation among responses to different ion beams is also reflected reasonably well for both systems by the displacement model, which is remarkable in view of the corresponding range of four orders of magnitude spanned by the ion doses.

A significant problem raised by this work is that, in general, each incident ion appears to be 5 to 10 times more effective in degrading H_c than is predicted on the basis of ballistic disordering alone, estimating the displacement rate with SRIM. This discrepancy has also been noted by Devolder et al. [15,16], who concluded that the major magnetic changes occur after irradiation leading, in principle, to only about 0.2 displacements per atom. The origin of the discrepancy is open to speculation.

Does SRIM underestimate the rate of ballistic displacements in these systems by an order of magnitude? The values of μ_{SRIM} are somewhat dependent on the value assumed for displacement energy E_d. The calculation used 25 eV for Fe, Co and Pt, and other values between 20 and 40 eV might be more accurate. However, only by imposing $E_d \approx 3$ eV would μ_{SRIM} increase by the factor required — a most unlikely situation. By its nature, SRIM is not constructed to take into account lattice structure of the sample. Yet here we are discussing ordered metals, for which *replacement collision sequences* are expected to occur. These processes have been modeled extensively by molecular dynamics simulations [17–20]. In these events, an atom displaced within a few degrees of a major crystal axis may produce a knock-on chain of typically 4 to 20 displacements, depending on the orientation of the chain. Such events can dominate the disordering process in ordered systems such as FePt or Co/Pt, and might serve to increase the effective value of μ_{SRIM} for ballistic collisions by a large factor. That factor would be expected to be largely independent of incident ion species (because most displacements result from cascaded recoiling species near their end of range), consistent with the observation that the missing scale factors are rather similar among different ion species and energies.

An apparent problem with the replacement collision sequences solution arises, however. While the factor $\beta_{expt.}/\beta_{SRIM}$ is in the general range 6–10 for both $[Co/Pt]_n$ and FePt, one would expect that the alternating monolayer FePt structure would offer a greater scope for cascaded displacements causing loss of anisotropy in many layers than for the Co/Pt system where a collision sequence would not be likely to reach more than one Co/Pt sandwich layer.

A different mechanism for coherent displacement of atoms in a single crystal has been modeled by Nordlund et al. [5], in which transient thermal excursions in the cascade lead to shear-induced slip motion. How relevant that would be for a multilayer system is not clear.

The question remains essentially open, with a need to incorporate consequences of a single bond-breaking event that extend beyond the destruction of a single pair bond. Zhang et al. [21] have proposed that for Co/Pt multilayers, a substantial contribution to uniaxial anisotropy K_u originates from the internal stress (compressive for Pt, tensile for Co, with approximately 6% strain) required to sustain the ordered Co/Pt interface. K_u would then be expressed as

$$K_u = -2\pi M_s^2 + K - (\text{const. } \lambda\sigma) + 2K_s/t_{Co} \qquad (3.9)$$

where σ = stress and λ is the related magneto-striction, and K = anisotropy energy of bulk Co. Ion beam disordering of a highly strained Co configuration may well promote relaxation of the interface in a very non-linear way (each Co atom displacement being analogous to excising one knot in a taut fishnet). Such a mechanism could account for the observed heightened magnetic sensitivity of the system to each atom displacement. It is not known at present whether a similar argument could be applied for the conceptual "atomic planes" of the fct phase of FePt.

3.5.6 Other Systems

Modification of magnetic anisotropy using ion or electron beams has also been reported for other systems that incorporate physically layered structures.

- Electron beam irradiation at 10 keV of Co films, 3 to 8 monolayers thick, deposited by MBE on Pt(111) has been reported by Allenspach et al. [22]. The easy-axis orientation was found to switch from in-plane to the perpendicular direction after exposure to 2.4×10^{18} electrons/cm^2, concurrently with a tenfold growth of domain size. The mechanisms responsible remain open to speculation.
- In a report of ion-irradiation control of the easy-axis orientation in Co/Pt films by Chang et al. [23], the authors found that in-plane easy-axis orientation rotated to match the direction of movement of the swept 80 keV Ar$^+$ beam across the sample. Application of an external magnetic field during irradiation produced a different orientation. The mechanism of these processes also remains subject to speculation.
- [Co/Pd]$_n$ multilayers have been found to display kinetics similar to those of [Co/Pt]$_n$ multilayers.
- CoPt$_3$ is a clustered alloy [24] whose chemically ordered phase displays strong anisotropy, and whose easy axis perpendicular to the substrate finally changes to the in-plane orientation after disordering by irradiation depletes the anisotropy, as in the case of FePt.

3.6 MODIFICATION OF MAGNETIZATION

In a different class of materials, it is possible for ion irradiation to induce a transition from an initial ferrimagnetic phase to a disordered non-magnetic phase, as in the case of the CrPt$_3$

($L1_2$ phase) [24]. Chemically ordered $CrPt_3$ films display strain-induced uniaxial magnetic anisotropy. Ion irradiation of such films converts the ferrimagnetic phase into the chemically disordered non-magnetic fcc phase, apparently as a result of the same ballistic mixing process that produces transformations in FePt or in $[Co/Pt]_n$ multilayers, at similar ion doses (e.g., ~ 10^{15} ions/cm^2 for 700 keV N$^+$). Because the irradiated region of a $CrPt_3$ film becomes completely non-magnetic, this material could be attractive as a patternable medium (or device structure), in which magnetic effects are completely absent from the irradiated area (e.g., the space separating the active bit patterns), in contrast to the multilayers of Co/Pt or FePt, where the moment of the irradiated material remains large (and possibly intrusive), even though its uniaxial anisotropy has been removed.

Another interesting alloy is $FePt_3$ [25], whose chemically ordered $L1_2$ phase is non-magnetic, but whose irradiated, chemically disordered phase displays ferromagnetism, with a uniaxial anisotropy of $K_u = 6 \times 10^4$ erg/cm^3, and a Curie temperature of ~ 400 K.

3.7 MAGNETORESISTANCE AND REMANENT MAGNETIZATION RATIO

Cai et al. [26] investigated the effects of 1 MeV Si$^+$ irradiation on [Co(16 Å)/Cu(20 Å)]$_{30}$ immiscible multilayers, with the concept of using the ion interaction as a sensitive means of uniformly altering the roughness of all the Co–Cu interfaces in the material, to observe the consequent changes in magnetoresistance (MR). They found that, for ion doses up to 10^{15}/cm^2, interface disorder was indeed produced, accompanied by large changes in resistivity ρ and MR (Figure 3.9). Furthermore, the remanent magnetization ratio (M_r/M_s) was reduced by 80%. The authors note that resistivities of comparable pure films of Cu and of Co were not affected by the irradiation, thus discounting the possible effects of bulk defects induced by the ion beam. It is proposed that even the limited ballistic interface mixing expected for metals with such a high positive heat of mixing [27] would be sufficient

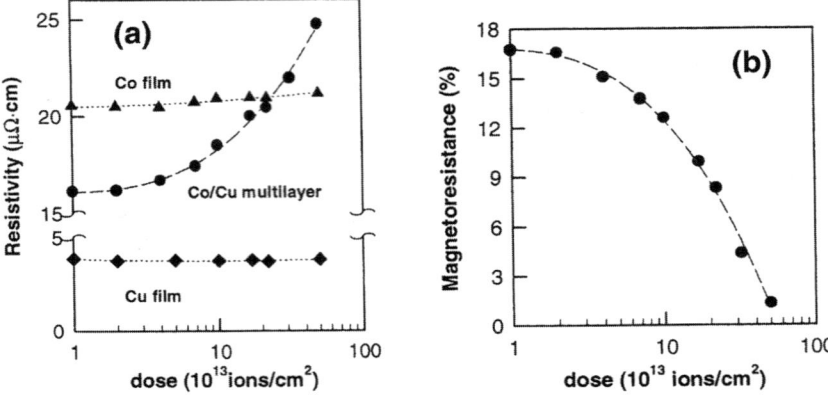

Figure 3.9 (a) Resistivities of a $[Co(15\text{Å})/Cu(20\text{Å})]_{30}$ multilayer, as well as of 100 nm Cu and Co films, as functions of dose of 1 MeV Si$^+$ ions. (b) Magnetoresistance ratio of this multilayer, as a function of ion dose. (From M. Cai, T. Veres, S. Roorda, R.W. Cochrane, R. Abdouche and M. Sutton, *J. Appl. Phys.* **81**, 5200–5202, 1997.)

to give rise to a region having sufficient spin-independent interface scattering, consistent with the rise in multilayer resistivity. With increasing ion dose, the magnetic interlayer coupling is believed to change from mainly antiferromagnetic (AF) coupling to ferromagnetic coupling, with the giant magnetoresistance (GMR) being progressively degraded.

In a later study of Co/Ag multilayer systems, Veres et al. [28,29] demonstrated the structural and magnetic consequences of irradiation, as a function of thickness of the deposited Co layers. In addition to interface mixing at the level of one or two atom layers, irradiation promotes segregation of the Co as islands within the host Ag, whose size and history govern the final magnetoresistive properties.

3.8 POTENTIAL APPLICATIONS IN MAGNETIC DISK STORAGE TECHNOLOGY

In the pursuit of magnetic disk storage densities above 100 Gb/in.^{2}, the disk drive industry needs to exploit new technologies for defining the "bits" on a disk, and for reading and writing information with spatial resolution of a few tens of nanometers. Such new systems must maintain high reliability, speed, signal-to-noise ratio and physical durability, while being manufacturable at very low cost, with robust and reliable processing steps.

The boldness of the challenge is highlighted by the trend plot of Figure 3.10, which shows the projected feature size required for the semiconductor industry roadmap based on Moore's Law. A comparative plot shows feature sizes required to define one storage "bit," as the projected storage densities are increased in the imminent future. The message of this figure is clear: the limits in spatial resolution acknowledged by the semiconductor industry (chiefly lithographic issues) must not be the determinants of magnetic bit resolution, if

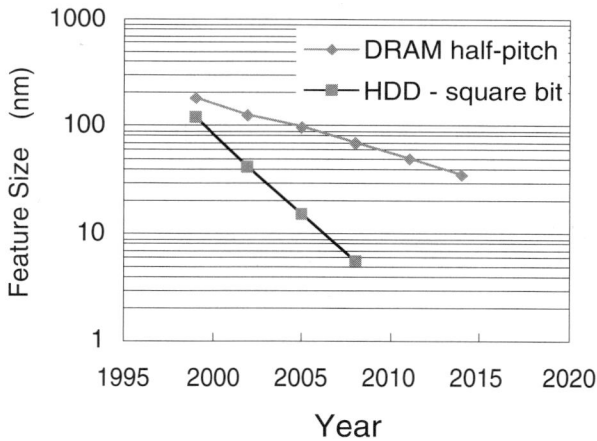

Figure 3.10 Roadmap projections for future increased bit density of magnetic disk storage (solid line). For comparison, the plot also displays the projected decrease in feature size for lithographic technology for the semiconductor industry (dashed line). Since the latter is likely to represent the frontier of lithographic resolution, it is evident that the disk storage ambitions will need to be fulfilled by a different viable process that is not subject to optical interference limitations.

the disk drive ambitions are to be realized. In addition to defining bit areas, the storage technology must include strategies for maintaining read/write sensitivity while the energy required to switch a bit becomes smaller and smaller as bit area decreases. In fact, the use of pre-patterned storage media seems essential, and the use of layered AF-coupled media becomes necessary for overcoming the nominal limits of superparamagnetism for bits of nanometer dimensions at room temperature.

The use of ion irradiation (rather than standard subtractive-etch lithography) for high-resolution pre-patterning of magnetic disks is now evidently feasible, as we shall see below. It offers the advantage of custom-tailoring the magnetic response of the irradiated regions of a continuous disk coating, without physical roughening of the critical disk surface, in a fast, one-step vacuum process. The use of a physically patterned stencil mask to enable convenient processing of the entire disk surface would offer, in principle, a fundamental advantage in spatial resolution over optical lithography with its major diffraction limitations, since particle wavelengths are negligible. As we shall see, projection ion beam tools now available can enable the use of robust, long-lived masks whose features are one or two orders of magnitude larger than the features desired in the projected pattern.

The sharp spatial definition of an ion beam may also find unique application in the fabrication of GMR read-heads and also write-heads, offering improved effective track width compared with those currently produced by means of standard lithography.

3.8.1 Fundamental Resolution Issues for Ion Beam Processing

Although ion beam lithography is free from diffraction limitations like those inherent in optical lithography, its spatial resolution is subject to other physical limitations.

Consider the schematic cartoon of a stencil mask in Figure 3.11. A stencil mask, placed in contact with the workpiece, must have a thickness exceeding the projected range R_p of

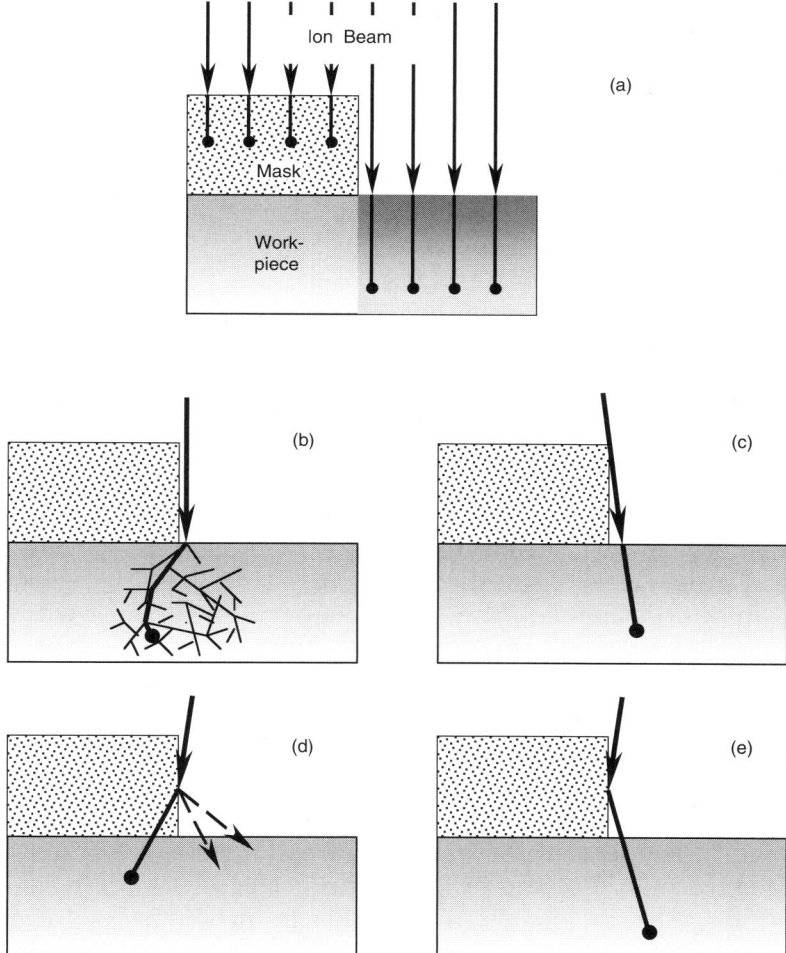

Figure 3.11 Schematic picture of a stencil mask edge used to pattern the exposure of a workpiece to a continuous ion beam. Some potentially resolution-limiting ion scattering or transmission phenomena are illustrated. (a) The idealized irradiation of the non-masked area. (b) Lateral straggling of ionization, recoils and possibly the incident ion, in the cascade within the workpiece. (c) Slight angular dispersion of the incident ions may cause a mask shadow. (d) Incident ions may enter the mask wall and penetrate through its corner. Mask atoms may also be sputtered. (e) Incident ions may scatter into the exposed field, as also may sputtered atoms.

the ions, in order to stop all the ions everywhere except in the open sections. Depending on the mask material, and the angular dispersion of the incident beam, some ions will strike the walls of the transmission holes, and scattered primary ions or sputtered ions of the mask material may contaminate the intended beam image and blur its edges. Of additional concern, energetic ions scattered in the beam hole may well be transmitted through the corners of the mask profile, again making a blurred boundary. Use of a thick mask with high aspect ratio features, together with a highly collimated or directed ion beam, can minimize these problems.

Even if we succeed in illuminating the work surface with a perfectly defined, parallel beam of ions, we must consider the broadening of the irradiation feature due to lateral straggling of the ionization, or collision cascades that represent the interaction of each ion as soon as it enters the workpiece. Such lateral straggling can, in general, be minimized in the case of irradiation of very thin films, by selecting an energetic (e.g., MeV He$^+$) beam of low-Z ions, whose interaction in the thin surface coating is mainly that of ionization produced by the projectile ion itself. This could apply, for example, for patterned cross-linking of organic materials resulting from ionization. However, in order to modify the magnetic materials that we have discussed above, ballistic displacement interactions are apparently needed. Typically, the lateral straggling radius for primary and secondary recoils is of the order of the projected range R_p (for low-energy heavy ions), which implies an intrinsic resolution limitation of the order of the thickness of the magnetic film being treated. The problem is made more acute by the stochastic nature of the recoil cascade process, in which finite probabilities exist for occasional laterally directed recoil paths exceeding R_p.

Despite the apparent limits of resolution cited for ion beam lithography, we emphasize that the technique holds more promise than most alternative approaches for producing patterned storage media at very high densities. However, this technique is still at the stage of exploratory development.

3.8.2 Recording Media

In a pioneering experiment, Chappert et al. [30,31] used 30 keV He+ ions, with a polymer resist mask in contact with the sample, to produce clean magnetic patterns in Co/Pt multilayers, with micron-sized features. Devolder et al. [32] used a Ga+ focused ion beam to produce similar patterning of Co/Pt multilayers.

Subsequently, Terris et al. [33,34] produced sub-micron scale magnetic patterns on Co/Pt multilayers [Co(4 Å)/Pt(1 0Å)]$_{7-14}$, using silicon contact masks 10 μm thick, each covering an area of 1×1 mm with a graded test pattern of arrays of 1 μm holes, at various separations. Successful patterns were obtained with He+ at 2.3 MeV, or N+ at 700 keV, or Ar+ at 1.2 MeV. A schematic diagram of the experimental setup is shown in Figure 3.12, together with an MFM (magnetic force microscope) image showing the magnetic contrast obtained from a patterned region. An SEM (scanning electron microscope) image of the mask profile is also shown. The MFM images represent the remanent state obtained after placing the sample briefly in a 20 kOe saturating field normal to the film surface. Within the circular dots, the easy magnetization direction was found to be rotated from out-of-plane (as-deposited) to in-plane after irradiation. Well-defined 100 nm dots were also achieved. AFM (atomic force microscope) images revealed topographic changes of only a few nanometers, attributable to density changes caused by the penetrating ions in the silica substrate.

3.8.2.1 Projection Ion Beams

A recently developed technique [35–38] of ion projection imaging has been applied to high-resolution patterning of Co/Pt multilayers prepared on full standard Microdrive™ disks, using 4×10^{13} ions/cm^2 of 45 keV Ar+ ions. The ion projection system is based on precision engineered electrostatic ion optics. It serves to reduce the image provided by a stencil mask placed in the optical condenser plane by one order of magnitude, or greater, at the surface of the disk, thus enabling the use of a robust mask (125 mm diameter Si membrane, 3 μm

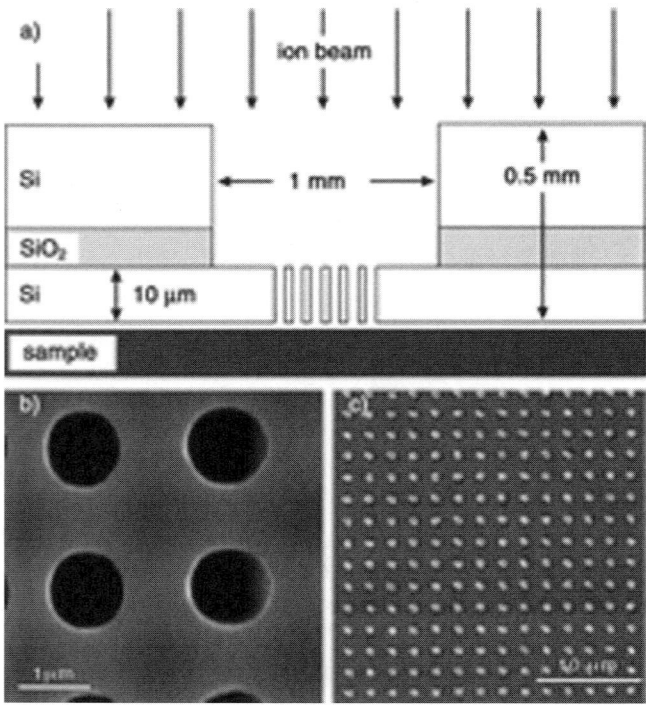

Figure 3.12 (a) Schematic diagram of ion irradiation through a stencil mask. (b) SEM image of a mask region having 1 μm holes on a 2 μm pitch. (c) MFM image, showing the magnetic contrast introduced by irradiation of the $[Co/Pt]_{10}$ multilayer with 700 keV N^+ at 5×10^{15} ions/cm². (From B.D. Terris, L. Folks, D. Weller, J.E.E. Baglin, A.J. Kellock, H. Rothuizen and P. Vettiger, *Appl. Phys. Lett.* 75, 403–405, 1999.)

thickness, patterned by an e-beam writer), thick enough to permit the construction of retrograde stencil-mask openings, and large enough to pattern a full disk in a single exposure, without step-and-repeat procedures. The optical precision and freedom from aberrations enables the generation of well-defined features at a scale beyond that of the e-beam mask writer.

Figure 3.13 shows an MFM image of a portion of the ion-patterned disk (using only 4× reduction), with magnetic bit

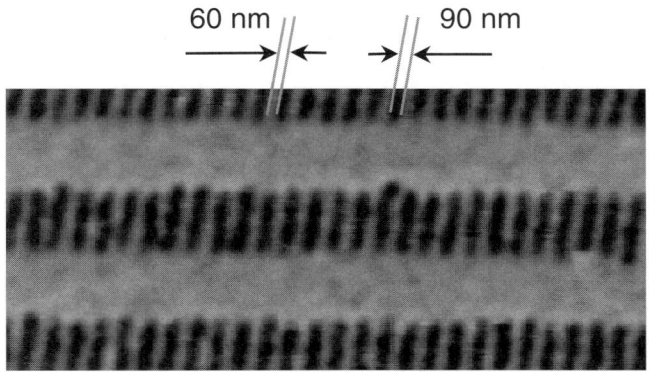

Figure 3.13 MFM image of a small area (4.5 × 2.0 μm) of a Microdrive™ [Co/Pt]$_n$ coated disk, after exposure to a stencil-masked ion projection beam of 45 keV Ar⁺ (4 × 10¹³ ions/cm²), showing 90 nm magnetic bit regions at 60 nm spacing. The stencil pattern was subject to 4× reduction in this case [35–38].

regions definable at 60 nm separation. Such feature dimensions would already be consistent with storage densities approaching 100 Gb/in². Newer versions of the system offer much greater image-reduction factors.

Parenthetically, we note that in Co/Pt, the bars of the image are the irradiated ferromagnetic regions. For such systems, this stencil mask produces the reverse image from that needed to produce recordable dots surrounded by a disabled medium. The ion projection system may, in principle, provide a useful option in this case, with the possibility of making double-irradiation exposures with a slightly shifted image, to create unirradiated bit islands surrounded by irradiated (soft) material.

3.8.2.2 Soft Bit in Hard Matrix

As proposed by Devolder et al. [15], the ion beam process for highly anisotropic materials such as FePt or Co/Pt multilayers suggests the patterning of arbitrarily "soft" bits into a high coercivity matrix. Such a medium would be characterized by a relatively small nucleation field, with both writing and reversal occurring through domain wall motion at the same modest propagation field.

The resolution of the bit reversal process has been studied by Devolder et al. [13,39], Kusinski et al. [40,41], Rettner et al. [42] and others. Key issues are illustrated by Devolder in the magneto-optic images (Figure 3.14) of the progressive switching of $[Co/Pt]_n$ at a border between an unirradiated area (lower right corner) and a region irradiated with a low dose of He^+ ions. Reverse magnetized areas appear in black. Images were made in remanence, following brief application of an external electric field that was higher for each successive cycle. The border between masked and irradiated zones clearly becomes the nucleation center for reversed domains, which expand preferentially along that border, eventually defining a sharp border between regions of antiparallel magnetization. At higher fields, marked by de-pinning at the border, an irregular boundary develops for several of the magnetically harder, non-irradiated areas. Although the work of Figure 3.14 refers to an original medium of quite low coercivity, similar behavior would be expected for magnetically harder media. The strong border nucleation effects have been attributed [13] to enhanced softening of the medium at the mask edges, related to lateral straggling and ion scattering within the mask. It is possible that such local mask-edge phenomena may be reduced in an ion projection system, where the randomly scattered ions emerging from the mask can be dispersed by the subsequent ion optics.

3.8.2.3 Nanoparticle Islands

Sun et al. [43,44] have displayed self-assembled ordered arrays of 4 nm monodispersive FePt nanoparticles isolated by > 2 nm walls of organic binder, which suggest the prospect of future media with bit sizes of a few nanometers, all of uniform size and constituting independent magnetic domains, self-organized without the aid of complex lithography. (However, extended pattern coherence lengths have yet to be achieved.) Such a medium may be crafted to display good signal-to-noise

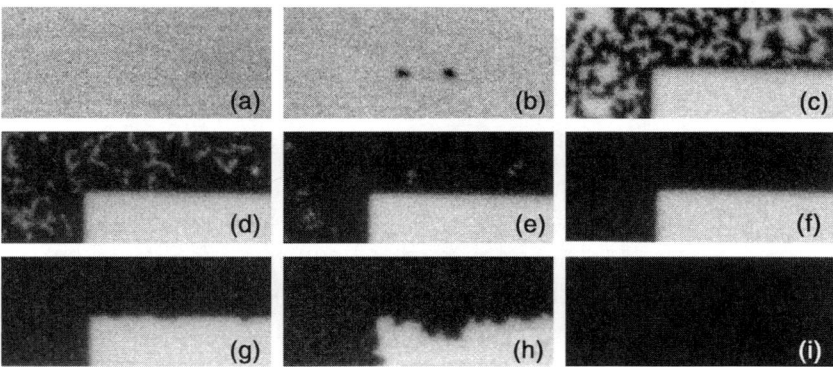

Figure 3.14 Magnetization reversal at the border between an area of $[Co/Pt]_n$ irradiated with 2×10^{15} He$^+$/cm^2, and a non-irradiated area (lower right hand corner). Reverse magnetized areas appear black. Images were made in remanence, after applied pulses of successively increasing field. (From T. Devolder, H. Bernas, D. Ravelosona, C. Chappert, S. Pizzini, J. Vogel, J. Ferré, J.-P. Jamet, Y. Chen and V. Mathet, *Nucl. Instrum. Methods Phys. Res.* B 175, 2001. With permission.)

properties. In the current absence of techniques for reading or writing on 4 nm features, it has been suggested [45] that islands or "rafts" of such nanoparticles, extending to diameters of 20–50 nm, could serve as highly uniform lithographically patterned single-bit areas, having accessible dimensions and magnetic performance. In the absence of magnetic coupling between the particles, the magnetic reversal process would exclude the problems of domain kinetics that were discussed above.

An elegant means of achieving such patterning of a surface coated with a single layer of FePt nanoparticles is proposed to be irradiation of the bit regions to locally cross-link the polymer embedment as a hard carbonaceous matrix, leaving the inter-bit regions to be completely removed by a solvent rinse. The patterned nanoparticle dots would then survive the high-temperature processing required to form the magnetically anisotropic fct phase of FePt without allowing thermal aggregation of the particles [44,45]. A means of producing a common orientation of the nanoparticles has yet to be developed, but would be essential for the concept to succeed.

3.8.3   Track-Width Definition of GMR Sensors

The conventional method of fabricating giant magnetoresistance sensors is schematically shown in Figure 3.15. The sensor area is shaped from a continuous multilayer via ion milling with a resist mask. The remaining surface is then coated with a hard bias layer and the lead, adjacent to the remaining sensor layer. This process leads to poorly defined track edge profiles, and a significantly broadened effective track width (TW). An alternative process is described by Folks et al. [46] (Figure 3.16), in which a highly collimated beam of energetic ions is shaped by the mask structure, and passes through the sensor layer, locally destroying its magnetic performance, yet leaving the masked region unaffected. The process does not involve physical etching of material, and is expected to produce a sharp sensor-edge profile (typically ~ 2 nm).

Experiments using 700 keV N^+ ions with a representative spin-valve sensor structure indicated total extinction of magnetoresistance and exchange field after an ion dose of $10^{15}/cm^2$, all of which is apparently due to ion beam mixing at the many interfaces and thin layers in the sensor.

3.8.4   Track Width for Magnetic Write Heads

Folks et al. [9] have proposed a method for replacing the standard ion-milling process for creating a "notch" in the larger of the two poles of a magnetic write head — a process that leads to degraded performance due to the removal of part of the smaller pole. In the new process, Cr^+ ions are implanted into the continuous $Ni_{80}Fe_{20}$ films to be notched after masking the areas that are to remain ferromagnetic. At high Cr^+ doses, the Curie temperature of $Ni_{80}Fe_{20}$ is reduced, and it becomes paramagnetic at room temperature. As in the previous example, good spatial definition is achieved and irregular removal of material by milling is avoided.

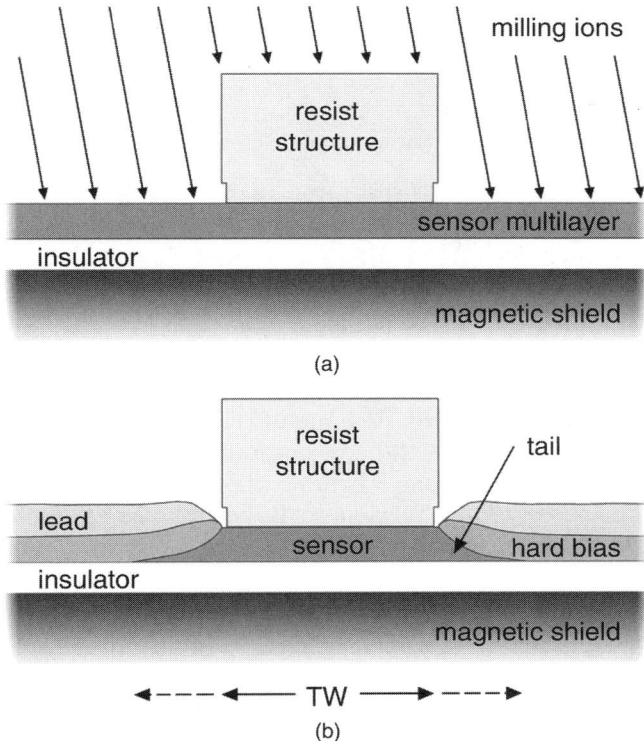

Figure 3.15 Schematic depiction of conventional processing of magnetic read sensors. (a) Shaping of the sensor mesa by ion milling, using a resist structure; followed by (b) deposition of the hard bias layer and the lead adjacent to the sensor mesa. The resulting poorly defined track edge profiles are indicated by the dashed lines. (From L. Folks, J.E.E. Baglin, A.J. Kellock, M.J. Carey, B.D. Terris and B. Gurney, *IEEE Trans. Mag.* 37, 2001. With permission.)

3.9 CONCLUDING REMARK

Ion beam modification of the magnetic performance of multi-layers and alloy structures offers many new possibilities for increasing the bit density in future disk storage. It also offers new options for the design of innovative devices of the future.

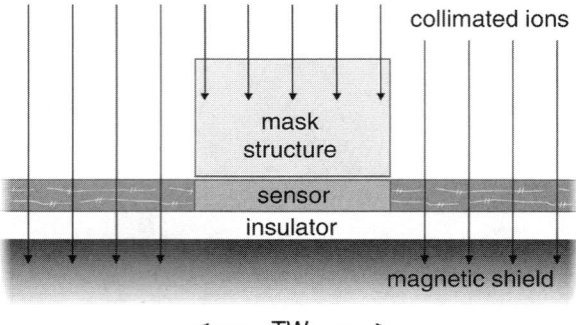

Figure 3.16 Definition of the GMR sensor by masked irradiation with collimated light energetic ions, which disorder the interfaces in the exposed regions of the sensor multilayer, reducing magnetoresistance in those areas. (From L. Folks, J.E.E. Baglin, A.J. Kellock, M.J. Carey, B.D. Terris and B. Gurney, *IEEE Trans. Mag.* 37, 2001. With permission.)

REFERENCES

1. J.P. Biersack and L. Haggmark, *Nucl. Instrum. Methods Phys. Res.* 174, 257 (1980); see also www.srim.org.

2. R.S. Averbach and T. Diaz de la Rubia, in *Solid State Physics,* Eds. H. Ehrenreich and F. Spaepen, Academic Press, New York, 51, 281–402 (1998).

3. T. Diaz de la Rubia, R.S. Averback, R. Benedek and I.M. Robertson, *Radiation Effects and Defects in Solids* 113, 39–52 (1990).

4. K. Nordlund, M. Ghaly and R.S. Averback, *J. Appl. Phys.* 83, 1238–1246 (1998).

5. K. Nordlund, J. Keinonen, M. Ghaly and R.S. Averback, *Nature*, 398, 49 (1999).

6. J.E.E. Baglin, M.H. Tabacniks, R. Fontana, A.J. Kellock and T.T. Bardin, *Mater. Sci. Forum* 248–249, 87–93 (1997).

7. J.P. Rivière, P. Bouillard and J.P. Eymery, *Phys. Lett.* A 138, 223 (1989).

8. D. Kurowski, R. Mechenstock, K. Brand and J. Pelzl, *J. Magn. Magn. Mater.*148, 99 (1995).

9. L. Folks, R. Fontana, B. Gurney, J. Childress, S. Maat, J. Katine, J.E.E. Baglin, A. Kellock and P. Saunders, *J. Phys. D: Appl. Phys.* 36, 2601–2604 (2003).

10. D. Weller, J.E.E. Baglin, A.J. Kellock, K.A. Hannibal, M.F. Toney, G. Kusinski, S. Lang, L. Folks, M.E. Best and B.D. Terris, *J. Appl. Phys.* 87, 5768 (2000).

11. C.T. Rettner, S. Anders, J.E.E. Baglin, T. Thomson and B.D. Terris, *Appl. Phys. Lett.* 80, 279 (2002).

12. J.E.E. Baglin, C.T. Rettner, B.D. Terris, D.K. Weller, J.-U. Thiele, A.J. Kellock, S. Anders and T. Thomson, *Proc. SPIE 4468*, 1–7 (2001).

13. J.M. MacLaren and R.H. Victora, *IEEE Trans. Magn.* 29, 3034 (1993).

14. J.M. MacLaren and R.H. Victora, *Appl. Phys. Lett.* 66, 3377 (1995).

15. T. Devolder, H. Bernas, D. Ravelosona, C. Chappert, S. Pizzini, J. Vogel, J. Ferré, J.-P. Jamet, Y. Chen and V. Mathet, *Nucl. Instrum. Methods Phys. Res.* B 175, 375–381 (2001).

16. T. Devolder, *Phys. Rev.* B 62, 5784 (2000).

17. G.J. Ackland, D.J. Bacon, A.F. Calder and T. Harry, *Phil. Mag. A* 75, 713–732 (1997).

18. H.F. Deng and D.J. Bacon, *Phys. Rev.* B 53, 11376–11387 (1996).

19. H.F. Deng and D.J. Bacon, *Phys. Rev.* B 48, 10022–10030 (1993).

20. H.F. Deng and D.J. Bacon, *Rad. Eff. Def. Solids* 130–131, 507–512 (1994).

21. B. Zhang, K.M. Krishnan, C.H. Lee and R.F.C. Farrow, *J. Appl. Phys.* 73, 6198–6200 (1993).

22. R. Allenspach, A. Bischof, U. Dürig and P. Grütter, *Appl. Phys. Lett.* 73, 3598–3600 (1998).

23. G.S. Chang, T.A. Callcott, G.P. Zhang, G.T. Woods, S.H. Kim, S.W. Shim, K. Jeong, C.N. Whang and A. Moeves, *Appl. Phys. Lett.* 81, 3016–3018 (2002).

24. O. Hellwig, D. Weller, A.J. Kellock, J.E.E. Baglin and E.E. Fullerton, *Appl. Phys. Lett.* 79, 1151–1153 (2001).

25. S. Maat, A.J. Kellock, D. Weller, J.E.E. Baglin and E.E. Fullerton, *J. Magn. Magn. Mater.* 265, 1–6 (2003).

26. M. Cai, T. Veres, S. Roorda, R.W. Cochrane, R. Abdouche and M. Sutton, *J. Appl. Phys.* 81, 5200–5202 (1997).

27. W.L. Johnson, Y.T. Cheng, M. Van Rossum and M.-A. Nicolet, *Nucl. Instrum. Methods Phys. Res.* B 7/8, 657 (1985).

28. T. Veres, M. Cai, R.W. Cochrane and S. Roorda, *J. Appl. Phys.* 87, 8504–8512 (2000).

29. T. Veres. M. Cai, M. Rouabhi, F. Schiettekatte, S. Roorda and R.W. Cochrane, *J. Appl. Phys.* 87, 8513–8521 (2000).

30. C. Chappert, H. Bernas, J. Ferré, V. Kottler, J.-P. Jamet, Y. Chen, E. Cambril, T. Devolder, F. Rousseaux, V. Mathet and H. Launois, *Science* 280, 1919–1921 (1998).

31. T. Devolder, C. Chappert, Y. Chen, E. Cambril, J.-P. Jamet and J. Ferré, *Appl. Phys. Lett.* 74, 3383–3384 (1999).

32. T. Devolder, C. Vieu, H. Bernas, J. Ferré, C. Chappert, J. Gierak, J.-P. Jamet, T. Aign, P. Meyer, Y. Chen, F. Rousseaux, V. Mathet, H. Launois and O. Kaitasov, *Comptes Rendus Acad. Sci. Paris* 327, 915–923 (1999).

33. B.D. Terris, L. Folks, D. Weller, J.E.E. Baglin, A.J. Kellock, H. Rothuizen and P. Vettiger, *Appl. Phys. Lett.* 75, 403–405 (1999).

34. B.D. Terris, D. Weller, L. Folks, J.E.E. Baglin, A.J. Kellock, H. Rothuizen and P. Vettiger, *J. Appl. Phys.* 87, 7004–7008 (2000).

35. H. Loeschner, E.J. Fantner, R. Korntner, E. Platzgummer, G. Stengl, M. Zeininger, J.E.E. Baglin, R. Berger, W.H. Brünger, A. Dietzel, M.I. Baraton and L. Merhari, *Mater. Res. Soc. Symp. Proc.* 739, H1.3.1 (2003).

36. A. Dietzel, R. Berger, H. Grimm, W.H. Brünger, C. Djionk, F. Letzkus, R. Springer, H. Loechsner, E. Platzgummer, G. Stengl, Z.Z. Bandic and B. Terris, *IEEE Trans. Mag.* 38, 1952 (2002).

37. H. Loechsner, G. Stengl, H. Buschbeck, A. Chalupka, G. Lammer, E. Platzgummer, H. Vonach, P. de Jager, P. Kaesmaier, A. Ehrmann, S. Hirschner, A. Wolter, A. Dietzel, R. Berger, H. Grimm, B. Terris, W. Brünger, G. Gross, O. Fortagne, D. Adam, M. Boehm, H. Eichhorn, R. Springer, J. Butschke, F. Letzkus, P. Ruckhoeft and J. Wolfe, *Proc. SPIE 4688,* 595 (2002).

38. A. Dietzel, R. Berger, H. Loechsner, E. Platzgummer, G. Stengl, W.H. Brünger and F. Letzkus, *Adv. Mater.* 15, 1152–1155 (2003).

39. T. Devolder, C. Chappert, V. Mathet, H. Bernas, Y. Chen, J.-P. Jamet and J. Ferré, *J. Appl. Phys.* 87, 8671–8681 (2000).

40. G.J. Kusinski, K. Krishnan, G. Denbeaux, G. Thomas, B.D. Terris and D. Weller, *Appl. Phys. Lett.* 79, 2211–2213 (2001).

41. G.J. Kusinski, G. Thomas, G. Denbeaux, K.M. Krishnan and B.D. Terris, *J. Appl. Phys.* 91, 7541–7543 (2002).

42. C.T. Rettner, M.E. Best and B.D. Terris, *IEEE Trans. Mag.* 37, 1649–1651 (2001).

43. S. Sun, C.B. Murray, D. Weller, L. Folks and A. Moser, *Science* 287, 1989–1992 (2000).

44. S. Sun, S. Anders, T. Thomson, J.E.E. Baglin, M.F. Toney, H.F. Hamann, C.B. Murray and B.D. Terris, *J. Phys. Chem.* B 107, 5419–5425 (2003).

45. J.E.E. Baglin, S. Sun, A.J. Kellock, T. Thomson, M.F. Toney, B.D. Terris and C.B. Murray, *Mater. Res. Soc. Symp. Proc.* 777, T6.5.1 (2003).

46. L. Folks, J.E.E. Baglin, A.J. Kellock, M.J. Carey, B.D. Terris and B. Gurney, *IEEE Trans. Mag.* 37, 1730–1732 (2001).

Chapter 4

Selected Topics in Ion Beam Surface Engineering

D.B. FENNER, J.K. HIRVONEN,
AND J.D. DEMAREE

CONTENTS

ABSTRACT

This chapter covers three distinct areas of ion beam research and their application to the study or modification of surface-mediated properties. These vacuum-based techniques include directed-beam ion implantation, ion beam-assisted deposition, and gas-cluster ion beams. Discussion of the general capabilities of these techniques and their application to surfaces are described at a non-specialist level to allow the assessment of their applicability for the study or modification of other surface properties. The ion implantation of directed beams of energetic ions beneath the surface of materials is discussed first from the context of non-semiconductor substrates, i.e., metals, ceramics, and polymers. The extreme versatility of surface alloying virtually any substrate with any ion (element) at low temperatures makes it a powerful technique for surface studies and surface engineering. The technique, however, is subject to its intrinsic limitations of shallow alloying region and line-of-sight treatment requirements. Although best known for its predominant role in semiconductor fabrication since the 1970s, commercially driven uses of this unique near-surface modification technique in the non-semiconductor arena (as covered here) continue to evolve, and several examples are described herein. Ion beam-assisted deposition (IBAD) involves the ion bombardment with energetic ions of a vacuum-deposited film during film growth. This hybrid coating process allows control of several important film properties (e.g., adhesion, density, stress, texture, and composition), and the non-equilibrium energetic bombardment can provide the possibility of low substrate temperature processing. Historically, this process has been most widely used for producing robust optical coatings, but other scale-up uses are emerging, including the production of template layers for high-temperature superconducting coatings. Gas-cluster ion beams (GCIB) have recently emerged in applications for surface modification, film deposition and secondary-ion analysis. Gas-cluster beam-generation apparatus and methods are briefly summarized along with a short history of the development of GCIB. More satisfactorily treated as a nanoscale phenomenon, the impact of these weakly bound condensed-gas ions can induce fine-scale smoothing of slightly rough surfaces, or low ion-energy reactive deposition of thin films. The ion–surface physics is quite distinct from that of conventional ion beams and the current understanding of this is described here. The early stage commercial relevance of GCIB to microelectronics and photonics manufacturing is sketched, and a few specific areas of technology application are presented. The potential for GCIB-assisted film deposition of metal oxides and metal nitrides is also reviewed, as is the use of cluster ions as a primary beam to stimulate secondary ions whose mass spectroscopy provides chemical analysis of the surface.

4.1 INTRODUCTION

Ion implantation evolved from the ion–solid interaction studies of the 1960s to an indispensable tool for semiconductor doping in the 1970s. There are several thousand ion implantation machines in the world today processing microprocessors for computers, industrial components and household appliances. The immense success of doping semiconductors by this athermal method of forcing dopant atoms in the form of an energetic ion beam beneath the Si surface led several R&D groups worldwide to explore ion implantation as a means of beneficially modifying the properties of non-semiconductor materials by the same process. This research phase was most active in the 1980s. It necessitated much higher ion doses to achieve atomic percent level doping (alloying) concentrations in materials typically required to affect the many mechanical and chemical surface mediated properties that were found to be beneficially modified by ion implantation. Research on implantation effects on polymers (mainly conductivity and hardness changes) and on ceramics (fracture toughness) was also explored in this time period. To date, most of the commercial non-semiconductor properties that are being used on a commercial level involve wear of precision tooling or treatment of high value components such as surgically implanted artificial knees and hips in the medical field where discrete coatings are to be avoided because of concern for delamination within the human body. The intrinsic line-of-sight limitations of beam ion implantation have been addressed by utilizing a pulsed plasma mode of implantation that allows high coverage of complex shapes (i.e., high throwing power) and this technique will be covered in a separate chapter (Misra and Nastasi: Chapter 8) of this volume. It is anticipated that advances of this nature in processing technology and in ion source development will foster new applications or allow benefits previously demonstrated in the laboratory to be exploited on a commercial basis using economies of scale.

During the 1980s, many researchers in the particle–solid interaction community turned their attention to the hybrid process of simultaneous vapor deposition and ion bombardment (ion beam-assisted deposition) to overcome the intrinsic depth/thickness limitations of the ion implantation process. Optical coating developers were amongst the first users of this technology and it remains a standard process for producing robust, high-density optical coatings for critical applications as well as for fabricating thin film filters for telecom applications.

The gas cluster ion beam (GCIB) technique has also had a rich history of development. The present thrust of applications for GCIB in industrial technology is toward surface treatment of high value-added microelectronic and optical components, and especially surface smoothing. GCIB processing is less well known and distinctly different from the processes of ion implantation and ion beam-assisted deposition and therefore, more attention will be given to describing some of its basics in Section 4.4 in addition to discussing current applications.

This chapter will describe the general features of these ion beam based processes (see Table 4.1) on a level sufficient to understand their basics and will present illustrative examples of selected applications. The intention is to convey to the interested reader who is new to the subject how ion implantation, ion beam-assisted deposition, and gas cluster ion beams operate, and how they have been and can be utilized to enhance surface properties of materials with industrial importance, except for semiconductors. The selected examples, and the discussion of their potential advantages and intrinsic limitations, should serve the reader to better understand the potential of these ion beam processes for helping study or solving other surface-sensitive problems. References are cited to more detailed discussions of the topic areas discussed.

All of these processes are very surface oriented and limited to only the outermost few microns on the treated substrate, and in the case of GCIB to only the outermost few atomic layers. As such they are to be viewed as complementary to conventional surface treatments that normally provide much deeper alloyed zones or thicker coatings. Nevertheless, these techniques, by virtue of affording their unique combi-

TABLE 4.1   Features of Ion Beam Processes

	Affected Zone	Ion Species and Energy Regime	Benefits
Ion implantation (II)	nm to μm depth scale surface alloy region; no discrete layer	Any element into any substrate; keV to MeV energies	Alloying not dependent on thermodynamics
Ion beam-assisted deposition (IBAD)	1 nm–5 μm coatings	100 eV–10 keV ion assisting e-beam or sputter deposition	Highly adherent, pinhole-free coatings with controlled stress, composition, texture
Gas cluster ion beams (GCIB)	Nano-smooths and cleans outer surface to nanometer levels	Ar, CO and reactive gas clusters incident at keV energies (i.e., few to several eV per atom)	Atomic smoothing at room temperature without beam species incorporation or lattice damage

nations of near room temperature processing with highly adherent and controllable surface alloy or coating structures, continue to find niches for the processing of critical surfaces including those of micro- or nanostructured components.

4.2   ION IMPLANTATION

Ion implantation is the forcible injection of energetic particles in the form of ionized atoms beneath the surface of a material. Due to this brute force means of near surface alloying virtually any elemental species can be alloyed into the near surface region of any substrate. This technology evolved from basic particle–solid interaction physics studies and rapidly gained a foothold in the semiconductor industry starting in the 1970s. Its relatively rapid acceptance in that industry was prompted by the fact that it gave a controllable and reproducible means of doping semiconductor substrates over a wide range of impurity levels, displacing thermal diffusion as a doping method. Its acceptance was accelerated by the fact that new and improved device structures could be devised that conventional diffusion technology would not allow. Accounts of the early history of ion implantation in semiconductor processing are given by Fair (R.B. Fair, *Proc. IEEE*, 86, 111, 1998) and Rose (P.R. Rose, *Nucl. Instrum. Methods Phys. Res. B*, 6, 1–8, 1985). The extensive materials R&D conducted to find processing variables for silicon technology paved the way for investigations at several laboratories worldwide into other materials systems that might benefit from implantation alloying. Papers appeared as early as the 1960s on non-semiconductor applications of ion implantation requiring considerably higher ion doses than semiconductors. Expectations of achieving these high doses in reasonable times was not too far-fetched, considering that the ion source technology developed for the high-current, large-scale isotope separators used during the 1940s and 1950s was available.

4.2.1 Fundamentals of Ion Implantation

The penetration of energetic ions into the surface of a material results in the ion interacting with both electrons and the core atoms in their slowing down process. Interaction with the electrons, analogous to a viscous drag, increases with increasing ion velocity (i.e., with higher energy and lower mass ions) while elastic (billiard ball type) collisions with target nuclei increase with higher mass projectile/substrate combinations and predominate at low energies. These elastic collisions produce displacements of target atoms from their lattice positions and produce either highly damaged lattices following implantation or a net of damaged structures whose density depends on lattice reordering (annealing) during the implantation process. During the slowing down period, each ion will undergo multiple collisions as depicted in Figure 4.1, producing a "trail" of lattice atoms displaced from their equilibrium sites. Metallic lattices undergo a very high degree of lattice self-annealing

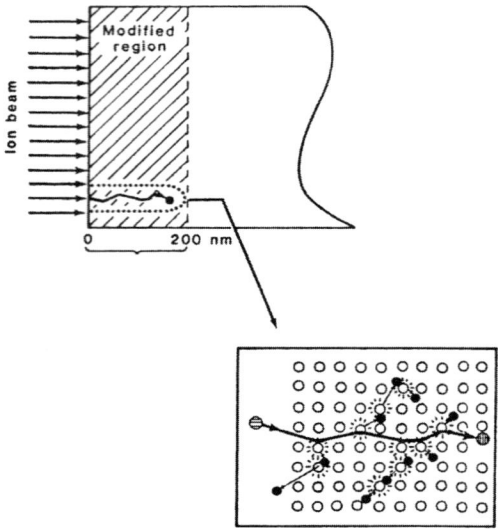

Figure 4.1 Schematic view of ion implantation process (top) and depiction of the ion interactions with the substrate atoms in a single collision cascade (bottom). (From G.K. Hubler, *Ion Beam Processing*, NRL Memorandum Report 5828, Naval Research Laboratory, Washington, DC, 1987. With permission.)

due to their non-directional bonding whereas covalent and ionic lattices may suffer total loss of near crystal order due to this radiation damage effect. In almost all cases the modified region is within the outermost micrometer of the substrate, often only within the first few hundred angstroms (i.e., microinches) of the surface. The modified depth scales with the ion energy, typically tens to hundreds of kilovolts.

During implantation, ions come to rest beneath the surface in typically less than 10^{-12} sec, producing a modified near-surface region with up to 50 atomic percent concentration of the implanted element. Because of this rapid quench time, many novel surface alloys or compounds unattainable by any other technique can be produced at room temperature, including highly metastable and amorphous alloys with unique physical and chemical properties. These include substitutional solid solutions of normally immiscible or low solubility elements.

At low doses, ion implantation generally results in a Gaussian impurity element distribution centered about an average range (R_p) with an associated width (ΔR_p). At the energies normally employed for ion implantation (typically 10–200 keV) projected ranges (i.e., penetration depths normal to the surface) are on the order of nanometers to hundreds of nanometers, depending on the energy and ion–substrate combination. The ion energy directly affects both the range and the distribution of the implanted ions in a given substrate. At higher energies a greater spread in the ion distribution is realized for a given dose, which is the number of implanted ions per unit area of the surface of the material (i.e., areal density), usually expressed as ions/cm^2. This areal dose (also called fluence) is a natural unit of ion implantation since the atomic depth concentration profile achieved for a particular implantation is a function of ion energy, species and substrate material as they influence the ion range and range straggling. The determination of atomic concentration (atoms/cm^3) requires relating the areal density of implanted atoms to their spatial extent. At low ion doses, the implant profile can normally be approximated as a Gaussian distribution centered about the projected range. Accordingly, the width of the distribution can be expressed by the standard deviation of the Gaussian distribution, i.e., ΔR_p. The expression for the peak concentration (N_p) of a Gaussian distribution is given in terms of the applied dose (N_A) and range straggling as:

$$N\left(\text{atoms/cm}^3\right) \equiv \frac{\left[N_A\left(\text{atoms/cm}^3\right)\right]}{\left[\Delta R_p\left(\text{cm}\right)\right]\sqrt{2\Pi}} \tag{4.1}$$

For non-semiconductor materials, beneficial doses can span the range of 10^{15} to 10^{18} ions/cm^2, depending on the application. These areal doses correspond to peak concentration values of up to several atomic percent using Equation (4.1).

There are now commercially available computer programs available (TRIM, PROFILE) that can calculate range energy curves for any ion into any substrate of interest. As the dose increases other processes, such as sputtering, normally alter the net shape of the distribution and limit the total number of delivered atoms to an equilibrium value. The ratio of the number of substrate atoms ejected per incident ion is commonly termed the sputtering coefficient, S. As a general rule, for a given substrate material, S will increase with the increasing ion mass. For a given ion, S depends inversely on the surface binding energy of the material, which can be related to its heat of sublimation. Therefore, for a given ion, S will decrease with increasing heat of sublimation for the substrate material. Values for S can range from less than 1 for the case of a light ion incident on a heavy substrate (e.g., nitrogen implanted into iron) to greater than 10 for very heavy ions in a lighter substrate (e.g., tantalum ions implanted into iron). For example, if the sputtering coefficient for an ion on a substrate is unity, the ultimate atomic concentration attainable is $50\% = 1/(1 + S)$ where S is the sputtering coefficient. A schematic of a low-dose implant profile and a sputter limited profile is shown in Figure 4.2. The sputter-limited depth distribution can be seen as evolving from consecutive low dose implantations, each removing a portion of the surface until an equilibrium distribution is attained. Sputtering effects ultimately limit the achievable concentration one may attain by implantation alloying to several tens of atomic percent over a depth comparable to the range.

In addition, it should be noted that more recent high ion current implantation effects have allowed enhanced diffusion of species into substrates overcoming the normal ballistic based ranges described above.

The following comments elaborate on the advantages/limitation bullet statements described in Table 4.2.

The ability to use ion implantation to forcibly inject alloying species into the near surface region of virtually any material, independent of equilibrium thermodynamic criteria such as solid solubility and diffusivity, is an important feature in tailoring the surface alloy to meet particular environmental operating conditions (i.e., wear or corrosion related). It has also been applied for basic metallurgical studies where one can prepare supersaturated or metastable systems and study

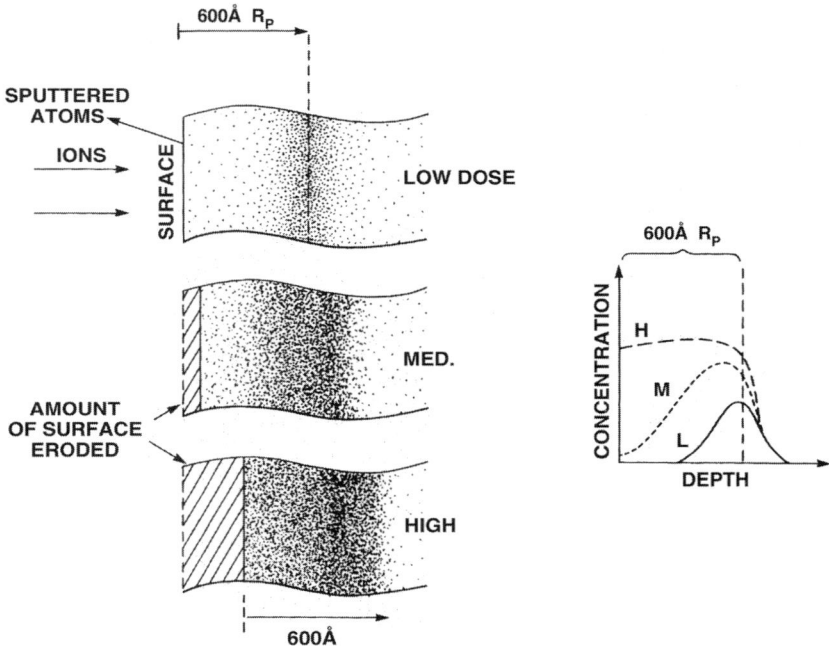

Figure 4.2 Depiction of how sputtering affects the ultimate depth distribution of high dose implantation profiles, evolving from a buried Gaussian distribution to a surface enriched region. (From G.K. Hubler, *Ion Beam Processing*, NRL Memorandum Report 5828, Naval Research Laboratory, Washington, DC, 1987. With permission.)

their return to equilibrium by suitable thermal annealing and observation of their microstructure. This application of ion implantation technology has been exploited to measure low temperature diffusion, solid solubility, and trapping phenomena in alloys and is discussed briefly later in this chapter. Several features set ion implantation apart from conventional surface processing techniques. First, the alloying is an extreme athermal technique enabling virtually any ion species to be introduced into any substrate desired, irrespective of normal thermodynamic constraints. The process is analogous to rapid quench processes since the time scale associated

TABLE 4.2 Advantages and Limitations of the Ion Implantation Process

Advantages	Limitations
Produces surface alloys independent of equilibrium thermodynamic constraints	Limited thickness of treated zone
No delamination concerns	High vacuum process
No significant dimensional changes	Line-of-sight process
Ambient temperature processing possible	Alloy concentrations dependent on sputtering
Enhance surface properties while retaining bulk properties	High value-added process
High degree of control and reproducibility	
Allows fast screening of different alloy properties	

with an ion entering and stopping is in the sub-picosecond regime. If one invokes a "local" lattice temperature as being in the 10^3 K range, since many atoms in the ion slowing-down path are displaced from their lattice sites; this would lead to "local" quench rates in the range of circa 10^{15} degrees per second. Although crude, this type of description is consistent with the high degree of metastability observed for many implanted alloy systems.

The fact that implanted near-surface layers are an integral part of the substrate, without an abrupt interface, affords the possibility of sustaining very high stresses without concern for the delamination that can plague certain types of coatings, especially those deposited at low temperatures.

Ion implanted surfaces experience dimensional changes due to microscopic surface erosion caused by sputtering that are usually less than 0.1 μm. Also, at doses greater than 10^{17} ions/cm^2 some degree of surface roughening may usually occur due to sputtering effects and this may be exploited for surface patterning. However, on a practical engineering scale, it is normally accurate to portray implantation as a process that does not adversely affect substrate dimensions or topography.

In contrast, it will be shown in Section 4.4 that gas cluster ion bombardment can have very pronounced surface smoothing effects in the nanometer scale surface regime.

Implantation processing is often touted as being a room temperature process. However, this depends on the thermal mass of the substrate, the relative ion beam power (W/cm^2) being delivered to the substrate, and ability to conduct heat away from the substrate, since radiative cooling is negligible at desired processing temperatures. These require adequate thermal management practices and sometimes require batch processing to average the thermal load over many components per treatment.

The ability to surface alloy a component (at low temperatures) without changing bulk properties is another attractive feature that has found interest in areas such as improved corrosion behavior of alloys (e.g., precision bearings) that are chosen principally on the basis of their strength and fatigue properties.

Perhaps one of the most under-utilized capabilities of the ion implantation process is its use as a materials research tool, such as a controllable surface alloying method for screening surface alloying content on properties. This attribute has been exploited for studies of oxidation resistance as a function of alloying content, for example. Another powerful use of the technique as a research tool has been the controlled alloying (doping) of selected elements into a narrow region to a controlled concentration and observing the evolution of the implanted layer after thermal annealing; thus giving information on solid solubilities, diffusivities, and trapping in a depth/temperature regime not accessible by other means. The ability to alloy impurity atoms in intimate contact with host atoms without any interfaces present allows these low temperature measurements to be accurately made.

The limited thickness of the implanted layer (typically less than a micrometer) is an intrinsic limitation dictated by the physics of the energy loss processes. This feature is obviously a doubled-sided issue: allowing for no significant dimensional changes but at the same time limiting the surface region that can be alloyed.

Ion implantation is intrinsically a vacuum process requiring background pressures of ~ 10^{-6} torr or less to avoid incorporation of background vacuum species. Component handling through the use of vacuum compatible fixtures, and heat sinking is often required to ensure both uniform dosage and adequate cooling to dissipate the imposed heat load on components due to the energetic ion beam. Directed beam ion implantation is a line-of-sight process. It lacks the "throwing power" of certain other plasma-based coating processes that can better treat irregular surface features, due to multiple scattering of atoms (ions) within the (higher pressure) plasma. Pulsed-plasma ion implantation (see Chapter 8) overcomes this line-of-sight limitation and allows treatment without as much manipulation of components to achieve adequate uniformity for complicated shapes.

The sputter-limited concentration issue was discussed previously as limiting attainable near-surface implanted element concentrations to normally less than 50 at.%., i.e., max. conc. = $1/(1 + S)$, where S is the sputtering coefficient, representing the ratio of the number of surface atoms ejected per incoming energetic ion.

Unlike conventional coatings, implanted atoms are normally in an intimate mixture with atoms of the original substrate. Their relative concentration will be determined by their energy and the sputtering rates of the substrate atoms. Variations of the basic implantation process allow tailoring the energies and fluences of energetic species so as to produce a graded region consisting of an "ion mixed" surface alloyed region with a discrete overlayer composed of very low energy impinging (ions) atoms.

With these many features and potential advantages in mind, the research community has explored many materials science and engineering applications for this process. Table 4.3 lists several of the topical areas have been documented in the open literature. The purpose of the discussion here, however, is to give the reader a flavor of where it has been applied, what has been found, and references to where more detailed descriptions of particular topic areas may be found.

TABLE 4.3 Application Areas of Ion Implantation

Mechanical	Chemical	Other
Tribology: Friction and wear	Aqueous corrosion: Electrochemistry	Electrical conductivity: polymers & ceramics
Surface hardening: metals/ceramics/ polymers	Thermal oxidation	Optical coatings; eliminate pinholes, control refractive indices, control stress & adhesion
Fatigue	Catalysis	Metallurgical research alloying tool for diffusion, solubility, and trapping studies
Fracture toughness: Ceramics	Atmospheric: Tarnishment	Create/modify nanostructures and MEMS

4.2.2 Tribological and Mechanical Property Changes

The wear and friction of ion implanted materials was one of the first non-semiconductor topics widely investigated. Much of this original work was conducted at AERE Harwell, U.K. by Dearnaley and colleagues who were pioneers in this field (G. Dearnaley, Historical perspective of metal implantation, *Surf. Coat. Technol.*, 65, 1–6, 1994). Harwell scientists developed many of the nitrogen ion implantation protocols and were among the first to transition the technology into industrial applications, such as the treatment of expensive plastic molding components as shown in Figure 4.3. Their work was key to several other laboratories worldwide initiating programs in this area in the 1970s and 1980s (Table 4.4). Figure 4.3 also shows an early published result of wear reduction in metals under sliding wear conditions following high-dose nitrogen ion implantation. It is seen that a sufficiently high N dose is required to initiate wear reduction. Such studies

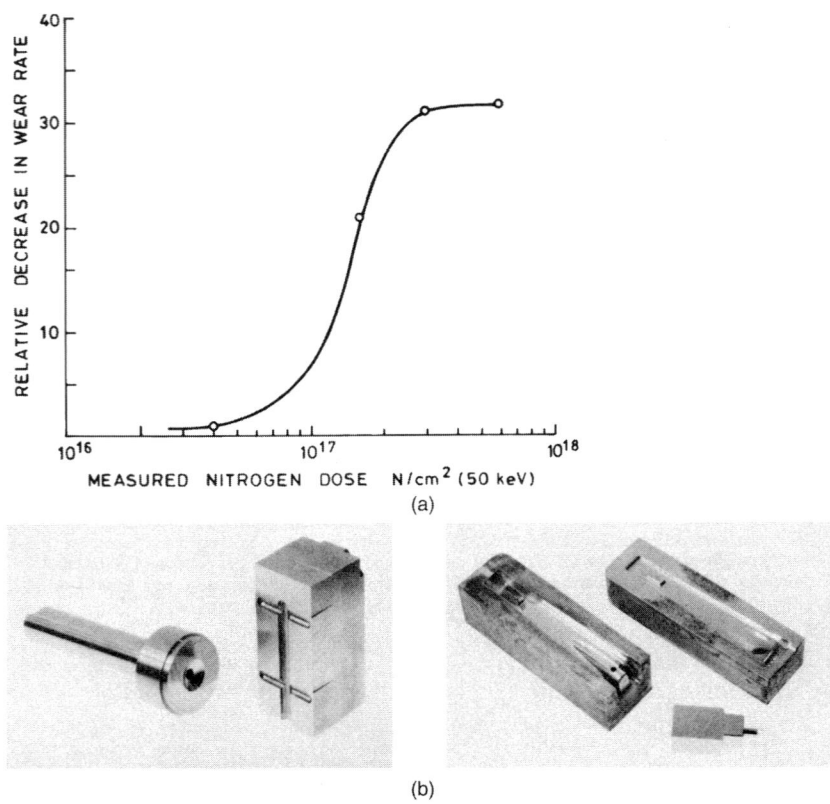

(a)

(b)

Figure 4.3 (a) One of earliest published accounts on the effect of N implantation dose for increasing wear resistance under lubricated sliding wear conditions. (Adapted from N.E.W. Hartley, *Inst. Metall. London* 1101-78-Y, 197–209, 1978.) (b) Examples of plastic molding components benefiting from N implantation conducted at Harwell in the 1970s and 1980s. Sprue bush and runner block (left) and impression mold for injection molding of thermosetting plastics (right). (G. Dearnaley, Practical applications of ion implantation, in *Ion Implantation Metallurgy*, AIME, Warrendale, PA, 1980. With permission.)

formed the basis of applications to compatible industrial wear problems such as are found in the injection molding business. Here, abrasive fillers, such as TiO_2 added for coloring, can be lifetime limiting factors of these expensive molds, in both initial and rework costs as well as in terms of production downtime. Harwell and subsequent commercial vendors in the U.K. treated numerous plastics molding tools and were able to substantially improve the lifetimes of these expensive, precision tools that could not be treated by any other process without risking thermal distortion or compromising the mirror-like finishes typical on such tools. A side benefit of nitrogen implantation for such tools can be enhanced mold release, a feature that directly relates to the quality of finished parts.

The mechanisms of wear are notoriously complicated even without the introduction of a highly non-equilibrium process such as ion implantation that creates both high concentrations of alloy species (N in Fe) and lattice defects that can interact with the implanted species. The types of wear that have been best ameliorated by ion implantation include metal-to-metal adhesive wear and mild abrasive wear. The mechanism(s) responsible for wear reduction by ion implantation are still being elucidated. Besides the normal surface hardening mechanisms deemed important for wear reduction, other less obvious mechanisms appear operative, especially regarding the extended depth range over which N implantation sometimes appears to impart wear resistance. Nitrogen implantation has been attributed to pinning defect structures such as dislocation networks that can directly inhibit wear. This type of treatment was found effective during adhesive wear situations such as lubricated steel-on-steel and for mild abrasive wear situations such as for abrasive fillers in plastic molding operations. Wear reductions of as much as 5- to 30-fold have been observed following high dose N implantation as seen in Figure 4.3. Surface nitriding by nitrogen implantation typically requires doses of 10^{17} ions/cm^2 at implantation energies of tens to hundreds of kilovolts energy, yielding surface N concentrations of 20–30 at.%.

Another early demonstration of implanted N effects on mechanical properties was shown by Hu et al. (W.W. Hu, H. Herman, C.R. Clayton, J. Kozubowski, R.A. Kant, J.K. Hirvonen, and R.K. MacCrone, in *Ion Implantation Metallurgy*, AIME, Warrendale, PA, 1980), who showed a marked decrease in cavitation wear and internal friction of mild steel following N implantation.

Other surface-hardening implantation protocols (see Table 4.4) have been found that also provide wear resistance of steels based on an implantation-induced surface carburization. This accompanies high-dose implantations of carbon-gathering species such as Ti or Ta (G.K. Hubler, *Ion Beam Processing*, NRL Memorandum Report 5828, Naval Research Laboratory, Washington, DC, 1987). These ion beam treatments have been found to be effective in martensitic alloys that had most of their bulk-hardening mechanisms used up, and were thus promising for these bearing type steels. Other wear-reducing implanted metal species (e.g., low-dose Y and Ce) appear to increase the adhesion of surface oxides and thereby stave off deleterious nascent metal-to-metal contact wear modes (G. Dearnaley, in *Ion Implantation Metallurgy*, AIME, 1980; N.E.W. Hartley and J.K. Hirvonen, *Nucl. Instrum. Methods,* 209/210, 933, 1983).

The possibility of improved fatigue performance (Table 4.5) was a natural follow-on to wear improvements based on common mechanisms, including dislocation pinning. Early experiments in this area included Ti/6Al/4V alloy implanted with either N or C ions by Vardiman (R.G. Vardiman, The improvement of fatigue life in Ti-6Al-4V by ion implantation, *J. Appl. Phys.*, 53(1), 1982) and low carbon (1018) steel by Hu et al. (W.W. Hu, H. Herman, C.R. Clayton, J. Kozubowski, R.A. Kant, J.K. Hirvonen, and R.K. MacCrone, in *Ion Implantation Metallurgy*, AIME, Warrendale, PA, 1980). The third example given in Table 4.5 concerns fretting fatigue, closely coupled with the oxidative behavior of the surface. Ba implantation was shown to significantly reduce oxidation of Ti alloys by formation of perovskite structures and has been attributed with improving the fatigue behavior of Ti alloys by the same mechanism (G. Dearnaley and P. Goode, *Nucl. Instrum. Methods*, 189, 117–132, 1981).

TABLE 4.4 Examples of Ion Implantation Effects on Mechanical Properties

Wear Category	Substrate	Ion Species	Results	Ref.
Adhesive wear	Steels	N	5–30× wear reduction	Hartley (a)
Wear and fatigue	Steels, tools	N, B, C; Y, Ce	Improved lifetimes of tooling by N; 5–8× sliding wear reduction attributed to improved oxide adhesion	Dearnaley (b)
Adhesive wear	Steels	Ti, C	Amorphous surface alloy formed	Knapp et al. (c)
Abrasive wear	Steels	N, Ti	Review of mechanisms with qualitative model	Hubler and Smidt (d)
Fatigue and cavitation wear; internal friction	1018 steel	N	>10× lifetime increase seen for N-implanted and aged steel in low cycle fatigue; incubation period for cavitation wear prolonged under same conditions; internal friction reduced by ~ 2	Hu et al. (e)
Wear and friction	Production tooling	N	2–6× improvements found	Straede (f)

Sources: (a) N.E.W. Hartley, in *Treatise on Materials Science and Technology, vol. 18*, Academic Press, 1980; (b) G. Dearnaley, Practical applications of ion implantation, *J. Metals*, Sept., 18, 1982; (c) J.A. Knapp, D.M. Follstaedt, and B.L. Doyle, *Nucl. Instrum. Methods Phys. Res. B*, 7/8, 1985; (d) G.K. Hubler and F.A. Smidt, Application of ion implantation to wear protection of materials, *Nucl. Instrum. Methods Phys. Res. B*, 7/8, 1985; (e) W.W. Hu, H. Herman, C.R. Clayton, J. Kozubowski, R.A. Kant, J.K. Hirvonen, and R.K. MacCrone, in *Ion Implantation Metallurgy*, AIME, Warrendale, PA, 1980; (f) C.A. Straed, Practical applications of ion implantation for tribological modification of surfaces, *Wear*, e130, 1989.

TABLE 4.5 Ion Implantation Effects on Fatigue Lifetimes

Fatigue Category	Ion, Substrate	Results	Ref.
Low cycle fatigue	N,C in Ti/6Al/4V	C gives 20% increase in endurance limit and 4–5× lifetime increase	Vardiman (a)
Low cycle fatigue	1018 steel	2–3× improvements in fatigue lifetimes	Hu et al. (b)
Fretting fatigue	Ba implanted in Ti/Ti alloy plus shot peening	Duplex treatment	Dearnaley and Goode (c)

Sources: (a) R.G. Vardiman, The improvement of fatigue life in Ti-6Al-4V by ion implantation, *J. Appl. Phys.*, 53(1), 1982; (b) W.W. Hu, H. Herman, C.R. Clayton, J. Kozubowski, R.A. Kant, J.K. Hirvonen, and R.K. MacCrone, in *Ion Implantation Metallurgy*, AIME, Warrendale, PA, 1980; (c) G. Dearnaley and P. Goode, *Nucl. Instrum. Methods*, 189, 1981.

4.2.3 Chemical Property Modification

Historically, some of the first observations of improved surface resistance to environmental effects related to reduced atmospheric tarnishment following ion bombardment (see Table 4.6). Many of these laboratory curiosity observations were not pursued to a point of being understood but were merely a notable side issue in other investigations.

Ashworth et al. (V. Ashworth, R.P.M. Procter, and W.A. Grant, The application of ion implantation to aqueous corrosion, in *Ion Implantation, Treatise on Materials Science and Technology*, Academic Press, 1980) and Wolf (Chemical properties of ion implanted materials, in *Ion Implantation, Treatise on Materials Science and Technology*, Academic Press, 1980) were among the first to explore aqueous corrosion in the 1970s. Several research groups have since studied the basic mechanisms and potential application areas by which ion implantation can improve aqueous corrosion resistance. One application area that has been studied involves the corrosion protection of expensive aerospace bearing alloys for critical rolling element bearing components (F.A. Smidt, B.D. Sartwell, *Nucl. Instrum. Methods Phys. Res. B*, 6, 70–77, 1985). These components can suffer from aggravated localized pitting corrosion due to salt in the environment. These pits are the basis for bearing rejection due to the concern over rolling contact fatigue failures originating at such flaws. These efforts showed the technical feasibility of improving the surface by ion implantation alloying, albeit it to a very thin region, producing either Cr-rich stainless steel-like surfaces on corrosion-prone martensitic alloys.

These results and others (see Table 4.6) demonstrated that implantation alloying can yield surface alloys essentially equivalent to conventional alloy (e.g., stainless steels) and can also produce non-conventional (metastable/amorphous) alloys at near ambient processing temperatures without altering bulk properties. The usefulness of producing viable corrosion resistant surfaces by ion implantation will depend crucially on the absence of pinholes that can arise from surface debris during treatment or from having an initial surface finish of too rough a scale to allow adequate coverage by the line-of-sight ions. In this regard, the criteria for surface coverage can differ significantly for most wear situations where small defects are not normally deleterious.

4.2.4. Thermal Oxidation

The utility of ion implantation alloying has been used extensively in the study of thermal oxidation, both in terms of mechanisms and as a means of treating critical components (see Table 4.7). Here again, Harwell researchers have been leaders in studies of ion implantation on thermal oxidation. Their interests were influenced by potential reliability concerns in high-temperature power components. Advantages of ion implantation for this application include the high degree of control in surface alloying, the need for only small amounts of alloying elements, and applicability to a wide range of

TABLE 4.6   Chemical Effects of Ion Implantation

Corrosion System or Property	Substrate	Ions	Results, Comments	Ref.
Atmospheric tarnishment	Uranium	Ar	Atmospheric tarnishment significantly reduced	Trillat and Haymann (a)
Aqueous corrosion-equilibrium alloys	Fe	Cr	Ion implantation allows both study and production of corrosion resistant surface alloys	Ashworth et al. (b)
Corrosion (pitting) of bearing alloys	M50, 52100	Cr, P, Ta	Both equilibrium and amorphous alloys studies; effective against pitting corrosion	Clayton (c)
Aqueous corrosion, pitting corrosion	Martensitic bearing alloys	Cr, Ta, P	Reference reviews program to apply findings above to aerospace bearings	Smidt and Sartwell (d)
Corrosion/ wear	Ti-6Al-4V alloy used for ortho-pedic implants	N	Corrosion-wear tests to simulate surgically implanted alloys showed orders of magnitude improvement	Williams et al. (e)

Sources: (a) J.J. Trillat and A. Haymann, in J.J. Trillat (Ed.) *Le Bombardment Ionique*, CRNS, Paris, 1960; (b) V. Ashworth, R.P.M. Procter, and W.A. Grant, in *Treatise on Materials Science and Technology, vol. 18*, Academic Press, 1980; (c) C.R. Clayton, in L. Rehn, S.T. Picraux, H. Wiedersich (Eds.), *Surface Alloying by Ion, Electron, and Laser Beams*, Am. Soc. Met., Metals Park, Ohio, 1987; (d) F.A. Smidt and B.D. Sartwell, *Nucl. Instrum. Methods Phys. Res. B*, 6, 1985; (e) J.M. Williams, G.M. Beardsley, R.A. Buchanan, and R.K. Bacon, *Mater. Res. Soc. Symp. Proc.*, 27, 1984.

TABLE 4.7 Thermal Oxidation Protection by Ion Implantation

Substrate	Ion Species	Results	Ref.
Ti	Ca, Ba	Implanted impurities appear to form diffusion barrier structures along defect routes	Dearnaley (a)
High Cr, Ni stainless steel in CO_2 coolant	Y, Ce, others	Time to initiation of oxide spallation prolonged by orders-of-magnitude: general review of ion implantation/oxidation studies	Bennett (b), Bennett and Tuson (c)

Sources: (a) G. Dearnaley, in *Treatise on Materials Science and Technology, vol. 18*, Academic Press, 1980; (b) M.J. Bennett, in G. Was and K.S. Grabowski (Eds.), *Environmental Degradation of Ion and Laser Beam Treated Surfaces*, AIME, Warrendale, PA, 1989; (c) M.J. Bennett and A.T. Tuson, *Mater. Sci. Eng. A*, 116, 1989.

substrates. The ability to use ion implantation as a development and research tool allowed high-temperature oxidation researchers the means to better understand the role of selected elements in the formation and growth of thermal protective oxide coatings.

Most metals when exposed to oxygen undergo a reaction of the following kind:

$$x\text{Me} + \frac{1}{2} y\text{O}_2 \rightarrow \text{Me}_x\text{O}_y$$

The reaction is normally driven by the change in free energy associated with the particular metal-oxygen interaction. Typical oxide growth is governed by diffusion-limited kinetics. A desirable protective oxide should therefore be very stable and dense.

Ion implantation has been demonstrated to be effective in improving factors important for oxidation resistance. These include improved oxide adhesion and cohesion within the growing oxide film. Both of these factors reduce the tendency of coatings to spall after repeated exposure to alternating thermal cycles. Oxide films formed on ion-implanted surfaces have often shown the tendency to form continuous, defect-free oxides more resistant to oxidation than non-implanted surfaces.

Another interesting observation has been made in the case of Ba or Ca implanted into Ti concerning the effective depth of the implanted species (G. Dearnaley, in *Treatise on Materials Science and Technology, vol. 18,* Academic Press, 1980). Here the proposed mechanism for observed oxidation resistance was the formation of perovskite structures (e.g., $BaTiO_4$) that blocked the defect structures (grain boundaries) that normally allow rapid ingress of O_2. Early work on this system showed that an oxidation inhibiting effect could be seen on the non-implanted side of 50 μm thick Ti foils that had been vacuum (10^{-7} torr) annealed at 850°C before being oxidized at 600°C in air. This strongly suggests rapid (grain boundary) migration of Ba along such defect routes at elevated anneal temperatures with a subsequent role in reducing oxidation.

The doses required to achieve these reductions were typically much lower (circa 10^{16} cm^{-2}) than required for tribological applications and would only require 1–5 sec treatment times for a 1000 μA (1 mA) ion beam current achievable with present ion beam technology.

The technical and economic feasibility of treating large-scale structures by ion implantation was actually demonstrated. The intended components of the project were the 50,000 fuel pins (1 m long) for all of the nuclear plants in the U.K. Although the U.K. government decided to cut back that nuclear program (Dearnaley), the technical feasibility of employing ion implantation treatment was convincingly demonstrated as described below. Figure 4.4 shows the effect of Ce and Y implantation on increasing the time to spallation for both ion implanted alloy surfaces and conventional

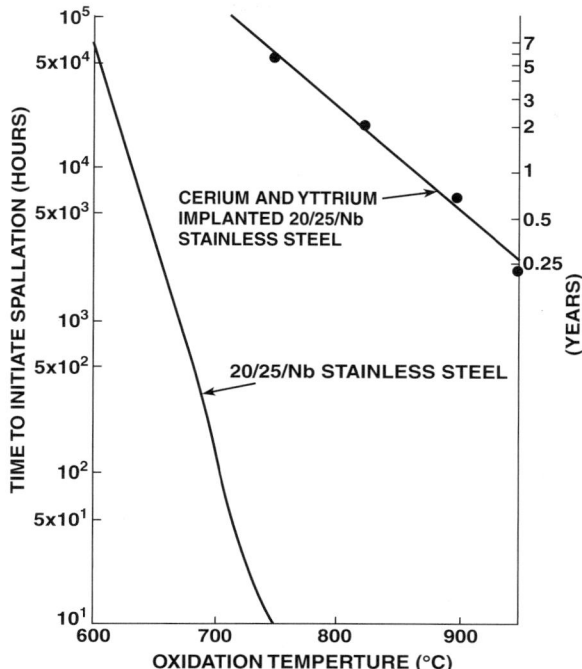

Figure 4.4 The time for oxide spallation to initiate for 20Cr–25Ni stainless steel when cooled from oxidizing temperature (600–950°C) to room temperature in CO_2, for both Y and Ce implanted samples and base alloy. (Adapted from M.J. Bennett, in G. Was and K.S. Grabowski (Eds.), *Environmental Degradation of Ion and Laser Beam Treated Surfaces*, AIME, Warrendale, PA, 1989, p. 261.

(untreated) alloy surfaces. It has been seen that these (implanted) elements segregate to the oxide grain boundaries and inhibit Cr outdiffusion and the oxide adheres better to the substrate (Dearnaley). The many orders-of-magnitude improvement seen here for implanted test coupons were unfortunately never tested under actual reactor conditions.

4.2.5 Implantation into Polymers and Ceramics

Although metals were the principal substrates of ion implantation studies, the use of implantation for altering polymer and ceramic properties, notably electrical conductivity for polymers and wear/fracture toughness for ceramics, became active research areas as well.

In comparison with metals, polymers and ceramics retain much more influence of the ion implantation lattice disorder in their behavior due to significantly less self-annealing during and after the ion implantation process. This, as stated earlier, is due to the more directional and fragile bonding configurations present in the polymer and ceramic substrates.

Early work (Venkatesan, *Nucl. Instrum. Methods B*, 7/8, 461–467, 1985) studied the effect of high-energy heavy (2 MeV Ar) ions on the various polymer–ion interactions as a function of dose. The effects of low dose ($< 10^{13}$/cm^2) ion bombardment encompass: (i) polymer cross-linking (10^{11} to 10^{13}/cm^2), (ii) beam-induced polymerization of monomers, and (iii) polymer dissociation. At somewhat higher fluences (10^{13} to 10^{17}/cm^2) carbonization, inorganic compound synthesis, and changes in the electronic transport are prevalent.

Novel material synthesis was demonstrated early on by the bombardment of an organo-silicon polymer with inert ions. Loss of hydrogen was observed in polymers during irradiation and enrichment of the Si that was attributed to reacting with the lattice C to produce (hard) SiC. Similar observations pertain to the controlled bombardment of silicated organic oils to produce silicated "diamond-like carbon" coatings of very high hardness and low friction that are discussed in a later section.

Dramatic decreases in electrical resistivity are also observed in many irradiated polymers, with specific resistivities approaching those of amorphous metals obtainable following implantation (Venkatesan, *Nucl. Instrum. Methods B*, 7/8, 461–467, 1985). Somewhat later work by the ORNL group shown in Table 4.8 showed that hardness improvements in irradiated Kapton could be attributed to the ionizing (electronic energy loss) constituent of the beam. This allowed hardness values comparable to mild steels to be attained for implanted Kapton.

Ion implantation into ceramics has been viewed as a potential means of improving surface-limiting features of

TABLE 4.8 Early Observations of Ion Implantation-Induced Changes in Polymers

Polymer	Ions, Energies	Hardness/Conductivity Property Change	Ref.
Photo resists	Inert species: Ar	Reactions ranging from scissioning to compound formation seen according to dose	Venkatesan (a)
Polystyrene, Kapton, Mylar	H, B, Si, Ar 200 keV to 2 MeV	Surface hardening up to 14 GPa observed; electronic stopping component; enhanced hardness increases while nuclear components appeared to reduce hardness	Lee et al. (b)

Sources: (a) T. Venkatesan, *Nucl. Instrum. Methods Phys. Res. B*, 7/8, 1985; (b) E.H. Lee, G.R. Rao, M.B. Lewis, and L.K. Mansur, Effects of electronic and recoil processes in polymers during ion implantation, *J. Mater. Res.*, 9(4), 1994.

ceramics, such as surface-initiated microcracking, and improved flexure strength. Some of the early reports on ion implantation effects in ceramics are listed in Table 4.9 along with reference to a more recent review of data in the field.

4.2.6   Ion Implantation as a Materials Science Tool

In an early, elegant series of experiments, S. Myers et al. of Sandia Laboratory used ion implantation as a metallurgical tool for the purpose of studying solubility, diffusion, and trapping (see Table 4.10). This was done by systematically alloying substrates by ion implantation to high doses (i.e., to a non-equilibrium state) and thermally annealing them to study their return to equilibrium using ion beam scattering techniques. Their studies allowed them to quantitatively measure solubility limits and diffusion coefficients in time/temperature regimes that were orders of magnitude lower than normally measured.

TABLE 4.9   Ion Implantation-Induced Changes in Ceramics

Ceramic Substrate	Ion Species	Result	Ref.
Alumina	Cr	30–40% increased hardness	McHargue et al. (a)
Al_2O_3	N, Ni	Flexure strength increased by 20% to 60%	Hioki et al. (b)
Al_2O_3, $SrTiO_3$, ZnO, $LiNbO_3$, and MgO	Self ions, Al, O, Fe, Cr, other	Review of ion implantation effects in ceramic substrates	White et al. (c)

Sources: (a) C.J. McHargue et al., in G.K. Hubler et al. (Eds.), *Ion Implantation and Ion Beam Processing of Materials*, Elsevier, New York, 1984; (b) T. Hioki et al., *Nucl. Instrum. Methods Phys. Res. B*, 7/8, 1985; (c) C.W. White et al., Ion implantation and annealing of crystalline oxides, *Mater. Sci. Rev.*, 4(2), 1989.

4.2.7   Emerging Research Areas: MEMS and Nano-Composite Structures

The advent of microstructured and nanostructured materials has ushered in a fresh wave of potential applications for ion implantation that are attractive because of the comparable length scales of component dimensions and the penetration distances associated with implantation. There are similarities with semiconductor usage in some cases, e.g., treatment of planar structures on a depth scale commensurate with ion ranges. Instead of electrical doping, however, ions are being used to alter tribological properties on a microscale, selectively etch micro-electric mechanical systems (MEMS) structures, alter optical properties by doping or damage creation, and form metallic nano-composites of selected elements in ceramic/polymer substrates for potential optical applications. Mazzoldi et al. (Chapter 6) review the area of nanoparticles for nonlinear optics, so no more will be said here on this topic.

TABLE 4.10 Ion Implantation as a Metallurgical Tool:
Metastability, Diffusion, Solubility, and Trapping Studies

Property Studied	System Studied	Result	Ref.
Alloy metastability	Several ion implanted elements in Cu, Fe, Be substrates	Early review of metastability of ion-implanted alloy systems as examined by ion channeling and transmission electron microscopy measurements and described in terms of Hume–Rothery and Miedema rules	Poate and Cullis (a)
	Several ion/ substrates	Review of topical area, discussion of models	Follstaedt (b)
Solubility	Fe, Cu in Be	Allowed extension of solid solubility data to much lower temperatures	Myers (c)
Low-temp. diffusion	Cu in Be, Sb in Fe	Enabled reliable diffusion data to be obtained at greatly lowered temperatures	Myers (c)
Solute trapping	Sb and Ti in Steel	Findings relevant to explaining impurity embrittlement phenomena	Myers (c)

Sources: (a) J.M. Poate and A.G. Cullis, in *Treatise on Materials Science and Technology, vol. 18*, Academic Press, 1980; (b) D.M. Follstaedt, *Nucl. Instrum. Methods Phys. Res. B*, 7/8, 1985; (c) S.M. Myers, in *Treatise on Materials Science and Technology, vol. 18*, Academic Press, 1980.

MEMS are structures that employ both mechanical and electrical inputs/ outputs as the name implies. The materials primarily used for MEMS fabrication are Si (wafers) and deposited polysilicon, with silicon nitride commonly used as a masking material along with silicon oxide and deposited metal films. One of the earliest uses for these type structures was as an accelerometer sensor for automobile airbags.

Electroplated Ni is commonly deposited in precision molds to produce mechanical components such as microgears that require metal-to-metal sliding contact, and untreated Ni-on-Ni is a poor tribological contact couple. Sandia Laboratory researchers have addressed this problem by applying Ti plus C ion implantation treatments, originally developed during the 1980s, to the surfaces of these component materials to improve their friction/wear properties for these high value microcomponents (see Table 4.11). Co-implantation of these elements has been shown previously to produce an amorphous surface alloy at sufficiently high doses, typically 20 at.% or greater. The amorphous surface alloy exhibits reduced friction (by a factor of two) and a wear rate reduced by an order of magnitude. In the case of Ti plus C implanted Ni, they saw an enhanced (2×) yield strength and increased elastic modulus for the implanted layer versus the untreated material and a significant hardening effect. These values were deduced from nano-indentation testing. Friction and wear testing of these surfaces showed the wear mode to be significantly reduced with a much lower friction coefficient as would be desired and

TABLE 4.11   Ion Implantation into MEMS Devices

Area	System	Ion Species	Result	Ref
Improved tribology of electroplated Ni (LIGA) for MEMS applications	Ni	Ti and C	Significant increase of hardness and lower friction of Ni (LIGA) MEMS surfaces	Meyers et al. (a)
Selective etching of Si for MEMs devices	Si	Au	Implanted Au areas reduce etching rate of selected area	Nakano et al. (b)

Sources: (a) S.M. Myers et al., *Surf. Coat. Technol.*, 1998; (b) S. Nakano et al., *Surf. Coat. Technol.*, 2000.

required for treatment of Ni MEMS surfaces (S.M. Myers, J.A. Knapp, D.M. Follstaedt, M.T. Dugger, and T. Christenson, *Surf. Coat. Technol.*, 103–104, 287–292, 1998).

D. Nagel (*Surf. Coat. Technol.*, 103–104, 138–145, 1998) discusses present and potential application of ion beams for MEMS fabrication. Besides the present implantation for inducing conductivity of poly-silicon, he points out their promise for controlled etching and deposition of materials for mechanical or RF tuning.

4.2.8 Commercial Usage and Applications

4.2.8.1 Precision Tooling

Ion implantation has been successfully used for a variety of commercial applications at a number of service centers in the U.S. and Europe for high-value-added components. For example, the Tribology Centre, Danish Technological Institute in Aarhus, Denmark is employing implantation as one surface processing tool among several to augment and complement conventional surface processes. The philosophy of implementation is to use ion implantation in concert with other surface treatment processes for a wide variety of surface related problems. Many of the center's most successful commercial examples are proprietary and thus not available for publication; however, two case examples are given in Figure 4.5.

Trion, a Danish company manufacturing ion implantation equipment, also works with potential users and issues case studies to customers for helping them identify markets and appropriate treatments. Table 4.12 is derived from its case study handouts.

4.2.8.2 Medical Materials and Applications

Implant Sciences Corp. is a U.S. provider of ion implantation services. Some of their commercial applications are described in more detail on their company Web site (http://www.implant-

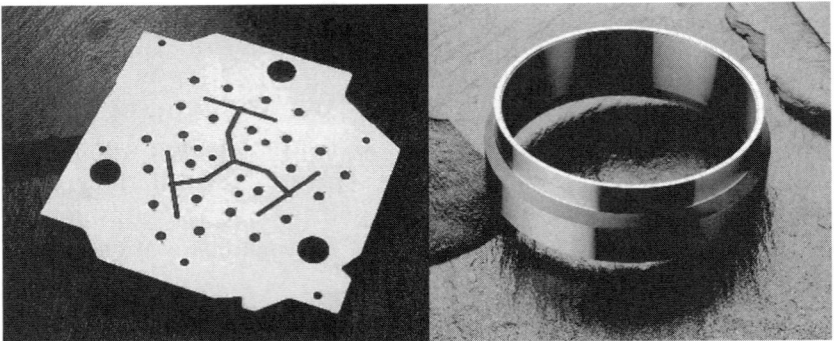

Figure 4.5 Case A (left) Corrosion of injection mold vents reme-
died by chromium ion implantation; Case B (right) Adhesive wear
of precision punching tools for manufacturing sheet metal contain-
ers reduced by nitrogen ion implantation. Case A: "A major manufacturer of sensor and control units has a large production of plastics
injection molding. For this particular form plate the company had
severe problems arising from corrosive attacks on the hardened steel
mold, localized to most of the hole edges at the vents of the form
plate. The corrosive attacks resulted from aggressive gases evaporating from the plastics during the molding process. However, the
company has now eliminated the corrosion problems by applying
chromium ion implantation on the attacked areas, with significantly
increased tool life, and the company has now solved a very significant and expensive production problem." (From C. Straede, Danish
Tribology Centre, Aarhus, Denmark. With permission.) Case B:
"DCT A/S in Denmark is a leading manufacturer of precision forming tools. Extremely long tool life is often demanded on bending,
drawing and punching tools for thin metal sheets and tin plates.
However, such tools are often subjected to adhesive wear, which
causes a micro-welding of the sheet metal to the tool surface, and
this will reduce the tool life significantly. By applying the nitrogen
ion implantation on the wear area at the cutting and bending edges
of the tools, the tool life is increased significantly. This delicate
precision punch, made of highly alloyed hardened tool steel by DCT,
has been nitrogen ion implanted at the cutting edge resulting in a
tool life extension by a factor of 3–5." (From N.J. Mikkelsen, Danish
Tribology Centre, Aarhus, Denmark. With permission.)

TABLE 4.12 Selected Commercial Nitrogen Ion Implantation Case Studies

Component	Problem	Result, Customer
WC step drill (20 mm) for SS4212 Al	Adhesive wear and sticking	Reduced adhesive wear, $3 \times$ lifetime; Unimerco AB, Sweden
M42 steel broach for Ti/6Al/4V	Heat buildup at tool corners, causing accelerated wear mode change	12-plus holes produced vs. 4–6 for untreated; for "leading aerospace company"
High-speed steel reamer for 42CrMo4 steel	Loss of tolerance due to wear	350 holes for implanted reamer; vs. 200 holes untreated; Guhring oHG, Germany

Source: B. Torp, Trion, Denmark.

sciences.com/products/surface/ion_implantation.html) and are directly quoted below with their permission (see Figure 4.6).

> Current applications include decreased polyethylene wear against CoCr bearing surfaces. Ti-6Al-4V is a softer material than CoCr or bone and fretting wear can be a significant problem on the stems of modular hip and knee prostheses. Ion implantation is used on the stem tapers to reduce fretting wear dramatically. Test results show an order of magnitude reduction in fretting of titanium alloy in micro-motion contact against CoCr. CoCr femoral hip and knee components cause their mating component of ultrahigh molecular weight polyethylene (UHMWPE) to wear. To reduce wear debris, the CoCr bearing surface is treated. Ion implantation will decrease the coefficient of friction of CoCr against polyethylene, and improve the material's wettability. Improved wettabiliy helps to retain a lubricating boundary layer and reduce wear of polyethylene components. Ion implanted CoCr on UHMWPE wear testing shows virtually no material loss after one million cycles.

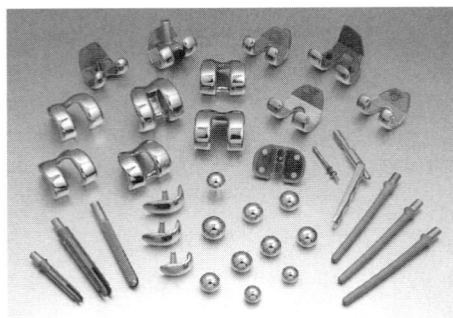

Figure 4.6 Orthopedic components treated with ion implantation and increased resistance of Ti-6Al-4V for fretting fatigue and wear. (From Implant Sciences Corp. Web site. With permission.)

Another application of ion implantation they market is for the predoping of brachytherapy seeds for surgical implants to treat cancerous growths such as prostate cancer. Xenon-124 is implanted into a Ti alloy "seed" and neutron activated to produce radioactive iodine-124. Their proprietary method of producing seeds in this manner avoids radioactive implantation and is a dry, reliable, and safe manufacturing method.

A third ion beam business area of theirs in biomedical products utilizes sputter etching to produce a patterned or roughened surface to improve adhesion, enhance tissue ingrowth, or increase electrical-charge transfer (Figure 4.7).

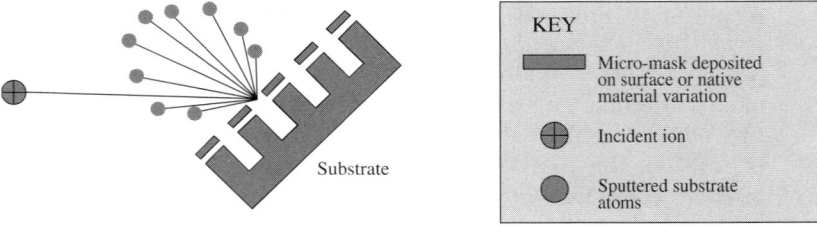

Figure 4.7 Schematic of ion beam texturing. (From Implant Sciences Corp. Web site. With permission.)

Ion beam texturing is accomplished using a high energy beam to selectively remove material from the substrate. Shown is a fine-mesh mask, which restricts the ablating of material to define (unmasked) areas. Natural and random seeding are alternative masking choices. Ion beam texturing results in surface morphologies with finer resolution than can be achieved with the use of laser beams. Current biomedical applications include peritoneal implants, soft tissue implants, hydrocephalic shunts, percutaneous connectors, dental implants, and orthopedic prostheses. One particular use is that of texturing pacemaker electrode tips to improve the threshold for pulsing and increase battery life.

4.2.8.3 Sensors

Another U.S. company employing ion implantation for microdevice fabrication is Ion Optics (http://www.ion-optics.com/ir.asp). The description quoted below is from their web site describing the gas sensors made possible by using ion beam textured surfaces (see Figure 4.8).

Ion Optics offers a new class of electrically pulsed, high intensity infrared radiators for gas analysis, spectroscopy and calibration. These radiators feature a low thermal-mass filament tailored for high emissivity in either the mid-wavelength (MWIR) (2–5 μm) or long-wavelength infrared (LWIR) (8–12 μm). Efficient in-band emission permits operation > 8 μm at temperatures several times cooler than tungsten bulbs and gives a radiator life of many years. This patented, high-efficiency device minimizes drive power, greatly reducing parasitic heating of detectors and optics; it also eliminates the mechanical choppers, permitting a sealed optical path.

Large Temperature Modulation: The high emissivity of the ion-beam-treated radiator surface enables it to efficiently and rapidly cool via thermal radiation. The infrared images below (left pair) show an Ion Optics source being electrically pulsed; the hot filament (left image) nearly cools to background temperature (right image) before the next pulse, thus providing several hundred degrees of temperature modulation. While the image pair

Figure 4.8 Electrically pulsed, high-intensity infrared radiators for gas analysis, spectroscopy, and calibration. These radiators feature a low thermal-mass filament tailored for high emissivity in either the MWIR (2–5 μm) or the LWIR (8–12 μm). (From Ion Optics. With permission.)

shows a traditional light source that must be run many times hotter than the Ion Optics source. The package heats up considerably and the apparent IR temperature difference in each pulse becomes only a few degrees.

The Technology in the Infrared Filaments: The emissivity of these metal filament infrared sources is enhanced and controlled by creating random surface texture (micron scale rods/cones). This texture modifies the reflection and absorption spectra relative to that for a flat filament of the same material. For wavelengths small compared to the feature sizes, the surface scatters most incoming light, therefore it has low reflectivity (the filaments appear visibly black), and by Kirchoff's law, it must also have high emissivity (>80%). For wavelengths long compared to the feature sizes, the surface still looks like flat metal and it therefore has low emissivity, characteristic of the flat metal (~ 0.1).

4.2.9 Ion Implantation Equipment

Ion implanters fall into two basic categories: mass-analyzed and non-mass-analyzed implanters. For applications where the ion beam purity and charge state (energy) distribution is critical (semiconductor processing), the use of a mass analyzed implanter is usually required. This is normally done by means of a beam-deflecting magnet, which adds to the size, complexity and cost of the equipment.

Simpler designs are possible when constraints on beams are not as critical, such as for nitrogen implantation of metals for wear reduction. Here, directed beam ion implanters have been built that enable high current (10 mA) non-mass analyzed beams of 20–100 kV energy to be achieved. The major challenges in utilizing these implanters on a commercial basis is in: (a) manipulating components so as to uniformly treat all important surfaces with ions, and (b) providing for heat removal from the components being treated in the high vacuum environment. The latter requirement for heat removal pertains to whatever design is used and is often the most vexing challenge for performing economic batch treatment of small parts.

The desire to circumvent the intrinsic line-of-sight limitations of beam-line implantation has prompted researchers to explore the use of pulsed plasma ion implantation to achieve highly uniform coverage of complicated shapes without the need of more complex manipulation jigging as often needed in beamline implantation. This approach is discussed in detail by Misra and Nastasi (Chapter 8) as to its features and applicability.

Continual evolution in ion source technology has allowed the efficient production of high beam currents of most metals using concepts from past fusion studies without the need for magnetic beam analysis. Estimates of treatment costs using such ion sources have been published and show that economy of scale is possible for even large sized objects (L $\propto$ 1 m) at a cost of cents (U.S.) per square inch. Other customized versions of ion sources are being configured for nominally difficult line-of-sight applications, such as the interiors of cylinders.

4.3 ION BEAM-ASSISTED DEPOSITION (IBAD)

The beneficial role that energetic ions play in thin film vacuum deposition techniques has long been realized by the coating community. Mattox (D.M. Mattox, *Handbook of Physical Vapor Deposition (PVD) Processing*, Noyes, 1998) showed as early as 1963 during his early development of ion plating that energetic ions within plasmas have an important influence on coating properties. Other plasma based deposition processes, such as activated reactive evaporation (ARE), developed by Bunshah and coworkers (P. Nath and R. Bunshah, *Thin Solid Films*, 69, 63, 1980), employ ionization to enhance film properties. Research on the role of ions in film deposition increased in the 1970s and early workers in the field include Weissmantel (C. Weissmantel, *Thin Solid Films*, 63, 315, 1979), Pranevicius (L. Pranevicius, Structure and properties of deposits grown by ion-beam-activated vacuum deposition techniques, *Thin Solid Films*, 63, 77, 1979), Pranevicius and Harper (J.M.E. Harper, J.J. Cuomo, R.J. Gambino, and H.E. Kaufman, in O. Auciello and R. Kelly (Eds.), *Ion Bombardment Modification of Surfaces: Fundamentals and Applications*, Elsevier, Amsterdam, 1984, pp. 127–162). It is difficult, however, in many of the plasma based coating techniques, to separate out the degree to which ion and neutral particle fluxes as well as ion energies affect resultant coating properties. The bombardment of growing films with energetic particles, coined IBAD (ion beam-assisted deposition) can produce beneficial modifications in a number of characteristics and properties critical to the performance of thin films and coatings such as improved adhesion, densification of films grown at low substrate temperatures, modification of residual stresses, control of texture (orientation), modification of grain size and morphology, modification of optical properties, and modification of hardness and ductility. The process has also been used in concert with other surface treatment directed beam processes such as those using lasers (R.P. Reade, S.R. Church, and R.E. Russo, Ion-assisted pulsed-laser deposition, *Rev. Sci. Instrum.*, 66, 3610, 1995).

This section will offer a generalized discussion of the IBAD process, illustrated by examples of particular coating properties affected without recourse to detailed mechanistic explanations. More detailed discussions of energetic ion/solid interactions are available elsewhere in this volume and in other treatises (M. Nastasi, J.W. Mayer, and J.K. Hirvonen, *Ion–Solid Interactions*, Cambridge University Press, 1993) for the reader interested in those specifics. The details of film growth without the assistance of ion bombardment will be briefly reviewed, followed by the influence of ion bombardment on thin film growth and on compound formation by IBAD processing. In general, the approach will be an empirical one, based on experimental results and observations. Analytical modeling in this field is limited, with most predictions obtained from computer simulations. Specific examples of results for IBAD processing including optical and electronic coatings, and the application areas of tribology and corrosion will be presented in a later section.

4.3.1 Film Growth Without Ion Beams

The microstructure of a thin film or coating is largely determined by the early stages of film nucleation and growth. Researchers such as Grovenor et al. (C.R.M. Grovenor, H.T.G. Henzell, and D.A. Smith, *Acta Metall.*, 32, 773, 1984) and Thornton (J. Thornton, *J. Vac. Sci. Technol. A*, 6, 3059, 1986) have examined the morphologies of metallic films grown by thermal evaporation and conclude that the microstructure developed in coatings is strongly influenced by the substrate temperature, T_S, expressed as a fraction (T_S/T_M) of the absolute melting point of the deposited material, in degrees Kelvin. Their analyses of the coating microstructures as a function of temperature showed that it is possible to classify the film morphology into four characteristic zones, shown schematically in Figure 4.9 and briefly described below.

Films grown on substrates held at temperatures T_S less than $0.15\ T_M$, are formed of equiaxial grains, typically 5–20 nm in diameter. In this temperature regime the deposited atom's mobility is low, and the atoms stick where they land.

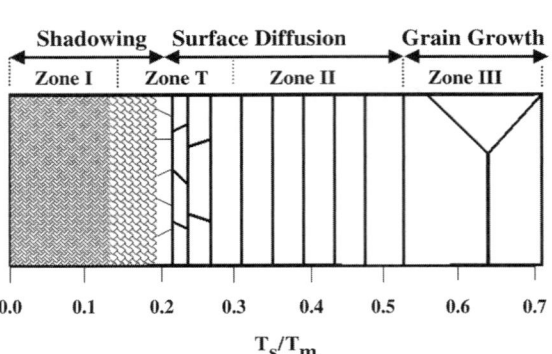

Figure 4.9 Zone model for the grain structure of vapor deposited metal films and mechanisms controlling microstructural development. (Adapted from C.R.M. Grovenor, H.T.G. Henzell, and D.A. Smith, *Acta Metall.*, 32, 773, 1984.)

In the next temperature regime, $0.15 \, T_M < T_S < 0.3 \, T_M$, a microstructure is observed that can be described as a transition microstructure between the equiaxed microstructure of Zone I and the columnar structure found in Zone II. This transition is attributed to the onset of surface diffusion, which allows the deposited atoms to surface migrate before being covered by the arrival of fresh material. Under these conditions, grain boundaries are becoming mobile, and the initiation of grain growth is expected.

In the third temperature regime, $0.3 \, T_M < T_S < 0.5 \, T_M$, the deposited atoms have sufficient mobility to diffuse and increase grain size before being covered by newly deposited atoms. In addition, grain boundaries become more mobile. Films grown in this temperature regime generally have a uniform columnar grain structure, with grain diameters less than the film thickness but increasing with both temperature and film thickness.

Finally, films grown in the fourth temperature regime, $T_S > 0.5 \, T_M$, have squat uniform columnar grains, with grain diameters typically larger than the film thickness. The grain structure in this zone is attributed to bulk rather than surface diffusion driven grain growth.

Several distinct modes of film growth have been observed in traditional (non-IBAD) thin film experiments. These include: (i) Frank–van der Merwe type in which the film grows layer by layer with complete coverage, (ii) Volmer–Weber growth in which islands form and grow before coalescing, and (iii) Stranski–Krastanov growth in which a monolayer first forms followed by island growth. Readers interested in further discussion of these different growth modes are referred to other sources, e.g., the excellent books by Smith (D.L. Smith, *Thin-Film Deposition*, McGraw-Hill, 1995), Venables (J.A. Venables, *Introduction to Surface and Thin Film Processes*, Cambridge University Press, 2000), and Ohring (M. Ohring, *The Materials Science of Thin Films*, Academic Press, 1992).

4.3.2   Ion Beam Effects on Film Growth

Ion bombardment during thin film deposition (i.e., IBAD) can beneficially influence several aspects of film growth and resultant thin film properties that will be described in the following sections (see Figure 4.10).

IBAD processing can be grouped into three different categories:

1. Nonreactive IBAD, where the main purpose of the ions, typically inert gas ions (Ar^+), is to influence the nucleation and growth of deposited elements or compounds.
2. Reactive IBAD, where the purpose of the ion beam is to both influence film growth as well as to provide reactant ion species for the growth of a chemical compound (e.g., Si or Ti deposition with nitrogen ion bombardment to produce Si_3N_4 or TiN).
3. A variation of this reactive technique is to provide reactant species in the form of backfilled molecular gas (e.g., N_2 or O_2), which is activated by the ion beam

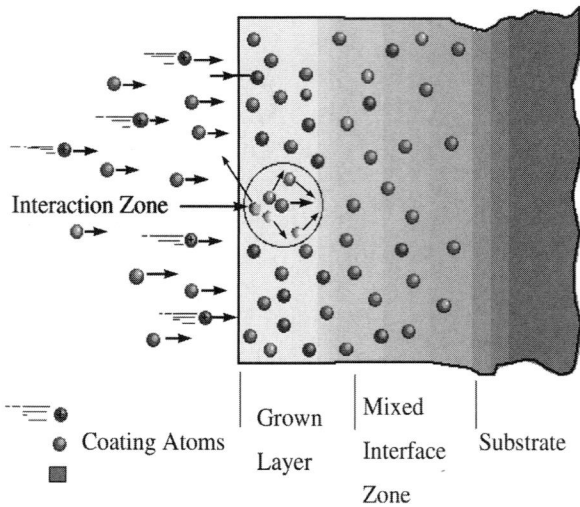

Figure 4.10 Schematic of film growth occurring in the presence of energetic ion bombardment. Although simplified, it depicts several of the phenomena occurring during assisted growth that are described in the text.

at the growth surface and then gets incorporated into the growing compound film.

This last variation can sometimes be used to create stoichiometric compounds if the evaporant is sufficiently reactive. It can also be used to make up for the loss of a constituent element (e.g., oxygen) when evaporating compounds (e.g., Al_2O_3 or SiO_2) that tend to decompose at high temperature giving a metal-rich coating in the absence of such an O_2 backfill.

4.3.3 Fundamentals of the IBAD Process

Several aspects of film growth and thin film properties are beneficially influenced by ion bombardment during thin film deposition including: (i) adhesion, (ii) nucleation or nucleation density, (iii) control of internal stress, (iv) morphology, (v)

TABLE 4.13 Advantages and Limitations of Ion Beam-Assisted Deposition

Advantages	Limitations
Control over many growth and coating properties	Line-of-sight processing
Stress, density, nucleation, texture, composition	Few micron thick scale coatings
Metastable film properties can be attained	Complexity of equipment

TABLE 4.14 Application Areas of Ion Beam-Assisted Deposition

Mechanical	Chemical	Other
Tribology: Friction and wear	Aqueous corrosion	Optical coatings; refractive index values
Rolling contact fatigue	Thermal oxidation	Buffer layers; magnetic thin films for sensors and memory
	Catalysis	Create/modify nanostructures and MEMS

density, (vi) composition, and (vii) the possibility of low-temperature deposition.

The IBAD process allows thicker alloyed regions to be attained than by either direct ion implantation or ion beam mixing, but still incorporates advantages attributed to ion beams, such as superior adhesion due to ion beam cleaning effects and ion mixing at the interface during the initial stages of deposition.

These potential advantages (Table 4.13) have encouraged exploration of this technique into several different areas as

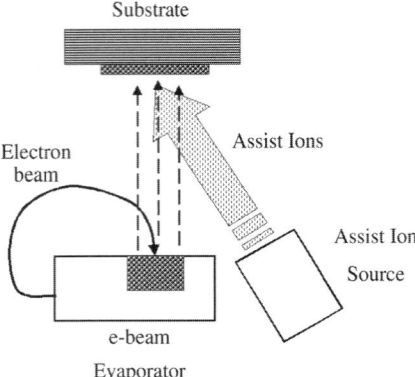

Figure 4.11   IBAD experimental configuration utilizing e-beam evaporation. (Adapted from J.M.E. Harper, J.J. Cuomo, R.J. Gambino, and H.E. Kaufman, in O. Auciello and R. Kelly, *Ion Bombardment Modification of Surfaces: Fundamentals and Applications*, Elsevier, Amsterdam, 1984, pp. 127–162.)

seen in Table 4.14, some of which are discussed in more detail below.

Figure 4.10 is a schematic of the IBAD process showing the interaction of the deposited coating atoms with energetic ions and the substrate. The figure depicts a physically mixed zone between the original substrate surface and coating produced by ion beam mixing (more prevalent at higher ion beam energies). Such a zone helps to avoid sharp interfaces and can improve adhesion if the coating and substrate elements have a mutual affinity in terms of solubility or compound formation. Other aspects of the ion/surface interactions pertaining to IBAD processing, such as ion beam neutralization and sputtering, are discussed below. During IBAD processing, the neutral species (deposited atoms) are normally delivered via physical vapor deposition via electron beam evaporation (Figure 4.11) or by sputter deposition (Figure 4.12). The ion species is typically produced by a low energy (0.2–2 keV) broad-beam ion source producing beam currents up to 1–2 mA/cm^2 (circa 10^{16} ions/sec per cm^2).

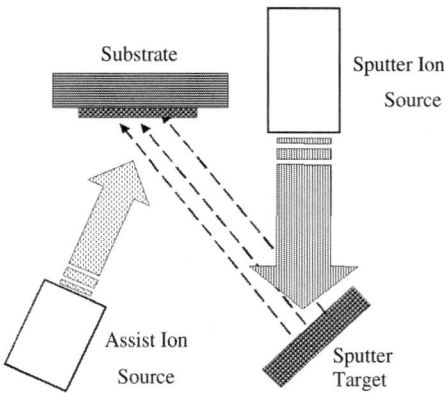

Figure 4.12 IBAD experimental configuration utilizing sputter deposition. (Adapted from J.M.E. Harper, J.J. Cuomo, R.J. Gambino, and H.E. Kaufman, *Ion Bombardment Modification of Surfaces: Fundamentals and Applications*, Elsevier, Amsterdam, 1984, pp. 127–162.)

The effect of an ion beam on film growth and its resultant physical properties will depend on the ion species, ion energy, and the relative flux ratio of the ions, J_I, and deposited atoms, J_A, customarily defined as R_i. Data showing the effect of ion bombardment on film properties is often expressed simply as a relative ion/metal atom flux ratio at the substrate (i.e., R_i) or as the average energy deposited per atom, E_{ave}, in eV/atom, which is simply the product of the relative ion/atom flux ratio and the average ion energy, E_{ion}.

In an IBAD system, deposition and ion bombardment usually operate at pressures higher (i.e., 10^{-5} to 10^{-4} torr) than typically employed for thermal deposition alone (10^{-6} to 10^{-7} torr). As the operating pressure increases, the probability of incorporating unwanted contamination from residual gases in the deposited film also increases. To minimize contamination from residual gases during deposition the following conditions and criteria must be considered, based on the relative flux rates of ions, neutrals, residual gases, and backfilled gases as indicated in Figure 4.13 (K. Miyake and T. Tokuyama,

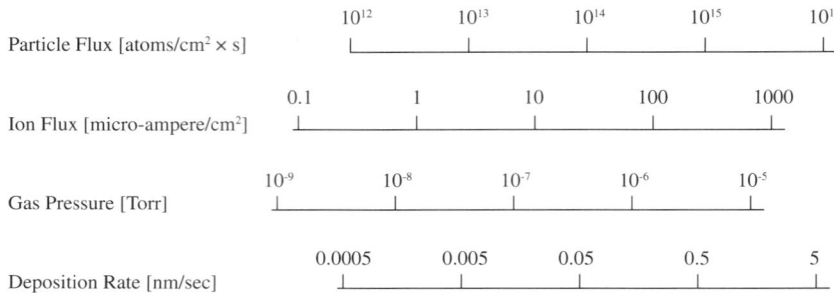

Figure 4.13   Chart of ion, gas, and neutral atom flux rates used to plan and compare deposition conditions in IBAD processing. (Adapted from K. Miyake and T. Tokuyama, Direct ion beam deposition, in T. Itoh (Ed.) *Ion Beam Assisted Film Growth,* Elsevier, Amsterdam, 1989, Chapter 8.)

Direct ion beam deposition, in T. Itoh (Ed.), *Ion Beam Assisted Film Growth,* Elsevier, Amsterdam, 1989, Chapter 8).

Figure 4.13 shows the relationship (i.e., the equivalency) of particle fluxes involved in IBAD processing, assuming that the sticking probability of impinging ions, residual gas atoms, and deposited atoms is unity. As an example of the use of Figure 4.13 consider the IBAD processing condition of $R_i = 1$. If the thermal deposition rate of atoms is set to 0.5 nm/sec, the atom arrival flux at the substrate will be circa 1×10^{15} atoms/cm^2/sec. For the condition $R_i = 1$, Figure 4.13 also shows that an ion flux at the substrate of 1×10^{15} ions/cm^2/sec which, assuming singly charged ions, translates to an ion current density of 160 µA/cm^2. Correspondingly, a low base pressure is required to avoid incorporation of contaminants from unwanted background species (e.g., H_2O), since a background (partial) pressure of 10^{-6} torr corresponds to an arrival rate of about 5×10^{14} atoms/cm^2/sec (i.e., about one-half a monolayer per second) of a particular constituent within the vacuum. When performing IBAD processing of highly reactive elements (e.g., Ti, Cr, Nb), the partial pressures of the vacuum species present should influence the choice of deposition rates or required background vacuum pressure to minimize impu-

TABLE **4.15**   Coating Properties Affected by Ion Bombardment

Coating Property	IBAD Parameter
Nucleation	Ion sputter cleaning, ion beam mixing
Adhesion	Ion sputter cleaning, ion beam mixing
Density	Ion flux
Stress	Ion flux and energy
Grain size	Substrate temperature, ion energy
Texture	Ion beam/substrate orientation; ion species and energy
Stoichiometry	Ion species and arrival ratio (R)

rity incorporation. Table 4.15 lists coating properties affected by ion bombardment that are described below.

4.3.3.1 Nucleation

An early deposition experiment by Pranevicius (L. Pranevicius, Structure and properties of deposits grown by ion-beam-activated vacuum deposition techniques, *Thin Solid Films*, 63, 77, 1979) aptly illustrates an effect of ions on initial film growth and coalescence. Aluminum was evaporated onto an insulator at a constant rate and the electrical conductivity between two separated electrodes on the surface was measured as a function of time by observing current flow between them. With no ion beam present during Al film growth at 10^{16} Al atoms/cm^2/sec, there was a considerable incubation time (40 s) required before the growing Al islands overlapped and provided electrical continuity. In contrast, the presence of an ion beam (5 keV Ar, 8 μA/cm^2 equivalent to 4.8×10^{13} ions/cm^2/sec, R = 0.005) served to significantly shorten the time period (to 15 sec) for overlap to occur. This was attributed to increased adatom mobility and nucleation sites produced for island formation and growth. Electron microscopy showed that there was a fourfold increase in nuclei density and a factor of 5 to 15 decrease in nuclei size in the presence of the ion beam. This early experiment demonstrates how the initial microstructural evolution of a thin film can be greatly influenced by ion bombardment.

4.3.3.2 Adhesion

Coating/substrate adhesion depends on a wide variety of factors including: interface chemistry, film stress, differential thermal expansion, contaminant levels at the interface, and surface morphology. Baglin has reviewed this area in some detail elsewhere (J. Baglin, Interface structure and thin film adhesion, in J.J. Cuomo, S.M. Rossnagel, and H.R. Kaufman (Eds.), *Handbook of Ion Beam Processing Technology*, Noyes, Park Ridge, NJ, 1989, Chapter 14) so only a cursory discussion of this technologically important subject will be given here.

Ion beam (*in situ* vacuum) cleaning offers an effective means of preparing substrates for coatings, via two mechanisms. First, this occurs by ion removal of contaminant layers such as adsorbed water, hydrocarbons and oxides. Second, ion bombardment can selectively remove surface material (texturing) to leave a favorable high bonding surface for either chemical or morphology (texture) influenced adhesion. For example, an initial surface that is chemically stable against bonding to deposited species can be decomposed to a depth of a few monolayers or so and thus render the surface highly reactive to subsequent deposition vapors. Metal-oxide compounds in ceramics can have the surface oxygen partially removed by the ion beam thus providing a higher density of metal atoms that can bond to another species, such as another metal or anion. Hence, deposition chambers designed for both ion cleaning and subsequent ion beam deposition are particularly attractive.

In addition to ion beam "precleaning," the use of high energy ions to inter-mix or "stitch" a thin metallization onto a substrate for improved (10- to 100-fold) adhesion has been recognized for many years. This technique, however, is limited to coating thicknesses the order of tens to hundreds of nanometers thick for medium mass ions of energies up to 200 keV to allow ion penetration of the film to be "stitched" with the substrate. Similar improvements in adhesion may also be realized at the much lower ion energies normally employed with IBAD processing (i.e., 0.1–10 keV) since ion bombardment is happening concurrently with the deposition of evaporated/sputtered atoms.

An early example of enhanced adhesion by Martin (P.J. Martin, Ion-enhanced adhesion of thin gold films, *Gold Bull.*,19, 102, 1986) following ion bombardment involves the adhesion of gold onto glass substrates for optical reflectors. They found that ion-assisted deposition of Au onto glass with an Ar$^+$ beam made only an insignificant improvement, whereas the use of an oxygen beam by itself or an oxygen plus argon ion beam made adhesion improvements of 100 times to over 400 times that of non-assisted or argon-assisted deposition. The improvement was attributed to some (unidentified) form of chemical bonding.

Optimal adhesion for film/substrate combinations not having chemical affinity between them, however, requires a broad interfacial region between the layers for good adhesion (G.K. Wolf, Modification of chemical properties by ion beam-assisted deposition, *Nucl. Instrum. Methods Phys. Res. B*, 46, 369, 1990). For IBAD processing, this would require high-energy beams. The substrate involved also has an important bearing on the appropriate ion energy to use for improving adhesion. For example, Ebe et al. (A. Ebe, N. Kuratani, S. Nishiyama, O. Imai, and K. Ogata, Metallization on polyimide film by ion and vapor deposition (IVD) method, *Jpn. J. Appl. Phys.*, 1993) found that the adhesion of Cu films on polyimide substrates was better for 0.5 keV Ar ions than for 5 or 10 keV ions. This was attributed to the (deleterious) carbonization of the polyimide at the higher energies. An alternative approach is to use an intermediate adhesive-enhancing layer between a chemically incompatible film and substrate, including metal-on-polymer systems. This adhesive interlayer layer would ideally form strong bonds with both the outermost film and the substrate.

4.3.3.3 Density

IBAD disrupts the porous columnar growth often found during unassisted deposition and stimulates bridging between neighboring grains, causing the closure of open voids. When an energetic ion interacts with a material target, the ion undergoes a collision cascade resulting in knock-on atoms, vacancies and interstitials, phonons, and electronic excitations. These processes are predominant in the collisional phase of the cascade and can be reasonably well modeled by Monte Carlo calculations (K.-H. Müller, Model for ion-assisted thin film densification, *J. Appl. Phys.*, 59, 2803, 1986). For ion bombardment in the low-energy regime of a few hundred electronvolts, most of the collision cascade processes are confined to the surface region of the target. For low ion energies the ions can be either backscattered or incorporated into the surface, while knock-on atoms may either leave the surface as sputtered atoms or be recoil implanted below the surface and become trapped as interstitials. The recoil implanted target atoms then produce an atomic density increase in the vicinity where they come to rest (i.e., at the end of their range) while sputtered atoms and the vacant sites left by the recoil implanted atoms result in a high concentration of vacancies at the surface and a reduction in surface density. The vacancies created near the surface by the bombarding ions can be partially filled by the vapor deposited atoms under IBAD conditions. At large ion-to-vapor-atom ratios (J_I/J_A) the IBAD process will result in an inward packing of the atoms in the film, eliminating the porous columnar network commonly observed in Zone I films, and favor the growth of a densely packed structure. Kuritano et al. describe such interactions for Cr growth under ion bombardment (N. Kuratani, A. Ebe, K. Ogata, I. Shimizu, Y. Setsuhara, and S. Miyake, Fundamental study of ion-irradiation effects on the columnar growth of chromium films prepared by ion beam and vapor deposition, *J. Vac. Sci. Technol. A*, 19, 153, 2001).

4.3.3.4 Stress

The intrinsic stresses in thin films can be related to their microstructure, as well as to the presence of incorporated impurities. Typically, thin films formed by thermal evaporation have a high void fraction (i.e., low atomic density) and are in a state of residual tensile stress. When thermally evaporated films are deposited with the aid of an ion beam, the microstructure begins to densify allowing even higher attractive interactions between adjacent atoms that can further increase the tensile stress to a maximum. As the ratio J_I/J_A increases, significant atomic compaction results in increased packing densities and the net film stress may be driven toward a state of compressive stress assisted by the forcible injection (i.e., ion implantation) of beam atoms. Figure 4.14 demonstrates this behavior for Nb and W deposition as a function of temperature with 100 eV and 400 eV Ar ion bombardment (R.A. Roy, D.S. Yee, and J.J. Cuomo, Control of metal film properties by ion-assisted deposition, *Mat. Res. Soc. Symp. Proc.*, 128, 23, 1989). A positive stress value indicates a tensile stress and a negative stress value indicates a compressive stress. The higher temperatures allow higher atomic mobilities so that the peak tensile stress and neutral (zero) stress states occur for lower beam currents. Analogous effects are seen as a function of ion energy, with lower critical current values needed to arrive at neutral stresses at higher ion energies.

The ratio of the (critical) ion current to the evaporant flux rate that produces a zero-stress state in the film is termed the critical arrival-rate ratio. Wolf has experimentally determined that the conditions required for this zero-stress state depend on the composition of the film as well as the energy of the ions (G.K. Wolf, Modification of chemical properties by ion beam-assisted deposition, *Nucl. Instrum. Methods Phys. Res. B*, 46, 369, 1990). For example, Cr requires 100 eV per atom for stress relief, whereas C or B require an order of magnitude less energy. The behavior of ionically bonded materials is even less well documented and understood thus far.

4.3.3.5 Grain Size

Thin film grain sizes show complicated dependencies on ion flux, ion energy, and substrate temperature. Many metals show a significant decrease in grain size with increasing ion-

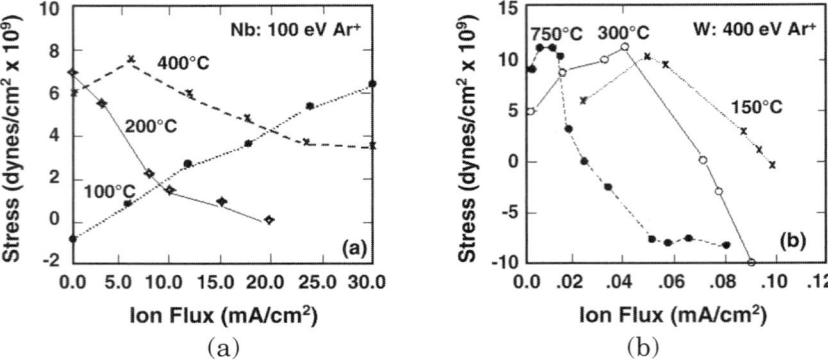

(a)   (b)

Figure 4.14 Intrinsic film stress as a function of substrate temperature and ion flux. (a) Nb films deposited (evaporated) with concurrent 100 eV Ar ion bombardment. (b) W films deposited (evaporated) with 400 eV Ar ion bombardment. (Adapted from R.A. Roy, J.J. Cuomo, and D.S. Yee, Control of microstructure and properties of copper films using ion-assisted deposition, *J. Vac. Sci. Technol. A*, 6, 1621, 1988.)

to-atom ratio, J_I/J_A, and increasing average energy deposited per atom, E_{ave}. Metals showing a significant decrease in grain size with increasing J_I/J_A values show an accompanying increase in electrical resistivity and an increase in microhardness. This may be understood qualitatively as follows: (a) as the grain size decreases, the grain boundary fraction increases, reducing the electron mean free path and thereby the electrical conductivity of the film; (b) similarly, grain boundaries impede dislocation motion during plastic deformation, which is also manifested as a hardness increase.

To date, no single mechanism has been confirmed to explain the grain size decrease with increasing ion bombardment. It has been argued that gas (Ar) becomes incorporated at the grain boundaries in growing (Cu) films and at sufficiently high concentrations may inhibit grain growth. Also at higher beam energies, irradiation-induced lattice disorder increases rapidly, which may also limit grain growth (C. Spinella, S. Lombardo, and F. Priolo, *J. Appl. Phys.*, 84, 5383. 1998).

4.3.3.6 Texture and Surface Roughness

The orientation and the size of grains in films deposited under IBAD conditions can also be a strong function of ion energy and the ion-to-atom flux ratio. For example, Yu et al. (L.S. Yu, J.M.E. Harper, J.J. Cuomo, and D.A. Smith, Alignment of thin films by glancing angle ion bombardment during deposition, *Appl. Phys. Lett.*, 47, 932, 1985; L.S. Yu, J.M.E. Harper, J.J. Cuomo, and D.A. Smith, Control of thin film orientation by glancing angle ion bombardment during growth, *J. Vac. Sci. Technol. A*, 4, 443, 1986) have shown that Nb films grown under 200 eV glancing angle (70° from normal incidence) Ar bombardment developed a (110) fiber texture with a restricted set of azimuthal orientations, whereas Nb films deposited without ion bombardment also show a (110) fiber texture but without azimuthal ordering. The degree of orientation (i.e., azimuthal ordering) increased with increasing ion-to-atom flux ratio. It was observed that the textured grains tended to be oriented so that the (110) planar (channeling) direction is aligned along the incident Ar beam direction. About half of the Nb grains are aligned to within 5° of the (110) planar (channeling) direction at a J_I/J_A ratio of 1.3. It was suggested that a low energy channeling effect was responsible for the development of a preferred orientation.

Dobrev (D. Dobrev, Ion-beam-induced texture formation in vacuum-condensed thin metal films, *Thin Solid Films*, 92, 41, 1982) has developed a model of texturing in IBAD films due to preferential channeling of bombarding ions along the open crystalline channeling directions in films growing under ion bombardment. Ions traveling in planar crystalline channels lose energy primarily through electronic energy loss processes, and as a result the ion's nuclear energy deposition near the film surface would be inversely proportional to the planar spacing. For face centered cubic crystals, the ease of planar channeling is in the order of most open planes (i.e., large planar spacing): i.e., <110>, <200>, <111>. Dobrev suggested that in a polycrystalline film, the crystallites with the most open channeling directions (easiest channeling direction) aligned with the ion beam would experience the least nuclear energy loss and thus experience the lowest lattice disorder (see

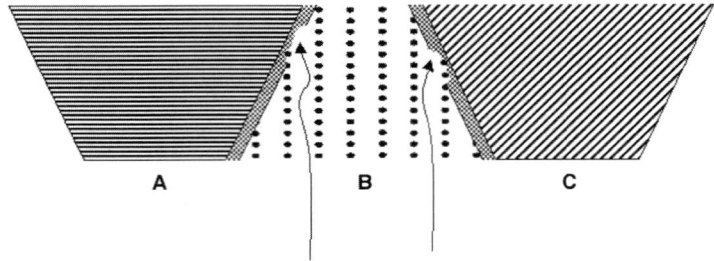

Figure 4.15 Schematic of texture development in thin film during ion bombardment. Figure indicates manner by which <110> texture develops in fcc films during normal-incidence ion bombardment. <110> grains (region B) provide easy channeling and hence serve as seeds for regrowth of grains not aligned with beam (i.e., regions A and C) that are preferentially dissolved at interfaces by ion beam. (Adapted from D. Dobrev, Ion-beam-induced texture formation in vacuum-condensed thin metal films, *Thin Solid Films*, 92, 41, 1982.)

Figure 4.15). These grains would then serve as the seeds for the regrowth of adjacent grains that were initially not well aligned with the ion beam and therefore would experience high levels of lattice disorder via nuclear energy deposition.

Bradley et al. (R.M. Bradley, J.M.E. Harper, and D.A. Smith, Theory of thin film orientation by ion bombardment during deposition, *J. Appl. Phys.*, 60, 4160, 1986) offer an alternate model to explain the development of texturing during ion bombardment based on the differences in sputtering yields at different crystallographic orientations. This model assumes that the film is deposited such that one crystal axis is normal to the film and that the azimuthal orientations are random. The orientation sputtering difference, which can be as a high as a factor of 5 in some materials, leads to the preferential removal of grains oriented with a high sputtering yield and the newly deposited material grows epitaxially on the low sputtering yield orientations. Calculations from this sputtering based model reproduced the general features of the Nb data from Yu et al. (L.S. Yu, J.M.E. Harper, J.J. Cuomo, and D.A. Smith, Alignment of thin films by glancing angle ion bombardment during deposition, *Appl. Phys. Lett.*, 47, 932, 1985).

Both of these IBAD texture models are based on the deposited energy density due to nuclear stopping varying inversely with the openness of the channeling orientations. The sputtering yield is directly proportional to the nuclear stopping. Thus, we expect both reduced sputtering and the reduction of dense radiation damage cascades when the bombarding ion is aligned along crystalline planes of those grains having large interplanar spacings. Based on these arguments, Smidt (F.A. Smidt, Use of ion beam-assisted deposition to modify the microstructure and properties of thin films, *Int. Mater. Rev.*, 35, 61, 1990) has summarized typical IBAD texture behavior. In general, thin films deposited by evaporation processes, without ion beam assistance, will normally be oriented with planes of highest atomic density parallel to the substrate surface. This crystallographic orientation corresponds to a fast growth geometry and will only be obtained when substrate temperatures are high enough for the depositing atoms to have surface mobility, i.e., $T_S > 0.15\ T_M$. Under such conditions face centered cubic (fcc) films have a <111> texture, body centered cubic (bcc) films have a <110> texture, and hexagonal close packed (hcp) films with an ideal c/a ratio will have a <0002> texture. During IBAD processing, the crystallographic orientation will be shifted so that widely spaced planes (i.e., planes of easiest planar channeling) will be aligned along the ion beam axis. Therefore, one might expect that the <111> texture in an evaporated fcc film will be shifted toward a <110> for an IBAD process with the ion beam aligned with the film normal. For an IBAD process with an off-normal incident ion beam, different textures will be produced. The evolution of texture, especially in technologically important oxide films has been studied in recent years (K.G. Ressler, N. Sonnenberg, and M.J. Cima, Mechanism of biaxial alignment of oxide thin films during ion-beam-assisted deposition, *J. Am. Ceram. Soc.*, 80(10), 2637–2648, 1997; G.I. Grigorov, and N. Savvides, Evolution of texture of CeO_2 thin film buffer layers prepared by ion-assisted deposition, *Thin Solid Films*, 350, 124, 1999).

The surface roughness of films can be critically important to their properties, especially in nanoscale electronic materials, where roughness can be on the same scale as the total feature dimensions. The energy deposited by ions during the IBAD process can create defects, remove coating atoms via sputtering, enhance surface diffusion, and densify voids, all leading to changes in surface morphology. These effects have been modeled using molecular dynamics simulations by Thijsse and others, including the effects of ion energy as well as the incident angle of both ion bombardment and vapor arrival (A. Robbemond and B.J. Thijsse, Ion beam assisted deposition of thin molybdenum films studied by molecular dynamics simulation, *Nucl. Instrum. Methods Phys. Res. B*, 127/128, 273–277, 1997). It is important to note that although ion assistance can reduce surface features via enhanced diffusion, especially for off-normal vapor incidence where large surface features can develop from shadowing effects, excessive ion energy can instead lead to significant surface roughness increases due to sputtering (P. Klaver, E. Haddeman, and B. Thijsse, Atomic-scale effects of sub-keV ions during growth and subsequent ion beam analysis of molybdenum thin films, *Nucl. Instrum. Methods Phys. Res. B*, 153(1–4), 228–235, 1999; B.S. Bunnik, C.D. Hoog, E.F.C. Haddeman, and B.J. Thijsse, Molecular dynamics study of Cu deposition on Mo and the effects of low-energy ion irradiation, *Nucl. Instrum. Methods Phys. Res. B*, 187(1), 57–65, 2002).

4.3.4   Application Areas of IBAD Coatings

4.3.4.1 Optical Coatings by IBAD

Optical coatings are becoming more prevalent in modern technology and as such carry particularly stringent requirements on their optical and mechanical properties regarding environmental durability and stability. In addition they pose additional challenges to deposition in thin film form, may they be transparent silica, metal-oxide compounds or layered combinations of films of these types. Fortunately, IBAD type processing has successfully addressed many of these problems. As a result, significant portions of the precision optical coating community have adopted ion beam-assisted processing because it offers several advantages over conventional e-beam evaporation with a minimum of additions to the equipment required.

One of the first demonstrations of this was by Martin et al. (J. Martin, R.P. Netterfield, and W.G. Sainty, *J. Appl. Phys.*, 55, 235, 1984). Optical coatings deposited at low substrate temperatures without ion assistance, frequently exhibit undesirable porosity that results in the pickup of water from the environment. This changes the optical properties of the coating in an unacceptable manner. Figure 4.16 shows the now classic demonstration of the transmission of ZrO_2 coatings (with and without ion assistance) both before and after exposure to (water containing) atmosphere. Exposure to humid air is seen to have a marked and undesired effect on the optical characteristics. The ability to avoid these types of environmental degradation by the use of ion-assisted deposition was soon embraced by makers of critical optical coatings and is commonly used today, mostly under proprietary shields for the several advantages it affords. This and some other advantages for employing ion assist on optical coatings are listed in Table 4.16.

Besides providing more dense, lower pinhole density coatings, ion beam-assisted deposition made it possible to coat optics at lowered temperatures and still get the same adherent, high quality coatings. For example, optical coaters could often control film stress by merely adjusting the ion energy and arrival rate during deposition. This feature has been especially valuable for controlling stress in thicker coatings required in the IR spectral region. In addition, IBAD optical coatings are often more robust than their non-ion-assisted counterparts. For metal-oxide compound films, the use of oxygen ions for IBAD processing can also aid in adjusting the stoichiometry of the oxide compound. These positive features combined with the possibility of converting existing optical (box coater) equipment at relatively low costs by simply adding ion source(s) made it attractive to the optical coating community.

The use of ions for precleaning a substrate is used extensively in plasma-based processes and is an important feature of ion beam-assisted processing of optical components for enhancing adhesion by removal of adsorbed contaminants.

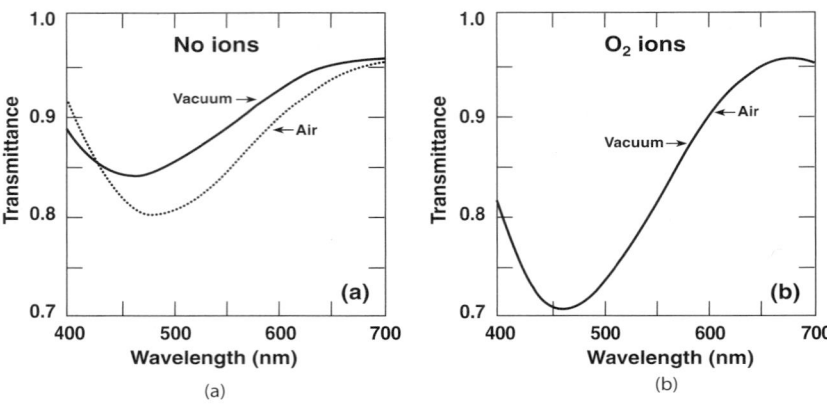

Figure 4.16 Optical transmittance of ZrO_2 film deposited (a) without and (b) with ion assistance. (Adapted from J. Martin, R.P. Netterfield, and W.G. Sainty, *J. Appl. Phys.*, 55, 235, 1984.)

This cleaning can be done at ambient temperatures, circumventing the requirement of employing elevated substrate temperatures normally required to gain adequate adhesion. In addition to aiding adhesion, ion cleaning allows the selective removal of (partially) deposited coatings whose deposition has been interrupted or improperly done. This alone can often economically justify the inclusion of an ion source due to the extremely high cost of precision optical substrates.

The enhanced reactivity afforded by energetic ion bombardment also allows for the production of stoichiometric optical coatings at lower temperatures than normally required, hence gives a greater choice of substrate materials, including temperature sensitive substrates.

Another important feature of ion-assisted processing is the ability to control stress levels in multilayered coatings. Ordinarily, stress can be the limiting factor in determining the maximum usable thickness, where it can lead to delamination. This is particularly important in multilayered coatings (at each of the layer interfaces) and for thicker coatings such as required for IR applications. An additional related factor promoting the acceptance of ion assistance within the

TABLE 4.16 IBAD Processing for Optical Coatings

Coating and/or Process Benefits	Benefits
Superior adhesion	Ions preclean substrate and promote interface bonding
Stress control	Better adhesion thus thicker (multilayered) structures feasible
Higher density films	Ions eliminate voids/open defects, and promote compaction
Higher refractive indexes	Improved spectral stability against humidity changes and aging
Lower physisorbed water pickup	Fewer layers required
Low temperature deposition	Energetic ions promote growth similar to elevated temperatures
Ions provide reactivity	Densified microstructures, more robust mechanically, greater cohesive strength
Lower reactive gas pressure required	Only minor changes in processing required
Higher durability coatings	Reduced rejection rate of very expensive optics
Additional equipment requires no significant reconfiguration	
Ion sources may be used in sputter-only mode to selectively remove coating material	

optical coating community is the promotion of superior abrasion resistance as a result of improved microstructure. Ion-assisted coatings have demonstrated higher resistance to the abrasion tests normally applied to coatings.

Several precision coating suppliers are presently using the method on a regular basis on some of their products, but specifics concerning most commercial uses are difficult to obtain because of proprietary concerns. The following is an account of scaling up ion beam treatment of large scale (meter plus in size) optics by Pawlewicz and coworkers in the early 1990s to produce both (a) larger scale optical coatings as well

TABLE 4.17   Optical Coatings by IBAD

Coating	IBAD Parameters	Results	Ref.
ZrO_2	O_2	High-density, low-pinhole coatings, No deleterious water pickup!	Martin et al. (a)
MgF		Improved stress and adhesion	Pawlewicz et al. (b)

Sources: (a) P.J. Martin, H.A. Macleod, R.P. Netterfield, C.G. Pacey, and W.G. Sainty, *Appl. Optics*, 22, 78, 1983; T.R. Culver, W.T. Pawlewicz, J.H. Zachistal, J.A. McCandless, M.W. Chiello, and S. Walters, *SPIE Proc.*, 1848, 192–199, 1992; (b) W.T. Pawlewicz, T.R. Culver, J.H. Zachistal, E.J. Prevost, J.D. Traylor, and C.E. Wheeler, *SPIE Proc.*, 1618, 1–16, 1991.

TABLE 4.18   Improved Mechanical Properties of MgF Optical Coatings Using IBAD

Coating Process	Moderate/Severe Abrasion	Intrinsic Stress ($\infty\ 10^6$ dyn/cm^2)	Maximum Thickness w/o Crazing (μm)
20°C No ions	Fail/fail	50 (tensile)	0.5
300°C No ions	Pass/pass	50 (tensile)	0.5
20°C With ions	Pass/pass	20 (tensile)	1

Source: W.T. Pawlewicz, T.R. Culver, J.H. Zachistal, E.J. Prevost, J.D. Traylor, and C.E. Wheeler, Ion assisted evaporative coating of 2-meter optics, *SPIE Proc.*, 1618, 1–16, 1991.

as (b) complex, multilayer optical coatings for use in the UV, visible, and IR spectral regions (T.R. Culver, W.T. Pawlewicz, J.H. Zachistal, J.A. McCandless, M.W. Chiello, and S. Walters, *SPIE Proc.*, 1848, 192–199, 1992; W.T. Pawlewicz, T.R. Culver, J.H. Zachistal, E.J. Prevost, J.D. Traylor, and C.E. Wheeler, *SPIE Proc.*, 1618, 1–16, 1991). They made use of the improved optical properties obtained using ion beams as discussed above. Their 2.4 m diameter chamber was 3.3 m tall, with 0.96 m top and 1.5 m bottom section diameters. It could handle optics up to 1.8 m in diameter and 2700 kg in weight. It was used for large-scale optics (> 1 m diameter) with *plus/minus 1% optical thickness uniformity* over the entire substrate. Electron beam evaporation was used for the deposition of oxides, fluorides, and sulfides at rates of 0.1–0.5 nm/s. It employed a high-current, low-voltage ion gun with ion currents selectable from zero up to 1 A at energies of 40 to 120 eV. The ion source ran Ar, O_2, or N_2 as required with peak currents of 70 µA/cm^2 at the optics surface. The ion to atom ratio used was typically 0.10 to 0.20 and the energy per deposited atom was in the range of 5–30 eV/atom. The ion current density profile was precalculated and optimized for the particular deposition geometry required. They used the same ion source for both precleaning and ion-assist during deposition, obviating the need for substrate heating. This greatly simplified and reduced production costs due to (i) shorter batch times, (ii) less expensive fixturing, and (iii) lowered substrate losses due to chipping and breakage. Although the optics described above are not typical for consumer items, it demonstrates how far advanced the production technique was already in the 1990 timescale.

Other optical IBAD references of possible interest to readers include the following:

J.R. McNeil, A.C. Barron, S.R. Wilson, and W.C. Herrmann, Ion-assisted deposition of optical thin films: low energy vs high energy bombardment, *Appl. Optics*, 23, 552, 1984

P.J. Martin, R.P. Netterfield, and W.G. Sainty, Modification of the optical and structural properties of dielectric ZrO2 films by ion-assisted deposition, *J. Appl. Phys.*, 55, 235, 1984

P.J. Martin, A. Bendavid, M.V. Swain, R.P. Netterfield, T.J. Kinder, W.G. Sainty, and D. Drage, Mechanical and Optical properties of thin films of tantalum oxide deposited by ion-assisted deposition, *Mater. Res. Soc. Symp. Proc.*, 308, 583, 1993

F.L. Williams, D.W. Reicher, C.-B. Juang, and J.R. McNeil, Metal oxides deposited using ion assisted deposition at low temperature, *J. Vac. Sci. Technol. A*, 7, 2286, 1989

F.L. Williams, L.L. Boyer, D.W. Reicher, J.J. McNally, G.A. Al-Jumaily, and J.R. McNeil, Ion beam processing of optical materials, *Mater. Res. Soc. Symp. Proc.*, 129, 17, 1989

Y. Yamada, H. Uyama, T. Murata, and H. Nozoye, Low-temperature deposition of titanium-oxide films with high refractive indices by oxygen-radical beam assisted evaporation combined with ion beams, *J. Vac. Sci. Technol. A*, 19, 2479, 2001

4.3.4.2 Wear Control Surfaces

The use of IBAD processing for low friction/wear coatings (Table 4.19) has been actively pursued in R&D laboratories since the early 1980s and is now being used commercially in niche applications including the production of wet/dry electric shaver heads (in Japan) as described below.

Refractory metal nitrides and carbides (e.g., Al_2O_3, TiN, TiC) are commonly used for wear-resistant or lowered friction coatings on tool steels or carbides. For these applications, a variety of plasma-based processes are typically used to economically coat tools in bulk lots at temperatures ranging from 400°C (activated reactive evaporation) to 1000–1100°C (chemical vapor deposition).

IBAD allows significantly lower temperature processing, in that the ion energy can activate chemical processes instead of using high temperatures. Titanium nitride coatings are evolving in commercial use faster than most other hard coatings made by conventional processing, and have served as a benchmark against which IBAD coatings can be compared.

4.3.4.2.1 TiN

Barth et al. (M. Barth, W. Ensinger, A. Schröer, and G.K. Wolf, *Proc. 3rd Int. Conf. Surface Modification Technologies*, Neuchatel, Switzerland, 1989) used reactive IBAD processing to produce both nitrides and carbides of titanium and chromium at low temperatures. They used nitrogen bombardment of titanium to form TiN as well as Ar ion bombardment of a growing titanium film in a N_2 background. They concluded that the energy per atom is a crucial factor for the film's orientation. The TiN coating microhardness was also found to depend strongly on deposition parameters and on coating stoichiometry. For a Ti/N = 1 layer grown with a 6 keV Ar^+ beam a Knoop microhardness of 3600 kg/mm^2 (50 mN load)

TABLE 4.19 Low Friction and Wear Coatings by IBAD

Coating/Substrate	IBAD Parameters	Results	Ref.
TiN	Ar ions activating reaction of Ti and N gas modeled	Microhardness and friction of films varies with deposition ratio (R)	Kant and Sartwell (a)
TiC,CrN		Low temperature production of TiC,CrN coatings demonstrated	Barth et al. (b)
Ag on ceramics	Ar onto Ag	Lowered μ plus 10 × lowered wear	Erdemir et al. (c)
Si-DLC on martensitic (bearing) steel	40 keV Ar onto silicated pump oil	High hardness and low friction	Fountzoulas et al. (d)
Cr_xN_y	Cr sublimation with N ions		Demaree et al. (e)

Sources: (a) R.A. Kant and B.D. Sartwell, *J. Vac. Sci.Technol. A*, 8, 861, 1990; (b) M. Barth, W. Ensinger, A. Schröer, and G.K. Wolf, *Proc. 3rd Int. Conf. Surface Modification Technologies*, Neuchatel, Switzerland, 1989; (c) A. Erdemir, B.R. Fenske, R.A. Erck, and C.C. Cheng, Ion beam assisted Ag solid lubricant coatings, *STLE Preprint*, No. 89-AM-5C-1, 1989; (d) C.G. Fountzoulas, J.D. Demaree, W.E. Kosik, W. Franzen, W. Croft, and J.K. Hirvonen, *Fall 1992 MRS Meeting, Symp. A*, Boston, MA, 1993; (e) J.D. Demaree, C.F. Fountzoulas, J.K. Hirvonen, M.E. Monserrat, G.P. Halada, and C.R. Clayton, in J.C. Barbour, S. Roorda, and D. Ila (Eds.), *Atomistic Mechanisms in Beam Synthesis and Irradiation of Materials*, *Material Research Society Proceedings, vol. 504*, 1998.

was obtained, and increasing or decreasing the Ti/N ratio or energy decreased the hardness.

They have also examined the effect of both nitrogen ion energy and substrate temperature on the critical load necessary to delaminate IBAD TiN coatings on stainless steel substrates. 300°C depositions were found to be superior to 130°C depositions in adhesion, and depositions at both temperatures improved monotonically with beam energy (from 5 to 35 keV).

4.3.4.2.2 Ag Solid Lubricants for Ceramics

Ceramics promise high temperature and oxidation resistance, but have poor intrinsic wear and friction properties and cannot be used with conventional lubricants at elevated temperatures. As such, suitable solid lubricants must be found. The three types of solid lubricants being used include: (i) lamellar solids — typically MoS_2 but also WS_2, WSe_2, $NbSe_2$, and other dichalcogenides; (ii) low adhesion polymers; and (iii) soft metals (i.e., In, Ag, Pb, and Au). Silver is attractive as a solid lubricant for high-temperature operations since it has (i) a high thermal conductivity required to dissipate frictional heat, (ii) a low shear strength, and (iii) is relatively chemically inert. However, it often suffers from inadequate adhesion to ceramic substrates. Erdemir et al. have used IBAD processing for depositing Ag onto Al_2O_3 substrates (A. Erdemir, B.R. Fenske, R.A. Erck, and C.C. Cheng, Ion beam assisted Ag solid lubricant coatings, *STLE Preprint*, No. 89-AM-5C-1, 1989). Their ion-assisted Ag coatings were denser than non-ion-assisted evaporated films, and showed nearly an order of magnitude decrease in the wear of a 2 μm thick IBAD Ag coating relative to the uncoated Al_2O_3. This was attributed to a much higher heat conduction away from the wear track by the extremely adherent IBAD silver coatings.

4.3.4.2.3 Silicon-Containing "Diamond-Like Carbon" Coatings

Hard amorphous carbon, sometimes called diamond-like carbon (DLC), appears promising for tribological and corrosion applications because of its high hardness, low friction, and chemical inertness. Conventional DLC films have unlubricated friction coefficients as low as 0.01 in dry atmospheres, but the friction coefficient increases considerably with increasing humidity (as high as 0.10 to 0.20 with 10% relative humidity). In the 1990s, researchers also examined Si-containing amorphous DLC films, and found that they exhibit extremely low friction coefficients (< 0.05) in both dry and humid atmospheres. Goode et al. (P.D. Goode, W. Hughes, and G.W. Proctor, Ion beam carbon layers, U.K. Patent No. GB2122224 B, 1986), used a polyphenyl ether liquid precursor and a large area ion beam from a bucket type ion source to produce Si-DLC. Carosella (NRL, private communication) used 150 keV Ti$^+$ ions and pentaphenyl-trimethyl-trisiloxane (Type 705 Dow-Corning silicone oil) as a precursor. Jones et al. (A.M. Jones, C.J. Bedell, G. Dearnaley, and C. Johnston, *Proc. Diamond and Diamond-like Carbon Films,* Nice, September 1991, 1992) used pentamethyl trisiloxane and polyphenyl ether as a precursor and a 50 keV nitrogen ion beam. Hioki et al. (T. Hioki, Y. Itoh, A. Itoh, S. Hibi, and J. Kawamoto, Tribology of carbonaceous films formed by ion-beam-assisted deposition of organic material, *Surf. Coat. Tech.,* 46(2), 233, 1991) combined vapor deposition of pentaphenyl-trimethyl-trisiloxane and simultaneous energetic ions of 1.5 MeV N$^+$ and 200 keV Ti$^+$. In each case the films showed coefficients of friction lower than those of conventional DLC films in both dry and humid atmospheres.

Diamond-like coatings with significant silicon content (Si-DLC) have also been prepared by Fountzoulas et al. (C.G. Fountzoulas, J.D. Demaree, W.E. Kosik, W. Franzen, W. Croft, and J.K. Hirvonen (i − $C_xSi_yO_zH_w$ films formed by ion beam assisted deposition, *Mat. Res. Soc. Symp. Proc.* 279, 645, 1993) using 40 keV Ar$^+$ ions to bombard tetraphenyl-tetramethyl-trisiloxane (Type 704 Dow-Corning silicone oil) vapor supplied from a heated reservoir. They found the relative amounts of C, Si, and O to be almost identical to those found in the precursor oil. The main effect of the ion beam appeared to be to drive off hydrogen and to add trace amounts of Ar. The fact that no oxygen appears lost implied that the siloxane "backbone" of the oil molecule was not destroyed by ion irradiation. Surface roughnesses appeared less than 10 nm as measured by photon tunneling microscopy, and transmission electron diffraction indicated that the films were amorphous. Figure 4.17 summarizes the microhardness, friction, and wear results of these Si-DLC coatings, normalized to the baseline properties of M50-bearing steel (a candidate to be coated with Si-DLC).

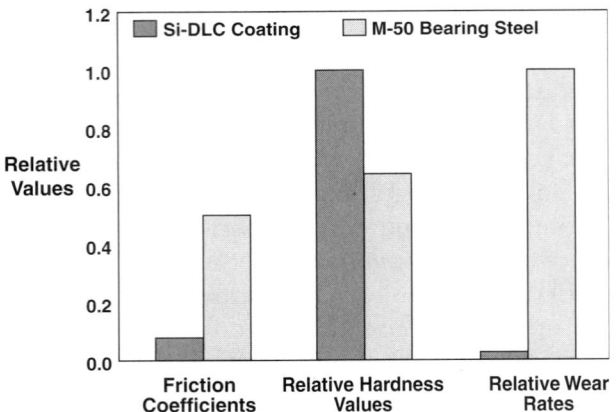

Wear rate of Si-DLC vs. M50 steel

Figure 4.17 Relative Knoop microhardness, friction, and wear rates of Si-DLC films compared to bare M50 steel. (Adapted from C.G. Fountzoulas, J.D. Demaree, W.E. Kosik, W. Franzen, W. Croft, and J.K. Hirvonen, *Mat. Res. Soc. Symp. Proc.* 279, 645, 1993.)

The combination of high hardness (Knoop microhardness from 1000 to 2100 at 15 g load), along with very low friction (unlubricated coefficients of friction as low as 0.03 against a steel pin, comparable to liquid lubricated systems) led to an extremely low wear rate in pin-on-disc tribological tests. The adherence of the coatings to the silicon and steel substrates was outstanding, as judged by automated scratch tests.

4.3.4.2.4 Chromium Nitride Hard Coatings

Environmental concerns regarding the disposal of toxic byproducts from the production of electroplated hard chromium (EHC), widely used as a wear- and corrosion-resistant tribological coating, have led many to consider IBAD and similar coating processes as potential replacements for electroplate technology. IBAD chromium nitride (Cr_xN_y) has been evaluated as a candidate for EHC replacement because of its good corrosion resistance, high hardness, and overall similarity to EHC. Several studies have investigated the deposition of Cr_xN_y coatings using reactive ion plating, reactive sputtering, and IBAD and have concluded that Cr_xN_y coatings can be formed with a hardness greater than that of ordinary Cr and a wear rate an order of magnitude lower. More-recent studies have focused on overcoming some of the difficulties often associated with IBAD and other physical vapor deposition (PVD) processes, namely, the optimization of coating adhesion to a given substrate, the optimization of coating hardness and wear resistance, and the management of residual internal coating stresses to allow the deposition of coatings more than a few microns thick.

Figure 4.18 shows some results from a study by Demaree et al. (J.D. Demaree, C.F. Fountzoulas, J.K. Hirvonen, M.E. Monserrat, G.P. Halada, and C.R. Clayton, in J.C. Barbour, S. Roorda, and D. Ila (Eds.), *Atomistic Mechanisms in Beam Synthesis and Irradiation of Materials, Mat. Res. Soc. Proc., Vol. 504*, 1998) in which Cr_xN_y coatings were deposited onto a variety of substrates using thermally evaporated Cr and a 1200 eV nitrogen ion beam from an RF-type ion source. As with most IBAD processes, the ion/atom arrival ratio was

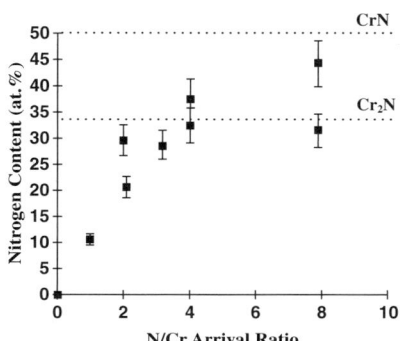 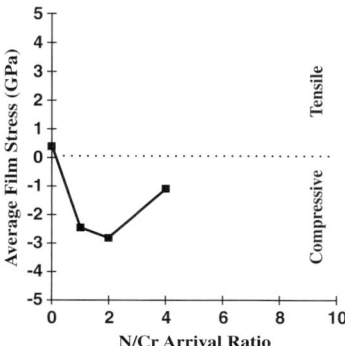

Figure 4.18 The effect of ion-to-atom arrival ratio on the composition and stress state of IBAD chromium nitride coatings. (Adapted from J.D. Demaree, C.F. Fountzoulas, J.K. Hirvonen, M.E. Monserrat, G.P. Halada, and C.R. Clayton, in J.C. Barbour, S. Roorda, and D. Ila (Eds.), *Atomistic Mechanisms in Beam Synthesis and Irradiation of Materials, Mat. Res. Soc. Proc., Vol. 504*, 1998.)

found to have a strong effect on the coating composition (N incorporation), film growth rate, and residual stress. As shown in the figure, the presence of a sufficiently intense nitrogen ion beam aided incorporation of nitrogen into the coatings during deposition (i.e., the formation of nitride phases), and produced coatings that approximated stoichiometric Cr_2N at high ion/atom ratios. This high nitrogen incorporation results in a hardening of the coating through a number of mechanisms, most likely the introduction of ion-induced dislocations and the effects of nitride precipitate growth, and induced a stress state that was highly compressive and less susceptible to cracking than ordinary evaporated chromium. Chromium nitride coatings produced by IBAD are extremely fine-grained, and do not exhibit the columnar structure often seen with PVD films without ion assistance, or the through-thickness cracks often found in electroplated coatings. Coatings produced by IBAD, therefore, can be more corrosion-resistant, since a columnar microstructure can act as a corrosion pathway to the underlying material.

The deposition parameters necessary for optimum performance as a wear-resistant hard coating (hardness, adhesion, scratch and scuff resistance, etc.) were identified using a small coating chamber and this technology was recently transferred to a large prototype production type ion beam facility constructed to demonstrate the efficacy of ion implantation and IBAD treatments at the NDCEE/CTC facility in Johnstown, PA, shown in Figure 4.19. Figure 4.19 also shows a selected number of DoD components treated at this facility that are currently undergoing testing by OEMs and military depots (M. Klingenberg, J. Arps, R. Wei, J. Demaree, and J. Hirvonen, *Surf. Coat. Technol.*, 158–159, 164–169, 2002). Other coating systems being pursued for tribological applications at the NDCEE/CTC facility include TiN, NbN, MoN, and Al_2O_3.

The ability to produce a nearly continuous range of stoichiometries makes IBAD a useful tool for investigating the wear- and corrosion-resistance of non-equilibrium phases and new alloy systems. In earlier studies of the effect on nitrogen on the aqueous corrosion of chromium nitride coatings (G.P.

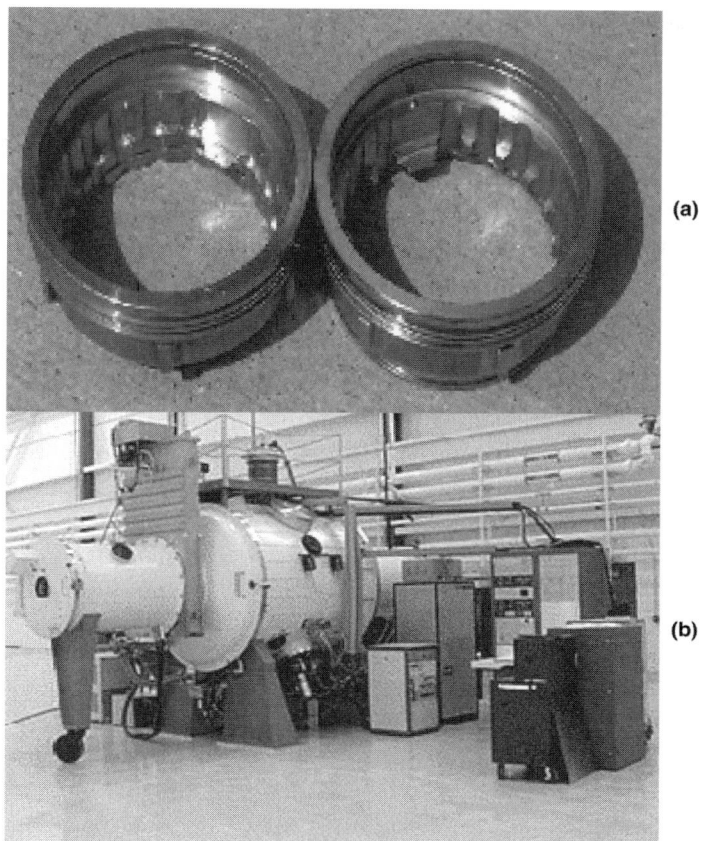

Figure 4.19 (a) Components coated with IBAD CrN for wear protection. (b) Ion implantation/IBAD demonstration facility at NDCEE/CTC in Johnstown, PA. (From M. Klingenberg, CTC. With permission.)

Halada, M.E. Monserrat, C.R. Clayton, and J.D. Demaree, in C.R. Clayton, J.K. Hirvonen, and A.R. Srivatsa (Eds.), *Advances in Coatings Technologies for Surface Engineering*, The Metallurgical Society, Warrendale, PA, 1996, pp. 339–351) researchers had only stoichiometric compounds to compare, viz. pure Cr metal and CrN pressed powder electrodes. In a subsequent study at the Army Research Laboratory (ARL),

Cr_xN_{1-x} coatings with $x = 0.0$ to 0.35 were synthesized with IBAD, and the passive oxides formed on these coatings during aqueous corrosion were examined with x-ray photoelectron spectroscopy (XPS) (J.D. Demaree, W.E. Kosik, C.R. Clayton, and G.P. Halada, in S. Seal, N. Dahotre, J. Moore, and B. Mishra (Eds.), *Surface Engineering in Materials Science I*, The Metallurgical Society, Warrendale, PA, 2000, pp. 335–345). The results of this study helped confirm the earlier hypothesis that nitrogen improves the corrosion behavior of chromium alloys by enhancing the number of chromate oxyanions in the protective oxide. In a similar manner, two e-beam evaporators were used to simultaneously evaporate Cr and Mo, with a nitrogen ion beam assist, to produce ternary Cr–Mo–N alloys. The corrosion resistance of these coatings was evaluated to assess the effectiveness of small amounts of Mo in improving the ability of CrN coatings to protect against corrosion (J.D. Demaree, in *Proceedings of the 2000 Conference on the Application of Accelerators in Research and Industry*, American Institute of Physics Conference Proceedings, 576, 2000, p. 915). In recent unpublished work, these coatings have also been evaluated to assess the effectiveness of small additions of Mo in improving the ability of CrN coatings to protect against corrosion through pinholes, scratches, and other coating flaws through a molybdate/chromate self-healing mechanism. These and other corrosion-resistant IBAD coatings are listed in Table 4.20.

4.3.4.2.5 Metastable Compound Formation

Metastable alloys are ordinarily formed under conditions involving a rapid thermal quench, such as in splat cooling where effective quenching rates of 10^6–$10^7\,°C/sec$ can be obtained. The stopping of ions can be viewed conceptually as similar on an atomistic scale. Individual atoms within a collision cascade, displaced in the wake of an ion, can have high kinetic energies (several eV) corresponding to extremely high equivalent kT values, and can dissipate this energy within time periods of $< 10^{-10}$ sec yielding very high local quenching

TABLE 4.20 Corrosion Control Coatings by IBAD

Coating	Substrate	IBAD Parameters	Results	Ref.
Cr/Mo/N	Steels, glass	2 μm thick	Enhanced passivation via N and Mo mechanisms	Demaree (a)
Si, Al, B, Cr	Steels	Up to 8 μm thick coatings deposited	Corrosion immersion tests and pull-off tests performed; best results shown by thickest (8 μm) coatings	Einsinger and Wolf (b)
Zn alloys	Low carbon, sheet steel 20–30 cm wide	2–8 μm thick Zn, Zn/Ti-, Zn/Cr-, and Zn/Mn-alloy coatings compared	All compared favorably with "optimized" electrogalvan-ized coating.	Wolf et al. (c)
TiN on shaver blades	Stainless steel shaver blades	N onto SS strip	Highly adhesive; decorative; wear/corrosion protection	Miyano and Kitamura (d)

Sources: (a) J.D. Demaree, Development of corrosion-resistant metal nitride coatings via ion beam assisted deposition, *Proceedings of the 2000 Conference on the Application of Accelerators in Research and Industry*, American Institute of Physics Conference Proceedings, 576, 2000, p. 915; (b) W. Einsinger and G.K. Wolf, Ion assisted coatings for corrosion protection studies, *Mater. Sci. Eng. A*, 116, 1–14, 1989; (c) G.K. Wolf, G. Preiss, R. Munz, and L. Guzman, Large area deposition of Zn-alloys in the coil coating mode, *Nucl Instrum. Methods Phys. Res. B*, 175–177, 756–761, 2001; (d) T. Miyano and H. Kitamura, Coating on the cutting edge of an electric shaver by ion beam assisted deposition, *Surf. Coat. Technol.*, 65, 179–183, 1994.

rates (10^{14}–10^{15}°C/sec). Such ultrafast "quenching" times associated with the decay of ion collision cascades have been utilized for producing various classes of metastable compounds, including cubic boron nitride. The quest for thin film cubic boron nitride is driven by its high hardness (second only to diamond), high temperature compatibility, high chemical inertness, and large band gap.

Several metastable compounds have also been prepared utilizing the high degree of control inherent to the ion beam process. Titanium nitride prepared by conventional CVD or reactive sputtering techniques often contains two or three different phases including alpha-Ti, Ti_2N, and TiN. However using the IBAD technique, single phase films of the Ti_2N phase, confirmed by x-ray diffraction, have been prepared for the first time by carefully controlling the ratio of the titanium deposition rate to the nitrogen ion beam (M. Kiuchi, K. Fuji, H. Miyamura, K. Kadona, and M. Satou, *Nucl. Instrum. Methods B*, 37/38, 701, 1989).

Another study involved low energy oxygen ion assistance for copper oxide formation (C.R. Guarnieri, S.D. Offsey, and J.J. Cuomo, in: J. Cuomo, S. Rossnagel, and H. Kaufman (Eds.), *Handbook of Ion Beam Processing Technology*, Noyes, Park Ridge, NJ, 1989). Oxygen-to-copper composition ratios were determined by Debye–Scherrer–Hull x-ray analysis for Cu–O films deposited by reactive evaporation of Cu with simultaneous oxygen ion bombardment at (a) 100 eV and (b) 200 eV energy. X-ray analysis of the 100 eV coatings confirmed the existence of polycrystalline Cu_2O and CuO, and the presence of another phase identified as Cu_2O_4, a new metastable compound. It is interesting to note that at a beam energy of 200 eV, neither the new metastable compound (Cu_2O_4) nor the Cu_2O phase is seen, presumably due to the higher degree of beam-induced lattice disorder.

4.3.4.2.6 Case Study of Commercial IBAD Processing: Electric Shaver Heads

The first known commercial use of an IBAD coating for a tribological/corrosion purpose is for TiN used on an electric shaver head for corrosion and wear resistance (Figure 4.20). It is currently being produced in Japan using reactive IBAD processing, and was introduced commercially in 1987 by Matsushita Electric Works, Ltd., Osaka, Japan. The development

TABLE 4.21 Metastable Compound Formation and Electronic Coatings by IBAD

Coating	IBAD Parameters	Results	Ref.
Metastable I-BN; c-BN	Defined zones for c-BN formation in ion beam space	N/Ar ion mixture and energy, current important to formation of cubic phase	Kester and Messier (a)
Metastable TiN$_2$ Compound	N beam onto evaporated Ti	Demonstrated ability to produce single TiN$_2$ phase film by suitable adjustment of IBAD parameters	Kiuchi et al. (b)
Metastable Cu oxides	Only for low-energy beam < 100 eV	See new binary Cu oxides; under controlled O ion to Cu atom arrival ratios	Guarnieri et al. (c)

Sources: (a) D. Kester and R. Messier, *J. Appl. Phys.*, 72(2), 504, 1992; *Mater. Res. Soc. Symp. Proc.*, 235, 721, 1992; (b) M. Kiuchi, K. Fuji, H. Miyamura, K. Kadona, and M. Satou, *Nucl. Instrum. Methods B*, 37/38, 701, 1989; (c) C.R. Guarnieri, S.D. Offsey, and J.J. Cuomo, in J. Cuomo, S. Rossnagel, and H. Kaufman (Eds.), *Handbook of Ion-Beam Processing Technology*, Noyes, Park Ridge, NJ, 1989.

of a hard and corrosion-resistant coating for the cutting blade of a Panasonic™ electric "wet and dry" shaver faced several obstacles to commercialization in the view of company researchers. Miyano and Kitamura (T. Miyano and H. Kitamura, Coating on the cutting edge of an electric shaver by ion beam assisted deposition, *Surf. Coatings Technol.*, 65, 179–183, 1994) have detailed their considerations of this particular application, which well demonstrates the technical concerns of employing an IBAD coating for a new application. Extensive electrochemical tests confirmed that the pitting potential of the IBAD TiN coatings (+150 to 200 mV) was superior to that of ion-plated coatings (−50 to 0 mV), due

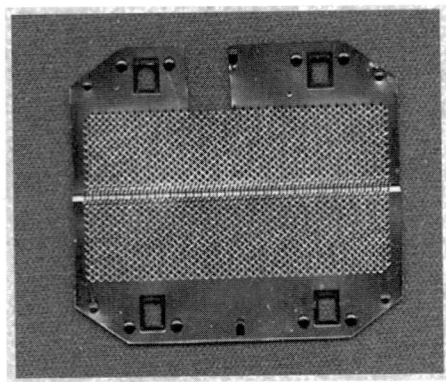

Figure 4.20 Stainless steel foil IBAD coated for electric shaver application. (As described by T. Miyano and H. Kitamura, Coating on the cutting edge of an electric shaver by ion beam assisted deposition, *Surf. Coat. Technol.*, 65, 179–183, 1994.)

presumably to fewer open defects (pores). Table 4.22 outlines the issues they considered important for this application.

Their success of applying this technique is ascribed to addressing or solving the issues listed in Table 4.23. Many of the issues listed in Table 4.23 are the same as for treating precision optics. However, the processing for this application is continuous versus batch processing of optics and involves treating sheet rolls of the thin martensitic stainless steel foil substrates that are used for the shaver heads. In the actual production roll coaters, each portion of the treated strip passes consecutively over the following components: from a source spool over (i) the cleaning ion source, (ii) the reaction ion source with evaporator, and (iii) a color monitor for quality control and process feedback, and finally to a take-up spool. Their process has been automated with feedback from the ion source and the color monitor, and runs unmanned 24 hours a day.

The examples shown above demonstrate the versatility and usefulness of IBAD processing for critical components. Successful commercialization of the IBAD technique requires the demonstration of a technically superior product, at com-

TABLE 4.22 Characteristics of Low-Energy IBAD Process for Commercialization

Film Formation	Equipment
ADVANTAGES:	
Superior adhesion	Substrate bias not needed
Arbitrary thickness films achievable by alleviating stress	Ceramic film formation enabled at ambient temperature
High film durability	Superior control and stability
	High current ions possible with compact power supplies
ISSUES:	
Lowering film losses by beam sputtering	Process limited to flat surfaces
	Large processing zone needed
	Higher deposition rates

Source: T. Miyano and H. Kitamura, Coating on the cutting edge of an electric shaver by ion beam assisted deposition, *Surf. Coat. Technol.*, 65, 179–183, 1994.

TABLE 4.23 Issues Considered During Equipment Preparation for Commercial IBAD Processing of Electric Razor Heads

Issues Considered	Requirements, Approach
Optimization of film formation conditions	Composition control and understanding of film formation mechanism
	High-speed film formation
	needed for large area, high current ion source
	Lower sputter losses→low energy operation
Equipment performance	Continuous operation, high stability→continuous 24-hour operation
Hardware issues	High current ion source
	Enlargement of process zone
Applications and product requirements	Products with small area and high added value
	Constant (high) production volume
	Continuous process capability (use of belt-shaped/hoop-shaped substrates)

Source: T. Miyano and H. Kitamura, Coating on the cutting edge of an electric shaver by ion beam assisted deposition, *Surf. Coat. Technol.*, 65, 179–183, 1994.

petitive or lower costs. Its acceptance in production will be greatly influenced by the relative ease of transitioning present equipment to accommodate IBAD processing.

4.4 GAS-CLUSTER ION BEAMS

The following sections deal with a distinctly different type of ion beam technology that has had its own evolution to its present status. Whereas much of the ion implantation and ion beam-assisted deposition work originated from nuclear, atomic, and plasma physics basics, the gas cluster ion beam (GCIB) technology started with basic interests in gas physics phenomena and was adopted into beam accelerator apparatus as described below. In spite of these disparate starting points, the techniques all do share common features including the efficient generation, transport and delivery of energetic charged particles in vacuum for beneficially modifying the surface properties of materials. For GCIB the predominant interest has been in its unique capability among ion techniques to smooth surfaces on a nanometer scale. The history, basic fundamentals, and present application areas of this technique are described in this final section of the chapter.

4.4.1. Gas-Cluster Ion Beam (GCIB) Techniques

In 1956 Becker, Bier and Henkes at the University of Marburg (and later of the University of Karlsruhe) reported the first vacuum transport properties of aerosols of condensed gases, which they referred to as a cluster beam (E.W. Becker, K. Bier, and W. Henkes, *Z. Phys.*, 146, 333, 1956). In that and subsequent work (E.W. Becker, *Z. Phys D*, 3, 101, 1986), they precooled various gases (Ar, CO_2, H_2, He, and N_2) then expanded them through a nozzle from a high-pressure reservoir into a vacuum chamber causing the gases to condense into very small droplets, i.e., essentially an aerosol. These neutral (unionized) "clusters" exit the nozzle as a narrow, high-velocity jet and are differentially pumped by passage through a pinhole, referred to as a skimmer. By use of a mechanical chopper wheel and an ionization manometer for detection of ion arrival, they were able to demonstrate time-of-flight (TOF) measurements for both the fast uncondensed gas and the more massive, slower gas clusters. The clusters were found to have a sharp appearance onset, an onset induced by cooling the inlet gas below a certain point or raising the pressure of the gas reservoir beyond a certain point, each characteristic of the particular gas in use (E.W. Becker, *Z. Phys D*, 3, 101, 1986). It was soon demonstrated (in 1961) that electron impact could be used to ionize the gas clusters and that subsequent acceleration could be utilized to form an energetic beam of the cluster ions, i.e., the gas-cluster ion beam (GCIB) (W. Henkes, *Z. Naturforsch.*, 16a, 842, 1961; E.W. Becker, *Z. Phys D*, 3, 101, 1986). Considerable refinement of the nozzle technique, especially for other applications, led to a thermodynamic and hydrodynamic theory of the nucleation and growth of gas clusters in nozzles and the subsequent formation of the cluster jet (O.F. Hagena and W. Obert, *J. Chem. Phys.*, 56, 1793, 1972; O.F. Hagena, *Surf. Sci.* 106, 101, 1981). Well-known equilibrium thermodynamic analysis of the free energy of a droplet reveals that there is a critical radius below which the droplet spontaneously and completely evaporates over a period of time (C. Kittel and H. Kroemer, *Thermal Physics*, Freeman, 1980). Typical critical droplet radii for these aerosols are 1–10 nm. Small-angle neutron scattering was reported for water aerosols and spherical shapes were observed with log-normal distributions of the droplet radii, which lead to highly asymmetric volume (mass) distributions (B.E. Wyslouzil, J.L. Cheung, G. Wilemski, and R. Strey, *Phys. Rev. Lett.*, 79, 431, 1997). As a consequence of the nonequilibrium thermodynamic nature of cluster formation in a nozzle and the statistical nature of ionization under electron impact, the mass-to-charge ratio (M/Q) for GCIB is observed to have a wide distribution, as measured by ion TOF.

Through the 1980s there was considerable effort to study the unusual state of matter of the cluster ion as well as to develop the practice of the cluster-beam technique. Much early effort was toward metal-vapor sources and generation of large metal-cluster ions for metal film and metal-oxide deposition (I. Brodie and J.J. Muray, *The Physics of*

Micro/Nano-Fabrication, Plenum Press, 1992; C.-Y. Ng, T. Baer, and I. Powis (Eds.), *Cluster Ions, Series in Ion Chemistry, and Physics,* Wiley, 1993; H. Haberland (Ed.), *Clusters of Atoms, and Molecules,* in *Springer Series of Chem. Phys., vol. 52,* Springer-Verlag, 1994; H. Haberland, M. Moseler, Y. Qiang, O. Rattunde, T. Reiners, and Y. Thurner, *Surf. Rev. Lett.,* 3, 887, 1996). The metal-cluster beam has not seen wide use so far (see Section 4.4.4.2), but the gas-cluster beam technique flourished in several research laboratories, including Kyoto, Karlsruhe, and Freiburg universities where cluster ions of Ar, CO_2, O_2, and N_2 were shown to have useful applications (J.A. Northby, T. Jiang, G.H. Takaoka, I. Yamada, W.L. Brown, and M. Sosnowski, *Nucl. Instrum. Methods Phys. Res. B,* 74, 336, 1993; P.R.W. Henkes and B. Krevet, *J. Vac. Sci. Technol. A,* 13, 2133, 1995; I. Yamada, J. Matsuo, Z. Insepov, D. Takeuchi, M. Akizuki, and N. Toyoda, *J. Vac. Sci. Technol. A,* 14, 781, 1996; I. Yamada and J. Matsuo, *Mater. Res. Soc. Symp. Proc.,* 396, 149, 1996; I. Yamada, U.S. Patent 5,459,326). Figure 4.21 illustrates the basic design of a present-day GCIB system and a configuration for beam processing of surfaces (D.B. Fenner and M.E. Mack, Advanced applications for ion implantation, in J.F. Ziegler (Ed.), *Ion Implantion Science and Technology, 7th Edition,* Ion Implant. Tech. Co., 2000, p. 125; B.K. Libby et al., U.S. Patent 6,486,478). Starting from the left side of the figure, the room-temperature high-pressure source gas (SG) is rapidly expanded in the nozzle (N) thus forming the cluster jet in the first vacuum chamber. Passage of the jet through the skimmer (S) removes most of the unclustered gas and allows differential pumping into the second vacuum chamber. The jet then passes through the core of a cylindrical screen where energetic electrons are provided by nearby filaments. This serves to ionize the gas clusters by electron impact, with one single positive charge per cluster predominating (i.e., cluster cations). The cluster ions are extracted, accelerated, and focused by various plates in the accelerator optics (AO), and the beam is then filtered by a transverse magnetic field (M) that serves to bend out of the beam any residual monomer ions. Passage through an aperture plate (AP) serves for a final differential

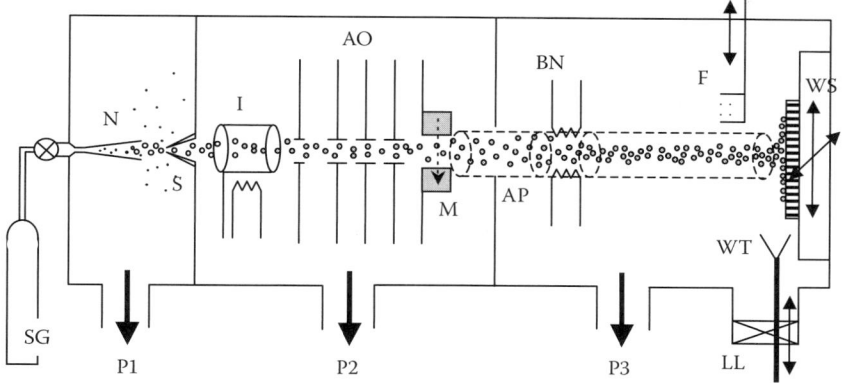

Figure 4.21 Schematic of basic GCIB system components, including: source gas (SG), nozzle (N), skimmer (S), ionizer (I), accelerator and optics (AO), magnet (M), aperture plate (AP), electron-flood beam neutralizer (BN), Faraday (F), wafer scanner (WS), wafer transfer (WT), load lock (LL) and vacuum pumps (P1, P2 and P3). The clusters are represented by the small circles and move from left to right. The open circles just after the nozzle are neutral clusters while the solid circles are ionized clusters. In the process chamber (P3) the cluster-ion beam has a sheath of electrons on its periphery, indicated by the dotted-line cylinder there.

pumping and entry into the process vacuum chamber. The cluster beam then passes near a filament that provides a flood of low-energy electrons to neutralize (EN) the beam. The beam current (ion charge flux, not mass flux) is monitored by insertion of a Faraday cup (F) into the beam. Processing is accomplished by allowing the beam to strike the work piece (e.g., wafer) at normal or near-normal incidence while mechanically scanning (WS) that work piece. Finally, efficient exchange of wafers is accomplished by a load lock (LL) and wafer transfer (WT) mechanism.

The design of GCIB apparatus requires attention to good vacuum practice and ion physics as well as several unique properties of cluster-ion beams (D.B. Fenner and M.E. Mack, Advanced applications for ion implantation, in J.F. Ziegler (Ed.), *Ion Implantion Science and Technology, 7th Edition,* Ion

Implant. Tech. Co., 2000, p. 125; B.K. Libby et al., U.S. Patent 6,486,478). As a consequence of the large M/Q of the clusters, the beam velocity is quite low and thus the charge repulsion within the beam causes lateral drift that can result in significant beam expansion. This quickly destroys the usefulness of a GCIB if attempts are made to focus the beam too tightly, or to transport the beam over relatively long distances, or to operate at too low an accelerating potential, or if a sufficient flux of neutralizing electrons is not provided. All of these issues become much more significant if the apparatus is designed to generate greater total beam current, which is required for most practical applications of GCIB. Available beam current for GCIB has increased considerably over the past few years. In 1993, the reported argon-cluster beam current was on the order of 1 nA (J.A. Northby, T. Jiang, G.H. Takaoka, I. Yamada, W.L. Brown, and M. Sosnowski, *Nucl. Instrum. Methods Phys. Res. B*, 74, 336, 1993), while by the end of 2000 commercial manufacture and sale of systems capable of 100 µA was announced (*Ultra-Smoother*® processing system from Epion Corp. of JDS Uniphase Corporation). Systems with even higher beam current have subsequently been reported and 400 µA is expected to be available soon from the same manufacturer.

The potential for applications of GCIB to surface processing was recognized in the early 1990s with the demonstration of the ability of argon clusters to reduce the roughness of (i.e., to smoothen) the surface of a gold film and remove surface contamination (clean) (J.A. Northby, T. Jiang, G.H. Takaoka, I. Yamada, W.L. Brown, and M. Sosnowski, *Nucl. Instrum. Methods Phys. Res. B*, 74, 336, 1993; I. Yamada, J. Matsuo, Z. Insepov, D. Takeuchi, M. Akizuki, and N. Toyoda, *J. Vac. Sci. Technol. A*, 14, 781, 1996; I. Yamada and J. Matsuo, *Mater. Res. Soc. Symp. Proc.*, 396, 149, 1996; I. Yamada, U.S. Patent 5,459,326). That same early study, however, reported that silicon crystal wafers after argon GCIB processing had sustained beam-induced lattice damage as measured with Rutherford backscattering spectroscopy (RBS) in the channeling mode. In time this was understood and overcome, making it possible to smoothen semiconductor crystal surfaces with little measurable damage (D.B. Fenner, V. DiFilippo, J. Bennett, J. Hirvonen, J.D. Demaree, T. Tetreault, L.C. Feldman, and A. Saigal, *Proc. SPIE,* 4468, 17, 2001). The majority of the early work that demonstrated the possibility of GCIB for smoothing and cleaning surfaces was carried out by research at Kyoto University and now also at Himeji Institute of Technology (I. Yamada, J. Matsuo, N. Toyoda, and A. Kirkpatrick, *Mater. Sci Eng. R*, 34, 231, 2001). The group at Freiburg University has reported surface smoothing by energetic cluster impact using large copper clusters in a deposition mode (H. Haberland (Ed.), *Clusters of Atoms, and Molecules,* in *Springer Series of Chem. Phys., vol. 52*, Springer-Verlag, 1994; H. Haberland, M. Moseler, Y. Qiang, O. Rattunde, T. Reiners, and Y. Thurner, *Surf. Rev. Lett.,* 3, 887, 1996; O. Rattunde, M. Moseler, A. Hafele, J. Kraft, D. Rieser, and H. Haberland, *J. Appl. Phys.,* 90, 3226, 2001). In this chapter, however, we emphasize the development of GCIB systems and processing for use as manufacturing tools in the microelectronics industry as well as establishment of a practical materials science of the interaction of GCIB with surfaces. (Major support for this development at Epion Corp. (Table 4.24) came from an award by the U.S. Department of Commerce through its Advanced Technology Program (ATP), #70NANB8H4011.) In this last regard, GCIB has grown out of the science and technology of ion implantation, of ion etch and of deposition yet is quite distinct from each of these. Assumptions of similarity without careful cognizance of the considerable basic differences can act to impede understanding and advancement of the field of GCIB and its applications.

The nature of a cluster-ion impact with the surface of a solid is now known to be quite distinct from that of a monomer ion (I. Yamada, J. Matsuo, Z. Insepov, D. Takeuchi, M. Akizuki, and N. Toyoda, *J. Vac. Sci. Technol. A,* 14, 781, 1996; I. Yamada and J. Matsuo, *Mater. Res. Soc. Symp. Proc.,* 396, 149, 1996; I. Yamada, U.S. Patent 5,459,326; D.B. Fenner and M.E. Mack, Advanced applications for ion implantation, in J.F. Ziegler (Ed.), *Ion Implantion Science and Technology, 7th Edition*, Ion Implant. Tech. Co., 2000, p. 125; B.K. Libby et al., U.S. Patent 6,486,478; D.B. Fenner, R.P. Torti, L.P. Allen, N. Toyoda, A.R.

Kirkpatrick, J.A. Greer, V. DiFilippo, and J. Hautala, *Mater. Res. Soc. Symp. Proc.,* 585, 27, 2000; D.B. Fenner, V. DiFilippo, J. Bennett, J. Hirvonen, J.D. Demaree, T. Tetreault, L.C. Feldman, and A. Saigal, *Proc. SPIE,* 4468, 17, 2001; I. Yamada, J. Matsuo, N. Toyoda, and A. Kirkpatrick, *Mater. Sci Eng. R,* 34, 231, 2001). The argon-cluster-size distribution spans the range from a few hundred to a few tens of thousands of argon atoms in each cluster, peaking at about 2000. Thus, average cluster diameters are ~ 5 nm. Basic surface smoothing is often done with argon GCIB at 10–20 kV of beam acceleration, and hence, the average kinetic energy per argon atom within the cluster is only ~ 10 eV. Although direct measurements are difficult, the charge of each cluster is thought to be single, so the cluster-ion collision is dominated by mass impact rather than charge interactions with the target. At practical beam currents, the frequency of impacts is very low compared with the duration of each impact (a few picoseconds, estimated). In addition, the bonding within a gas cluster is very weak, since the condensed gas is held together by van der Waals forces. For example, with Ar the bonding strength is ~ 0.08 eV per argon atom, from thermodynamic measurements of the bulk liquid and solid. This is much less than the cluster impact kinetic energy and hence the impact is highly inelastic, despite its low velocity. By contrast, low-energy monomer ion collisions are highly elastic. It is tempting to consider a cluster as a tiny projectile, i.e., a tiny meteorite. However, the latter are strongly bonded (a few electronvolts) and impacts are elastic except at extremely high velocity. Thus, the net result of impact of a gas-cluster ion is a very soft fragmentation with delivery of a localized pulse of kinetic energy followed by lateral exit of the now free argon atoms together with a small amount of the vaporized solid target.

The interaction volume of each cluster impact involves many atoms and bonds of the target surface, as well as a low average kinetic energy. Hence, penetration of any of the cluster atoms into the surface (i.e., implantation) is of low probability. Indeed, measurements have shown that under conditions that result in good surface smoothing, little or no argon is retained by the solid target (D.B. Fenner, J. Hautala,

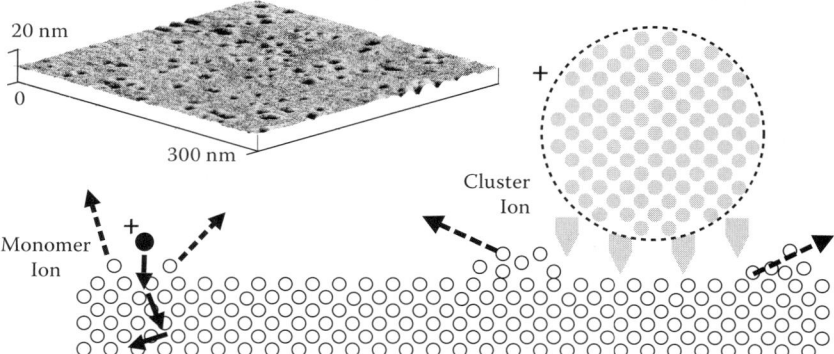

Figure 4.22 Schematic of monomer-ion (lower left) and cluster-ion (right) impacting the surface of a solid lattice (open circles). Both ions carry a single positive charge (Q=+1e) and are incident from the top down with velocity vectors indicated by the solid arrows. The sputtered surface material exits at a variety of angles indicated by dashed arrows. The inset (top left) is an AFM image of an initially-smooth SiO_2 film that has been exposed to a very low fluence GCIB.

L.P. Allen, J.A. Greer, W.J. Skinner, and J.I. Budnick, *Mater. Res. Soc. Symp. Proc.,* 614, F10.3.1, 2000; D.B. Fenner, R.P. Torti, L.P. Allen, N. Toyoda, A.R. Kirkpatrick, J.A. Greer, V. DiFilippo, and J. Hautala, *Mater. Res. Soc. Symp. Proc.,* 585, 27, 2000; D.B. Fenner, V. DiFilippo, J. Bennett, J. Hirvonen, J.D. Demaree, T. Tetreault, L.C. Feldman, and A. Saigal, *Proc. SPIE,* 4468, 17, 2001). Some of the features of impact are illustrated in Figure 4.22 along with an atomic force microscope (AFM) image of a few individual cluster impacts (L.P. Allen, Z. Insepov, D.B. Fenner, C. Santeufemio, W. Brooks, K.S. Jones, and I. Yamada, *J. Appl. Phys.,* 92, 3671, 2002). The cartoon in Figure 4.22 compares the impact of a monomer ion with that of a cluster ion. The former easily penetrates the surface since it carries a few 100 eV or more of kinetic energy, and when it is about one bond length from impact its size is comparable to a bond length of the solid, i.e., it has an interaction volume of atomic dimensions. The cluster ion, on the other hand, is more accurately thought of as a nanophase piece of condensed matter (argon snow) and acts cooperatively over a large interaction volume. The gas cluster is exceedingly soft and easily disintegrates upon impact. The AFM image in Figure 4.22 illustrates the individual impact features that are found if the surface is extremely smooth to start with (average roughness $R_a \sim 0.1$ nm, or less) and the clusters have relatively high acceleration (25 kV argon clusters in this case). The craters are ~ 4 nm deep and ~ 10 nm in diameter in this example, and have formed a small rim of ejecta (L.P. Allen, Z. Insepov, D.B. Fenner, C. Santeufemio, W. Brooks, K.S. Jones, and I. Yamada, *J. Appl. Phys.*, 92, 3671, 2002). For smoothing surfaces, typically less acceleration is used together with much higher fluence (D.B. Fenner, R.P. Torti, L.P. Allen, N. Toyoda, A.R. Kirkpatrick, J.A. Greer, V. DiFilippo, and J. Hautala, *Mater. Res. Soc. Symp. Proc.*, 585, 27, 2000; D.B. Fenner, V. DiFilippo, J. Bennett, J. Hirvonen, J.D. Demaree, T. Tetreault, L.C. Feldman, and A. Saigal, *Proc. SPIE*, 4468, 17, 2001), such that the impact features are much shallower and much denser, i.e., overlapping a great deal. (Since implantation does not occur with GCIB the term "dose" is inappropriate. Hence, we will refer to the time-integrated beam flux density as "fluence" with units of ions/cm^2.) Under those conditions, the many impact craters generate many small ejecta rims and these, in effect, constitute lateral motion of the surface material. As with thermally induced lateral motion (diffusion) on a surface, rough features are preferentially relaxed into less rough features, and thereby the surface is smoothed. Indeed, as with thermal diffusion, sharp features smooth much more readily than do broad features, i.e., those with in-plane dimensions a few orders of magnitude larger than the typical impact crater diameter (D.B. Fenner, V. DiFilippo, J. Bennett, J. Hirvonen, J.D. Demaree, T. Tetreault, L.C. Feldman, and A. Saigal, *Proc. SPIE*, 4468, 17, 2001).

There has been some effort to simulate by calculation the impact of gas clusters with solid targets. Individual impacts can be modeled with the methods of molecular dynamics (MD) and this provides information on the very short time behavior (I. Yamada, J. Matsuo, Z. Insepov, D. Takeuchi, M. Akizuki, and N. Toyoda, *J. Vac. Sci. Technol. A,* 14, 781, 1996; I. Yamada and J. Matsuo, *Mater. Res. Soc. Symp. Proc.,* 396, 149, 1996; I. Yamada, U.S. Patent 5,459,326; L.P. Allen, Z. Insepov, D.B. Fenner, C. Santeufemio, W. Brooks, K.S. Jones, and I. Yamada, *J. Appl. Phys.,* 92, 3671, 2002). Indeed, from this work it is predicted that the impact causes a crater with a small rim of ejecta as well as a flux of sputtered material and shock waves that propagate into the solid target. These simulations also suggest impact duration of a few picoseconds during which the local temperature and pressure reach extremely high values: 10^5 K and 10^6 bar, respectively. Direct experimental verification of these predictions has not been possible, and although it is undoubtedly true that fairly high temperatures and pressures occur, these numerical values would not seem to be supported by microscopic observations to date. The extremely nonlinear nature of the many-body collisions under highly nonequilibrium thermodynamic conditions greatly complicate interpretation. Detailed MD analysis of argon cluster etching of silicon crystal surfaces has also been reported (N.A. Kubota, D.J. Economou, and S.J. Plimpton, *J. Appl. Phys.,* 83, 4055, 1998). Other simulation approaches utilize continuum models and address issues of cluster impact such as the experimentally observed smoothing at high fluence (D.B. Fenner, D.W. Dean, V. DiFilippo, L.P. Allen, J. Hautala, and P.B. Mirkarimi, *Mater. Res. Soc. Symp. Proc.,* 647, O5.2.1, 2001; D.B. Fenner, *J. Appl. Phys.,* 95, 5408, 2004). Such phenomenological work has shown that a simple surface-diffusion model can predict key features of the evolution of surface morphology that compare favorably with AFM images of actual solids that were smoothed by GCIB.

4.4.2 GCIB Smoothing, Etching, and Cleaning Applications

The smoothing of tantalum metal surfaces with argon GCIB is illustrated in Figure 4.23 (D.B. Fenner, J. Hautala, L.P.

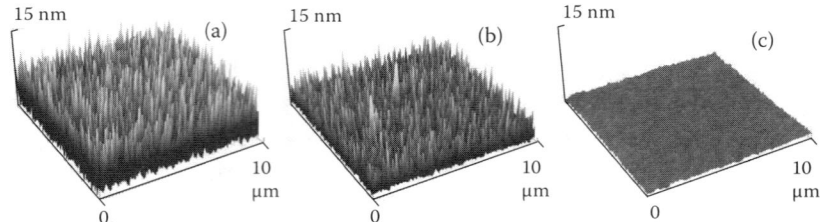

Figure 4.23   AFM images of the surface of a tantalum film at three stages of processing: (a) prior to exposure, (b) after 2×1014 ions/cm², and (c) after 3×1016 ions/cm² of argon GCIB.

Allen, J.A. Greer, W.J. Skinner, and J.I. Budnick, *Mater. Res. Soc. Symp. Proc.,* 614, F10.3.1, 2000). Shown is a sputter-deposited Ta thin film which, as deposited, is already very smooth, having an average roughness $R_a \sim 1.5$ nm. The typical columnar morphology can be seen in the AFM image of Figure 4.23a. After a fluence of 2×10^{14} ions/cm² the loss of roughness is already apparent, Figure 4.23b, and after a total of 3×10^{16} ions/cm² the roughness is only $R_a \sim 0.18$ nm, Figure 4.23c. Many such images were obtained for a sequence of argon GCIB fluences and at three accelerations of 14, 20, and 28 kV. The result is that the average (and the rms) roughness is reduced in approximately an exponential fashion with increasing fluence, as illustrated in Figure 4.24. Here it can be seen that the higher accelerations initially reduce the surface roughness rapidly, but that the lower accelerations trend toward a lower residual roughness at high fluence. Further, by fitting exponential decay functions to the data in Figure 4.24a–c it was found that the decay rates are linear with the beam acceleration, as shown in Figure 4.24d. The data in Figure 4.24a–c also suggest that a multistep smoothing process would have an improved efficiency (D.B. Fenner, U.S. Patents 6,375,790 and 6,805,807). That is, GCIB can be initiated at a relatively high acceleration, a small fluence applied and thus attain a rapid reduction of the initial roughness. At a high acceleration the residual roughness would be unacceptably high, so the apparatus is adjusted to lower acceler-

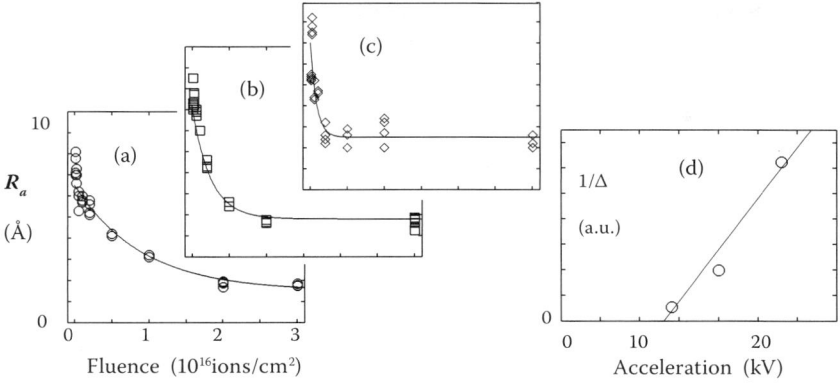

Figure 4.24 Average roughness (R_a) measured by AFM for Ta films exposed to argon GCIB at (a) 14, (b) 20, and (c) 28 kV acceleration. The solid lines are exponentials fit to the data. In (d) are plotted the decay rates ($1/\Delta$) from the curves in (a)–(c) along with a linear fit line.

ation and a second exposure made. With a judicious choice of accelerations and fluences a quicker overall reduction of the roughness to some residual roughness goal is possible. Indeed, today most industrial GCIB smoothing processes benefit from some variation on this progressive adaptation (i.e., multistep) of the basic process.

Surface smoothing by GCIB is accompanied by sputter cleaning and etching of the surface. The sputtering occurs at a constant rate during any constant GCIB flux, such as during the smoothing shown in Figure 4.24a–c. The sputter etch is of the directed-beam type rather than the plasma-etch type, and the rate has been measured by various means. Figure 4.25 shows the etch rate for argon GCIB incident upon a gold film that is the front electrode of a quartz crystal microbalance or monitor (QCM). The etch rate shown is the calculated rate effective for etching the entire surface of a wafer 150 mm in diameter with commercial apparatus (*Ultra-Smoother*®). The calculated sputter yield based on this data is ~ 12 atoms/cluster ion. Again, as in Figure 4.24d, a linear behavior with beam acceleration is observed along with a minimum acceleration

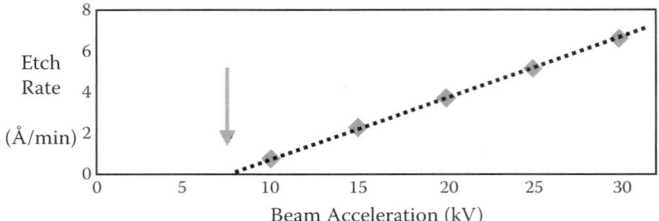

Figure 4.25   Sputter rate (mass etched) for a gold film under argon GCIB exposure. The arrow indicates the estimated threshold of cluster-ion acceleration required to initiate etching.

threshold. These features of GCIB were reported previously for sputtering yield, and it was proposed there that this is evidence that the cluster impact and etch mechanism is primarily one of kinetic energy delivery (I. Yamada, J. Matsuo, Z. Insepov, D. Takeuchi, M. Akizuki, and N. Toyoda, *J. Vac. Sci. Technol. A,* 14, 781, 1996; I. Yamada and J. Matsuo, *Mater. Res. Soc. Symp. Proc.*, 396, 149, 1996; I. Yamada, U.S. Patent 5,459,326; D.B. Fenner, J. Hautala, L.P. Allen, J.A. Greer, W.J. Skinner, and J.I. Budnick, *Mater. Res. Soc. Symp. Proc.*, 614, F10.3.1, 2000; I. Yamada, J. Matsuo, N. Toyoda, and A. Kirkpatrick, *Mater. Sci Eng. R*, 34, 231, 2001). Notice that higher accelerations give a more rapid etch but also have generally higher residual roughness, while the opposite is true of lower accelerations.

A wide range of materials has been investigated for potential application of GCIB as a tool to smoothen surfaces of various compositions (I. Yamada, J. Matsuo, Z. Insepov, D. Takeuchi, M. Akizuki, and N. Toyoda, *J. Vac. Sci. Technol. A,* 14, 781, 1996; I. Yamada and J. Matsuo, *Mater. Res. Soc. Symp. Proc.*, 396, 149, 1996; I. Yamada, U.S. Patent 5,459,326; D.B. Fenner and M.E. Mack, Advanced applications for ion implantation, in J.F. Ziegler (Ed.), *Ion Implantion Science and Technology, 7th Edition*, Ion Implant. Tech. Co., 2000, p. 125; B.K. Libby et al., U.S. Patent 6,486,478; J.A. Greer, D.B. Fenner, J. Hautala, L.P. Allen, V. DiFilippo, N. Toyoda, I. Yamada, J. Matsuo, E. Minami, and H. Katsumata, *Surf. Coat. Technol.,*

133–134, 273, 2000). Many of these materials are listed in Table 4.24. Especially for thin films, material properties can vary considerably, and depending on such factors as the method of film fabrication utilized, the effects of GCIB processing also vary. All of the materials in Table 4.24 are of current interest for fabrication into products for industry, and many for microelectronics and photonics. (Major support for the development at Epion Corp. came from an award by the U.S. Department of Commerce through its Advanced Technology Program (ATP), #70NANB8H4011.) Since the etch rate of GCIB is low and it smoothes at a fine scale only, surfaces with an initially high level of roughness or wide undulations are unsuited for GCIB smoothing unless preceded by another technique, such as chemical-mechanical polishing (CMP). For a given GCIB configuration, the rate of smoothing and the residual roughness attainable depend upon the nature of the material. Defects such as microvoids, nonabrupt grain boundaries, or film stress can cause the GCIB process to produce unsatisfactory results. However, all inorganic materials that are initially of very high quality appear at this point to be suitable for improvement of the surface morphology by GCIB. Even the complex metal-oxide compound of $YBa_2Cu_3O_{7-\delta}$ (YBCO) was smoothed by GCIB, although a brief anneal was required to regrow the outermost crystal layer of the film (W.-K. Chu, Y.O. Li, J.R. Liu, J.Z. Wu, S.C. Tidrow, N. Toyoda, J. Matsuo, and I. Yamada, *Appl. Phys. Lett.* 72, 246, 1998. W.-K. Chu et al., U.S. Patent 6,251,835). Experience with organic solids such as industrial polymers is limited so far, but some appear promising as was reported for polycarbonate and polypropylene (H. Biederman, D. Slavinska, H. Boldyreva, H. Lehmberg, G. Takaoka, J. Matsuo, H. Kinpara, and J. Zemek, *J. Vac. Sci. Technol. B*, 19, 2050, 2001; J.P. Dykstra et al., U.S. Patent 6,331,227).

Some surfaces may have relatively small average roughness (R_a) and yet much larger but occasional maximum deviations (Δz), such as spikes and pits or clumps of foreign material (contamination). Schemes to use cluster beams for surface cleaning have been proposed (I. Yamada, J. Matsuo, Z. Insepov, D. Takeuchi, M. Akizuki, and N. Toyoda, *J. Vac.*

TABLE 4.24	Materials Processed by GCIB at Epion Corporation

Alumina (Al2O3)–film	Alumina (Al2O3)–film
Alumina titanium carbide (AlTiC)–ceramic	Alumina titanium carbide (AlTiC)–ceramic
Aluminum–film	Aluminum–filmCalcium Fluoride–single crystalCopper–sputtered
Calcium Fluoride–single crystal	Copper–electroplated
Copper–sputtered	Cobalt Iron (CoFe)–film
Copper–electroplated	Cobalt chrome (CoCr)–bulk
Cobalt Iron (CoFe)–film	Diamond–polycrystalline film
Cobalt chrome (CoCr)–bulkDiamond–polycrystalline film	Gallium Arsenide (GaAs)–single crystal
Gallium Arsenide (GaAs)–single crystal	Gallium Antimonide (GaSb)–single crystal
Gallium Antimonide (GaSb)–single crystal	Gallium Nitride (GaN)–film
Gallium Nitride (GaN)–film	Gold–sputtered film
Gold–sputtered film	Hafnium oxide (HfO2)–film
Hafnium oxide (HfO2)–film	Indium phosphide (InP)–single crystal
Indium phosphide (InP)–single crystal	Indium tin oxide (ITO)–film
Indium tin oxide (ITO)–film	Iron oxide (Fe2O3)–film
Iron oxide (Fe2O3)–film	Lead zirconate titanate (PZT)–film
Lead zirconate titanate (PZT)–film	Lithium Tantalate (LiTaO3)–film
Lithium Tantalate (LiTaO3)–film	Nickel–bulk
Nickel–bulkPermalloy (Ni-Fe)–sputtered film	Permalloy (Ni-Fe)–sputtered film
Permalloy (Ni-Fe)–electroplated film	Permalloy (Ni-Fe)–electroplated film

Sci. Technol. A, 14, 781, 1996; I. Yamada and J. Matsuo, *Mater. Res. Soc. Symp. Proc.,* 396, 149, 1996; I. Yamada, U.S. Patent 5,459,326; I. Yamada, J. Matsuo, N. Toyoda, and A. Kirkpatrick, *Mater. Sci. Eng. R,* 34, 231, 2001; J.F. Mahoney, U.S. Patent 5,796,111). An example of GCIB used to clean and smooth surfaces is shown in Figure 4.26 (D.B. Fenner, J. Hautala, L.P. Allen, J.A. Greer, W.J. Skinner, and J.I. Budnick, *Mater. Res. Soc. Symp. Proc.,* 614, F10.3.1, 2000). In Figure 4.26a, AFM images at two different magnifications of the as-deposited surface of an alumina thin film are shown, while Figure 4.26b displays corresponding images of the surface after a GCIB process. The GCIB here was a multistep smoothing process consisting of three Ar-GCIB exposures, each at an

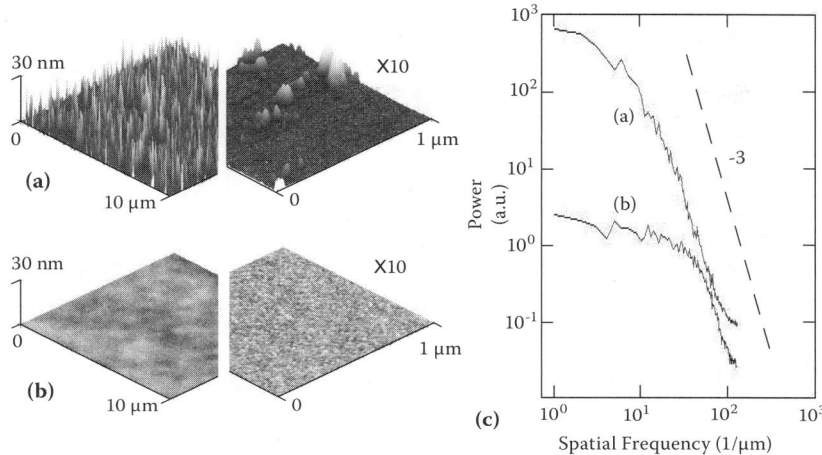

Figure 4.26 Smoothing of an alumina (Al_2O_3) thin film. The surface (a) before and (b) after GCIB exposure, at two AFM magnifications (X1 and X10) corresponding to scan length of 10 and 1 mm, respectively. In (c) are the PSD (log scales) calculated from the X10 images of (a) and (b). The dashed line is a power-law with a slope of –3.

acceleration incremented lower than the previous one, followed by an oxygen-GCIB exposure at a low acceleration. At the lower AFM resolution (scan length of 10 µm) the initial surface in Figure 4.26 is seen to have a high density of spikes. At the higher resolution (scan length of 1 µm) these spikes are seen to be lumps, of low areal density and set against an otherwise fairly smooth surface. By comparison of Figure 4.26a with b, it is seen that the GCIB exposure has removed the spikes (cleaned the surface) and left the underlying surface smooth.

The higher-resolution images shown in Figure 4.26a and b have been statistically analyzed by (one-dimensional) Fourier transform, and the squared Fourier components (averaged over the surface) are shown as power spectral density (PSD) curves in Figure 4.26c (D.B. Fenner, V. DiFilippo, J. Bennett, J. Hirvonen, J.D. Demaree, T. Tetreault, L.C. Feldman, and A. Saigal, *Proc. SPIE,* 4468, 17, 2001). It is seen here that the GCIB process has greatly reduced the strength of the spectral power of the roughness over the spatial frequency range of about 1–100 μm^{-1}. GCIB has a diminishing effect on much larger in-plane features sizes, and in fact the PSD measured at 10 μm and larger typically show little, if any, change upon GCIB processing. Thus, GCIB can be thought of as preferentially removing sharp features within this approximate lateral-size range. The roughness PSD can be compared with that of a random fractal. The dashed line at a power-law slope of –3 in Figure 4.26c is related to an effective fractal dimension (A.-L. Barabasi and H.E. Stanley, *Fractal Concepts in Surface Growth*, Cambridge Press, 1995; D.B. Fenner, *J. Appl. Phys.*, 95, 5408, 2004). The line is thus a gauge of the range of spatial wavelengths for which the surfaces show fractal behavior. By analysis of PSD from AFM images of a variety of surfaces after GCIB exposure, it was found that the PSD often show a "knee" shape (C. Santeufemio, *Microsc. Microanal.*, 8, Suppl. 2, 1138CD, 2002). Curve b in Figure 4.26 shows such a knee at ~ 50 μm^{-1}. Above the spatial frequency of the knee the PSD is a fractal-like, i.e., power-law slope. While below the knee and extending down to approximately 1–10 μm^{-1}, the PSD is nearly white-noise like, i.e., independent of spatial frequency.

Ion implantation causes copious amounts of lattice damage to single crystals and, especially with silicon, this can be successfully removed by thermal annealing after the implant. Highly accelerated argon GCIB was initially reported to cause lattice damage and some implantation into silicon (J.A. Northby, T. Jiang, G.H. Takaoka, I. Yamada, W.L. Brown, and M. Sosnowski, *Nucl. Instrum. Methods Phys. Res. B*, 74, 336, 1993; I. Yamada, J. Matsuo, Z. Insepov, D. Takeuchi, M. Akizuki, and N. Toyoda, *J. Vac. Sci. Technol. A*, 14, 781, 1996; I. Yamada and J. Matsuo, *Mater. Res. Soc. Symp. Proc.*, 396, 149, 1996; I. Yamada, U.S. Patent 5,459,326). Although processing is frequently done with argon, other source gases for GCIB sometimes provide particular advantages with certain applications. For example, argon-GCIB smoothing of SiC single-crystal wafers was found to be accompanied by a modest but significantly increased lattice-defect density near the wafer surface, presumably due to damage induced by cluster

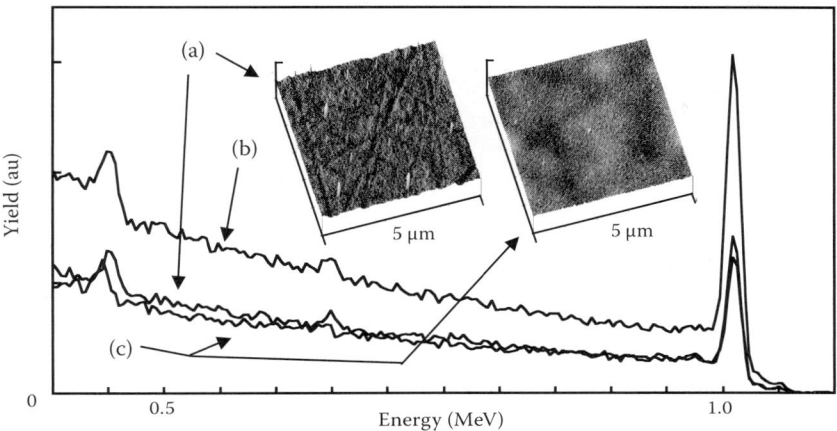

Figure 4.27 RBS channeling yield (return) from SiC wafers (a) before, and (b) after argon and (c) oxygen GCIB exposures. Insets show corresponding AFM images for (a) and (c).

impacts (D.B. Fenner, V. DiFilippo, J. Bennett, J. Hirvonen, J.D. Demaree, T. Tetreault, L.C. Feldman, and A. Saigal, *Proc. SPIE,* 4468, 17, 2001). Use of multiple argon-process steps afforded an improved final roughness, but some lattice damage persisted (smooth surfaces are, of course, not necessarily damage free). It was found that oxygen gas could be used instead of the argon for GCIB of SiC wafers that afforded smoothing but without damage. RBS channeling evaluation is particularly sensitive to damage in a single-crystal lattice and such an analysis demonstrated little if any increase in damage after oxygen-GCIB smoothing, as shown in Figure 4.27. Note the lack of an increase in the peak at ~ 1 MeV, the Si surface peak, for the oxygen-GCIB process (curve c in Figure 4.27), despite this being a process with no elevated temperatures (i.e., no anneal after the GCIB exposures).

The problem of near-surface lattice damage caused by argon GCIB, initially found with silicon wafers (J.A. Northby, T. Jiang, G.H. Takaoka, I. Yamada, W.L. Brown, and M. Sosnowski, *Nucl. Instrum. Methods Phys. Res. B*, 74, 336, 1993; I. Yamada, J. Matsuo, Z. Insepov, D. Takeuchi, M. Akizuki, and N. Toyoda, *J. Vac. Sci. Technol. A,* 14, 781, 1996; I. Yamada and J. Matsuo, *Mater. Res. Soc. Symp. Proc.,* 396, 149, 1996; I. Yamada, U.S. Patent 5,459,326), was corroborated by careful high-resolution microscopy (L.P. Allen, Z. Insepov, D.B. Fenner, C. Santeufemio, W. Brooks, K.S. Jones, and I. Yamada, *J. Appl. Phys.,* 92, 3671, 2002). Also, it was found that the silicon is oxidized under argon GCIB exposure (D.B. Fenner and M.E. Mack, Advanced applications for ion implantation, in J.F. Ziegler (Ed.), *Ion Implantion Science and Technology, 7th Edition,* Ion Implant. Tech. Co., 2000, p. 125; B.K. Libby et al., U.S. Patent 6,486,478; D.B. Fenner, R.P. Torti, L.P. Allen, N. Toyoda, A.R. Kirkpatrick, J.A. Greer, V. DiFilippo, and J. Hautala, *Mater. Res. Soc. Symp. Proc.,* 585, 27, 2000; L.P. Allen, Z. Insepov, D.B. Fenner, C. Santeufemio, W. Brooks, K.S. Jones, and I. Yamada, *J. Appl. Phys.,* 92, 3671, 2002). Argon GCIB acts as a physical sputter etch but at the same time the ions induce a chemical reaction between residual gases (e.g., water vapor) and the highly reactive silicon (freshly exposed by each cluster impact). Even individual argon-cluster impact features on silicon are found to be in the form of small hillocks of oxide rather than the small craters formed by the same beam on an oxide (SiO_2) film surface as in Figure 4.22 (L.P. Allen, Z. Insepov, D.B. Fenner, C. Santeufemio, W. Brooks, K.S. Jones, and I. Yamada, *J. Appl. Phys.,* 92, 3671, 2002). For Si wafers receiving a higher fluence of argon GCIB the surface oxidation was verified by x-ray photoelectron spectroscopy (XPS) after various GCIB exposures (D.B. Fenner and M.E. Mack, Advanced applications for ion implantation, in J.F. Ziegler (Ed.), *Ion Implantion Science and Technology, 7th Edition,* Ion Implant. Tech. Co., 2000, p. 125; B.K. Libby et al., U.S. Patent 6,486,478; L.P. Allen, Z. Insepov, D.B. Fenner, C. Santeufemio, W. Brooks, K.S. Jones, and I. Yamada, *J. Appl. Phys.,* 92, 3671, 2002). Figure 4.28 shows the Si $2p$ photoelectron emission by XPS, from surfaces that were exposed to argon and oxygen clusters of two accelerations. All of these produce a silicon-oxide film but the argon results in an oxygen-deficient oxide. Also, the higher acceleration oxygen produced the highest proportion of Si^{4+} indicative of full oxidation to SiO_2. Spectroscopic ellipsometry (SE) of the oxide on Si grown with

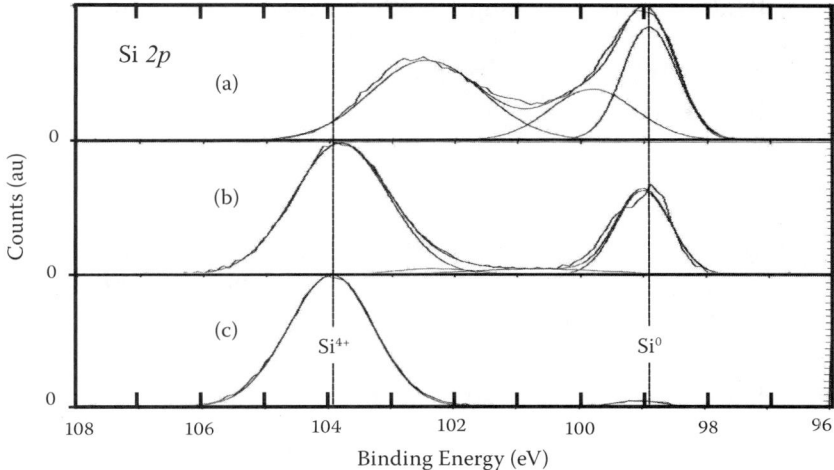

Figure 4.28 XPS Si $2p$ photoelectron emission from silicon wafer surfaces after 1×1015 ions/cm2 of GCIB fluence with the following cluster gas and acceleration: (a) argon at 7 kV, (b) oxygen at 7 kV, and (c) oxygen at 14 kV. Two of the valence states for silicon are indicated by the dashed lines. Deconvolution curves for (a) are also shown.

oxygen GCIB found that the oxide thickness is a saturating exponential growth with increasing fluences, and with ~ 15 nm typical after high fluence. A wet etch with dilute HF solution can be used to remove the oxide formed by oxygen GCIB. Indeed, this is practical for both Si and SiC, a wet etch having been done before the RBS channeling measurements shown as curves b and c in Figure 4.27. RBS return from oxygen occurs at ~ 0.65 MeV in Figure 4.27 but little if any is seen in curve c, demonstrating that oxygen is not implanted or otherwise retained after the wet etch.

Other source gases have found practical use as well (I. Yamada, J. Matsuo, Z. Insepov, D. Takeuchi, M. Akizuki, and N. Toyoda, *J. Vac. Sci. Technol. A,* 14, 781, 1996; I. Yamada and J. Matsuo, *Mater. Res. Soc. Symp. Proc.*, 396, 149, 1996; I. Yamada, U.S. Patent 5,459,326). Mixtures of SF_6 or CF_4 with O_2, for example, are found to be useful in some situations,

TABLE 4.25 Sputter Etching Depth and Induced Roughness of Silicon Wafers Resulting from GCIB Using Two Kinds of Source Gases: Pure Argon and a Mixture of 5% CF4 in Oxygen.

GCIB Gas	Ar	Ar	CF4 + O2	CF4 + O2
Wafer Orientation	Si (100)	Si (111)	Si (100)	Si (111)
Etch Depth d (nm)	30	47	370	360
Roughness Ra (nm)	1.1	1.1	0.72	0.71
Roughness Dz (nm)	12.5	12.6	8.6	8.5

Note: Silicon samples were of two crystal orientations, as indicated.

such as when more rapid etching of silicon is desired. Table 4.25 lists the measured etch depths for the CF_4 gas mixture in comparison with that of argon and for Si wafers of two crystal orientations, the (100) and the (111), all exposed to a fluence of 1×10^{16} ions/cm^2 (D.B. Fenner and Y. Shao, *J. Vac. Sci. Technol. A*, 21, 47–58, 2003). The mixture of 5% CF_4 in O_2 illustrated in the table etches about ten times faster than does argon but with little difference between the two wafer orientations. In addition, it was found that for these surfaces, both the R_a and Δz roughness were essentially independent of wafer orientation after the typical fast-etch processes being compared (i.e., at relatively high acceleration). Surprisingly the CF_4 mixture gas results in a somewhat lower roughness. Of course, if a maximally smooth surface is desired, this type of fast-etch process step may be followed by an additional fluence with perhaps oxygen GCIB at low acceleration. Orientation-independent ion etching of crystals is a clear indication that the etching is not limited by the rate of surface chemical reactions that produce the volatile species (M.A. Blauw, T. Zijlstra, R.A. Baker, and E. van der Drift, *J. Vac. Sci. Technol. B*, 18, 3453, 2000). Thus, the data of Table 4.25 indicate that both the physical etching with argon clusters and the chemical with CF_4 in O_2 are etch-rate limited by cluster-beam flux and not cluster-stimulated (chemical) vaporization of the target surface atoms.

Surface processing with clusters of CO_2 has also been reported and even used to demonstrate micromachining of silicon (I. Yamada, J. Matsuo, Z. Insepov, D. Takeuchi, M. Akizuki, and N. Toyoda, *J. Vac. Sci. Technol. A*, 14, 781, 1996; I. Yamada and J. Matsuo, *Mater. Res. Soc. Symp. Proc.*, 396, 149, 1996; I. Yamada, U.S. Patent 5,459,326; A. Gruber and J. Gspann, *J. Vac. Sci. Technol. B*, 15, 2362, 1997; I. Yamada, J. Matsuo, N. Toyoda, and A. Kirkpatrick, *Mater. Sci. Eng. R*, 34, 231, 2001). Thus, GCIB with reactive gases has some of the characteristics of a reactive-ion etch (RIE) or reactive-ion beam etch (RIBE). However, it is more accurately thought of as a directed-beam technique with a high-mass reactive gas, i.e., a cluster-ion beam etch (CIBE), thus incorporating important advantages of both RIBE and GCIB. An important advantage of CIBE with a reactive gas is afforded by the property that the gas of the cluster need not be in a reactive form as it is transported, since the cluster impact itself can release the reactive species by cracking the gas molecule and thereby render it highly reactive. If, as is often the case, the reactive species (radical or ion) has a short half-life, the cluster impact technique raises the probability that the reactive species survives to reach the target surface. Also, vacuum chamber design may be simplified in some cases by intentionally using a particular gas that is ordinarily unreactive with the chamber components but upon cluster impact is rendered reactive (with the target) for only a short time. However, residual gas in the chamber may become ionized and attack the components unless due care is taken.

4.4.3 Commercial Status of GCIB and Manufacturing Use

In 1993 an early research GCIB system was constructed at Kyoto University and attained a beam current of about 1 nA (J.A. Northby, T. Jiang, G.H. Takaoka, I. Yamada, W.L. Brown, and M. Sosnowski, *Nucl. Instrum. Methods Phys. Res. B*, 74, 336, 1993). Shortly afterwards, this research team began to demonstrate quite promising applications for their argon clusters, such as the first surface smoothing, which was done with a sputtered gold film (J.A. Northby, T. Jiang, G.H. Takaoka, I. Yamada, W.L. Brown, and M. Sosnowski, *Nucl. Instrum. Methods Phys. Res. B*, 74, 336, 1993; I. Yamada, J. Matsuo, Z. Insepov, D. Takeuchi, M. Akizuki, and N. Toyoda, *J. Vac. Sci. Technol. A*, 14, 781, 1996; I. Yamada and J. Matsuo, *Mater. Res. Soc. Symp. Proc.*, 396, 149, 1996; I. Yamada, U.S. Patent 5,459,326; I. Yamada, J. Matsuo, N. Toyoda, and A. Kirkpatrick, *Mater. Sci. Eng. R*, 34, 231, 2001). In early 1997, Epion Corp. fabricated a research GCIB system for the group at Kyoto University and this system delivered 100 nA of argon beam. By the middle of 1998, Epion had built and delivered an improved system that was capable of 1 μA of beam, and a photograph of this system is shown in Figure 4.29a. It was recognized that for widespread use of this technology, systems would need to be developed that could be used routinely in an industrial manufacturing environment and would provide sufficient beam current to allow rapid processing of wafers. At about this time, Epion received an award from the U.S. Department of Commerce through the Advanced Technology Program (ATP) and over the next three years Epion developed significant improvements in the GCIB toward its use as a manufacturing tool, supported by an award by the U.S. Department of Commerce through its Advanced Technology Program, #70NANB8H4011. By the end of 2000, Epion had systems that delivered 100 μA of beam with continuous operation over a week's time, and by the spring of 2002 they had demonstrated 1000 μA of beam. An example of these *Ultra-Smoother®* processing systems is shown in Figure 4.29b. Such systems are designed with semiconductor fabrication-line standards in mind and can be integrated into the "cluster tool" configuration common in microelectronics manufacture. The primary use is for smoothing and etching of whole wafers, and at present, systems are marketed that handle wafer diameter up to 200 mm and process with multiple source gases.

A wide variety of industrial applications are envisioned for GCIB. Some of these are listed in Table 4.26. To date, surface smoothing has been the primary application area.

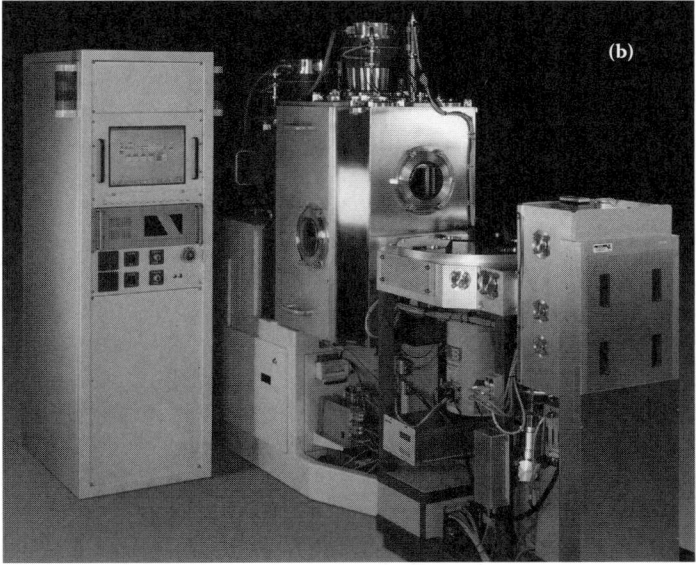

Figure 4.29 Photographs of GCIB systems: (a) is an early system fabricated for university research; (b) is a unit of the commercial type sold to microelectronic manufacturers.

TABLE 4.26   Potential Areas of Application for GCIB Technology

Nanoscale smoothing, planarization and cleaning of surfaces

Surfaces of all microelectronic materials, especially thin films

Final touch polish after chemical-mechanical polishing (CMP)

Multilayer magnetic and optical devices (GMR, MRAM, TMR, WDM, AWG)

Sputter etching with very little ion damage, mixing or roughening

Programmable thickness trimming (figuring) of films and wafers

Analytical instrumentation (end-point detection and high depth-resolution SIMS)

Cluster-assisted deposition (CAD), GCIB reactive gas coincident with other sources

Nitride-compound films (GaN, Si3N4, TaN) with nitrogen clusters

Oxide-compound films (SiO2, Ta2O5, ITO, Zr-Si-O) with oxygen clusters

Some films, such as the gate oxide (SiO_2) on silicon wafers, are fabricated at high temperatures where the film has a high enough in-plane diffusion to be smoothed simply by the growth, deposition, or anneal fabrication process itself. But other films are damaged by high temperatures or, in some cases, roughen when heated in a required annealing process such as for metal-oxide ferroelectrics. For these a final ion polish, at room temperature and in a controlled vacuum environment, is particularly attractive. The first significant markets have been use of GCIB for substrate and film smoothing within the manufacturing of multilayer films for fabrication of giant magnetoresistive (GMR) sensors for hard-drive data storage (D.B. Fenner, J. Hautala, L.P. Allen, J.A. Greer, W.J. Skinner, and J.I. Budnick, *Mater. Res. Soc. Symp. Proc.,* 614, F10.3.1, 2000; D.B. Fenner, J. Hautala, L.P. Allen, T. Tetreault, A. Al-Jibouri, J.I. Budnick, and K.S. Jones, *J. Vac. Sci. Technol. A*, 19, 1207, 2001). The thickness of some metal-film layers is as small as ~ 1 nm in GMR spin-valve devices and the newer tunneling-magnetoresistive (TMR) devices. Roughness of the substrate surface or any of the various buffer layers (underneath the actual sensor layers) will generally cause roughness in each overlayer, and hence, also in the sensor layers. The roughness is thought to limit performance and yield of the devices, and thus smooth surfaces are desired. Thermal smoothing is not possible since most of the film layers must be processed near room temperature and control of metal-film oxidation by the ambient is important. Progress with application of GCIB to GMR fabrication has been reported (J.J. Sun, K. Shimazawa, N. Kasahara, K. Sato, T. Kagami, S. Saruki, S. Araki, and M. Matsuzaki, *J. Appl. Phys.*, 89, 6653, 2001). The roughness constraints placed on TMR may well be even more difficult to meet with conventional fabrication and smoothing tools.

Other applications of GCIB are discussed in the next section.

4.4.4 Emerging Techniques and Applications

A variety of film deposition methods have been devised that utilize beam techniques. One method utilizes the neutral gas jet from a nozzle to crack gas molecules upon impact and deposit a film from that gas. In one of these, a high-pressure gas of He or H_2 is expanded through a nozzle under conditions such that it forms a supersonic-velocity jet but not such that it condenses into clusters (S.E. Roadman, N. Maity, J.N. Carter, and J.R. Engstrom, *J. Vac. Sci. Technol.* A, 16, 3423, 1998). The high-pressure gas is mixed with a small amount of SiH_4 or Si_2H_6 gas, and the jet is directed onto a silicon wafer at 550–750°C. The silane cracks and an epitaxial silicon film is formed. The beam kinetic energy of 0.5–2.2 eV serves to assist with reaction activation.

If the gas jet from the nozzle is chosen such that clusters form, and these are ionized and accelerated (i.e., GCIB), then even greater impact energy can be made available. This technique is also being utilized for film deposition. Additional features, useful for process manipulation and improvement, become available with ion acceleration.

4.4.4.1 Reactive-Gas GCIB Surface Treatment

Two surface treatments have been investigated for use of GCIB. In the first, a reactive gas is used to etch the surface of a target work piece. This was described briefly in the section above, and many areas of application for this new technique are under development. In the second, a reactive gas is used to treat a surface so as to form a thin film perhaps as a hard or chemically protective surface. For example, treatment of surfaces with a nitrogen-ion beam for improved properties has had a long and successful history; however, much thinner treated surface layers are now desired for some, more recent applications. In these cases, the nitrided surface layer must be as little as just a few monolayers thick and formed without significant ion damage. Three examples of this have been reported: (1) the reactive conversion of a GaAs crystal surface into a GaN layer, (2) the preparation of sapphire crystal surfaces (i.e., termination) to make them suitable for deposition and growth of crystalline semiconductors in the family of gallium nitride semiconductor materials, and (3) similarly thin layers of nitride compound formed on silicon-wafer surfaces for use as a barrier against dopant diffusion.

The first example was demonstrated by a collaboration of Kyoto University, using clusters formed from ammonia gas (H. Saito, T. Kato, M. Yoneta, M. Ohishi, J. Matsuo, and I. Yamada, *Phys. Stat. Sol. (a)*, 180, 251, 2000). The second and third examples have been demonstrated by Epion Corp. that has developed research apparatus consisting of a small GCIB system for generating a beam of nitrogen-cluster ions and a film-process vacuum chamber operating at ultrahigh vacuum (D.B. Fenner, U.S. Patent 6,498,107). With this system, nitrogen gas (N_2) is condensed into an aerosol (nitrogen snow), ionized, accelerated, and impacts a substrate crystal surface where the cluster vaporizes and some of the N_2 molecules are cracked and react with the substrate to form nitrogen compounds (D.B. Fenner and Y. Shao, *J. Vac. Sci. Technol. A*, 21, 47–58, 2003). Both silicon and sapphire crystals have been treated this way and both reveal a nitride-compound layer upon XPS analysis. Sapphire is the more difficult of the two, and in fact a sapphire wafer is ordinarily almost entirely unreactive with (static, neutral) N_2 gas for sustained lengths of exposure. A nitrogen plasma is required or use of ammonia gas if reaction with sapphire is to be practical, but both of these techniques have drawbacks. Figure 4.30

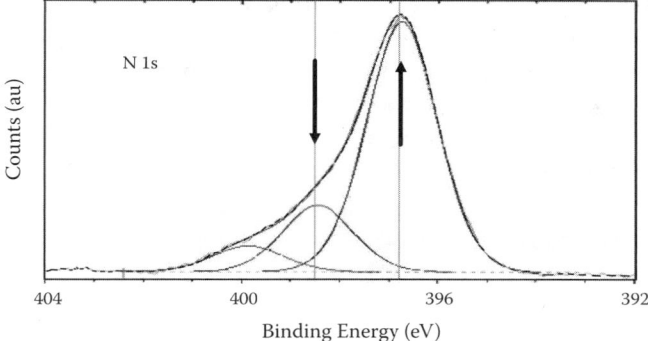

Figure 4.30 XPS N 1s peak of sapphire surface after exposure to a nitrogen cluster ion beam. The heavy dashed line is the measured data while the gray solid lines are deconvolution components of the data line. The arrow at 396.8 eV indicates the component expected for N bonded to Al. At 398.6 eV is the component expected for N bonded to O.

shows the XPS spectra for the N *1s* photoelectrons from a sapphire surface, heated to 400°C and exposed to nitrogen GCIB with 25 kV acceleration. Analysis of the binding energy observed indicates that the surface consisted primarily of Al bound to N, instead of the Al-O bonds that are found within the bulk of the sapphire (D.B. Fenner, U.S. Patent 6,498,107). The system is located in the Photonics Center laboratories at Boston University under the direction of Professor T.D. Moustakas.

4.4.4.2 GCIB-Assisted Thin-Film Deposition

As an extension of the reactive GCIB treatment of surfaces just described, thin films can be deposited by cluster-assisted deposition (CAD) (J.A. Northby, T. Jiang, G.H. Takaoka, I. Yamada, W.L. Brown, and M. Sosnowski, *Nucl. Instrum. Methods Phys. Res. B*, 74, 336, 1993; I. Yamada, J. Matsuo, Z. Insepov, D. Takeuchi, M. Akizuki, and N. Toyoda, *J. Vac. Sci. Technol. A*, 14, 781, 1996; I. Yamada and J. Matsuo, *Mater. Res. Soc. Symp. Proc.*, 396, 149, 1996; I. Yamada, U.S. Patent 5,459,326; I. Yamada, J. Matsuo, N. Toyoda, and A. Kirkpatrick, *Mater. Sci Eng. R*, 34, 231, 2001). To date, CAD requires that at least one other vapor source be provided within the process chamber together with the GCIB. The cluster-ion flux may consist of unreactive (e.g., argon) species and merely provide energetic assistance to the film deposition. Or the cluster gas may itself be reactive, such as a gas mixture with a reactive component (e.g., nitrogen or oxygen). Here, the deposited films consist of reaction products of the GCIB source gas and the constituents of the (additional) vapor source or sources. Many variations of CAD are possible, and in addition, the GCIB may be operated in more than one mode, such as initially using a beam to clean or smooth a substrate surface and then using a different beam to perform CAD. Assisted deposition on semiconductor wafers and films is of particular industrial significance (D.B. Fenner, U.S. Patent 6,498,107).

The nitrogen-GCIB research system at Boston University described in Section 4.4.4.1 is outfitted with a gallium thermal effusion cell so that the substrate can be simultaneously fluxed with both nitrogen clusters and gallium atoms while being heated and rotated. A photograph of this system appears in Figure 4.31. In this view, the nitrogen beam starts at the lower left and travels upward to strike the downward-facing sapphire substrate in about the middle of the photograph. In the top central region of the figure is the manipulator that rotates the sapphire wafer while heating it. Films of GaN have been grown on sapphire substrates with the CAD system of Figure 4.31. These films are as yet thin, but show good heteroepitaxial structure visible in cross-sections imaged by transmission electron microscopy (TEM) (Y. Shao, T.C. Chen, D.B. Fenner, and T.D. Moustakas, *Growth of GaN by Ion Beam Cluster Deposition, Nitride Semiconductor Workshop*, Richmond, VA, March 2002 (unpublished); Y. Shao, T.C. Chen, D.B. Fenner, T.D. Moustakas, and G. Chu, *Mater. Res. Soc. Symp. Proc.*, 743, L3.10, 2002). In addition, the cathodoluminescence (CL) of homoepitaxial films exhibit a strong emission at the expected band gap of GaN (363 nm) with a narrow linewidth (full width at half maximum ~ 9 nm).

Figure 4.31 Research GCIB system for nitrogen-cluster assisted film deposition.

As discussed elsewhere in this chapter, oxygen ions are known to be useful for assisting with deposition of optical films. Oxygen GCIB has been demonstrated as promising for CAD of Ta_2O_5, Nb_2O_5, TiO_2, SiO_2, and indium tin oxide (ITO) films (J.A. Greer, D.B. Fenner, J. Hautala, L.P. Allen, V. DiFilippo, N. Toyoda, I. Yamada, J. Matsuo, E. Minami, and H. Katsumata, *Surf. Coat. Technol.*, 133–134, 273, 2000; I. Yamada, J. Matsuo, N. Toyoda, and A. Kirkpatrick, *Mater. Sci. Eng. R*, 34, 231, 2001; N. Toyoda, K. Shirai, M. Terasawa, S. Matsui, and I. Yamada, *Mater. Res. Soc. Symp. Proc.*, 697, P5.20, 2002; N. Toyoda and I. Yamada, *Mater. Res. Soc. Symp.*

Proc., 749, W17.7, 2002). Much of that work was done with the system shown in Figure 4.29a. A flux of the oxide compound material is produced by electron beam evaporation and the oxygen clusters are accelerated to 7–11 kV, which is a level sufficient to obtain reaction of the oxygen with the condensing compound vapor, at the substrate. Both metal and metal-oxide e-beam sources have been explored, and deposition at 0.1 nm/s is achieved. With simultaneous GCIB assist the films are smoother (average roughness of 0.5 nm) and higher in refractive index (Ta_2O_5 up to 2.11) than deposition without GCIB assist (N. Toyoda, K. Shirai, M. Terasawa, S. Matsui, and I. Yamada, *Mater. Res. Soc. Symp. Proc.,* 697, P5.20, 2002; N. Toyoda and I. Yamada, *Mater. Res. Soc. Symp. Proc.,* 749, W17.7, 2002). Multilayers of Ta_2O_5/SiO_2 films fabricated by CAD show abrupt interfaces in cross-sectional TEM and little shift of the optical index after water exposure. Thus, the oxygen-cluster impacts serve to increase the oxygen content and density of the growing films as well as smooth the surface.

Diamond-like carbon (DLC) films have been reported as fabricated by CAD (T. Seki, M. Tanomura, T. Aoki, J. Matsuo, and I. Yamada, *Mater. Res. Soc. Symp. Proc.,* 504, 93, 1998; T. Kitagawa, I. Yamada, J. Matsuo, A. Kirkpatrick, and G.H. Takaoka, in J.L. Duggan and I.L. Morgan (Eds.), *Applications Accelerators Research Industry, Amer. Inst. Phys. Conf. Proc., vol. 576,* 2001, p. 963; I. Yamada et al., U.S. Patent 6,416,820), as well as directly from a (single source) beam of molecular hydrocarbon (small) clusters (D.A. Zeze, S.R.P. Silva, N.M.D. Brown, A.M. Joyce, and C.A. Anderson, *J. Appl. Phys.,* 91, 1819, 2002). In the former technique, a carbon vapor source such as a hydrocarbon or C_{60} is thermally evaporated and condensed onto a substrate while an argon GCIB is directed at the surface. The argon clusters crack the carbon-containing molecules and facilitate formation of the diamond-like film. Metal-film deposition has been reported from a single source of metal vapor from a magnetron that is condensed into clusters (0.4 to 10 nm in diameter) (O. Rattunde, M. Moseler, A. Hafele, J. Kraft, D. Rieser, and H. Haberland, *J. Appl. Phys.,* 90, 3226, 2001; G. Palasantzas, S.A. Koch, and J.Th.M. De

Hosson, *Appl. Phys. Lett.*, 81, 1089, 2002). Acceleration of these metal clusters is reported to be useful to form highly adherent films, a technique referred to as energetic cluster impact (ECI) and now marketed by Oxford Applied Research (Nanocluster source NC200K from Oxford Applied Research, Oxfordshire, UK). Sputter targets in the DC magnetron source can accommodate most metals, and oxygen gas can be added so that metal-oxide films are deposited.

4.4.4.3 SIMS and High-Resolution Depth Profiling with GCIB

The impact of accelerated gas-cluster ions vaporizes surfaces with some portion of the volatile products being in the form of ions, i.e., secondary ions, both atomic and molecular. If the target chamber is fitted with apparatus to measure ion mass, such as a quadrupole mass spectrometer (QMS), then the combination of this with a GCIB source (the primary ions) contains the basics of a secondary ion mass spectrometer (SIMS) instrument. This is to be distinguished from use of ion-mass analysis apparatus to monitor the cluster ions themselves. For example, with this latter technique the size distribution (or more accurately the M/Q distribution) of the clusters can be measured, as has been reported (I. Yamada, J. Matsuo, Z. Insepov, D. Takeuchi, M. Akizuki, and N. Toyoda, *J. Vac. Sci. Technol. A*, 14, 781, 1996; I. Yamada and J. Matsuo, *Mater. Res. Soc. Symp. Proc.*, 396, 149, 1996; I. Yamada, U.S. Patent 5,459,326; N. Toyoda, J. Matsuo, T. Aoki, S. Chiba, I. Yamada, D.B. Fenner, and R. Torti, *Mater. Res. Soc. Symp. Proc.*, 647, O5.1, 2001; I. Yamada, J. Matsuo, N. Toyoda, and A. Kirkpatrick, *Mater. Sci. Eng. R*, 34, 231, 2001). The new technique of GCIB for SIMS analysis of surfaces (GC-SIMS) has been investigated by Epion Corp. and has provided useful insight into the nature of cluster-ion interaction with surfaces (D.B. Fenner and Y. Shao, *Proc. SIMS Workshop,* Clearwater Beach, Florida, May 2, 2002 (unpublished); D.B. Fenner and Y. Shao, *J. Vac. Sci. Technol. A*, 21, 47–58, 2003). The research instrument constructed for that work is shown in Figure 4.32. In this image, the beam is generated on the right-hand side,

Figure 4.32   Research GCIB system for mass spectroscopy of secondary ions generated by cluster-ion impact.

travels horizontally, and strikes the target surface in the analytical chamber, which is situated underneath the left-most turbo pump. Differential pumping of the beam line is used along with baking of the analytical chamber to attain a base pressure in that chamber of about 1×10^{-10} torr. The beam impacts the target at normal incidence and the secondary ions are collected and mass analyzed by a QMS, located on the backside in Figure 4.32.

Initial results with the instrument have demonstrated both the utility of the GC-SIMS technique for analysis of surfaces and its value for advancing the understanding of cluster interactions with surfaces (D.B. Fenner and Y. Shao, *J. Vac. Sci. Technol.* A, 21, 47–58, 2003; D.B. Fenner and Y. Shao, *Proc. SIMS Workshop,* Clearwater Beach, Florida, May 2, 2002 (unpublished)). Elemental metal surfaces have been analyzed with this new instrument, and two typical mass spectra of the secondary ions emitted by gold and copper films while under exposure to an argon GCIB are shown in Figure 4.33. The mass peak at 12 amu (carbon ion) in both spectra is an indication of a small residual contamination of these surfaces that were not aggressively cleaned before the analysis. Also seen in these mass spectra are fragmentation pieces from the incident argon cluster in the form of Ar_2^+ ions, as well as reaction products of the metals with the incident argon, i.e., $AuAr^+$ and $CuAr^+$. These latter species are excimer compound ions that form by virtue of the weak molecular binding possible with an argon ion, unlike a neutral argon atom. It is of note also that the spectra do not show any multiply charged secondary ions. This is consistent with the expectation that the mechanism of cluster interaction depends little on the incident ion charge but rather its mass and energy, since the primary ions (clusters) have very high M/Q (N. Toyoda, J. Matsuo, T. Aoki, S. Chiba, I. Yamada, D.B. Fenner, and R. Torti, *Mater. Res. Soc. Symp. Proc.,* 647, O5.1, 2001; D.B. Fenner and Y. Shao, *Proc. SIMS Workshop,* Clearwater Beach, Florida, May 2, 2002 (unpublished)). The spectra for copper, the lower panel of Figure 4.33, was collected while the analytical chamber was not baked but at a base pressure of $\sim 1 \times 10^{-9}$ torr. Clearly residual water-vapor molecules (or

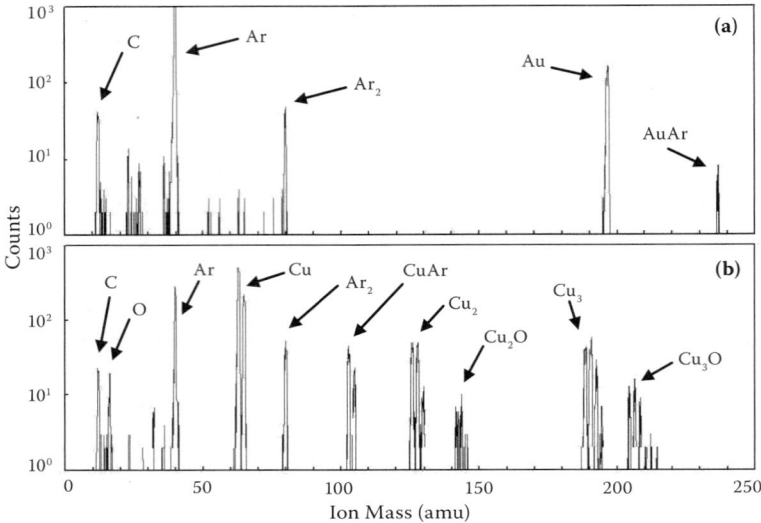

Figure 4.33 Secondary-ion mass spectrum of (a) gold and (b) copper films under argon-GCIB exposure. The secondary-ion flux is reported as "Counts" on a log scale with arbitrary units.

other oxygen containing species) are sufficiently numerous in the chamber as to cause oxidation of some of the sputtered copper thus forming the Cu–O secondary ions of various stoichiometries. This is direct evidence of cluster-ion stimulated surface reactions with gases not contained in the cluster beam itself, such as that utilized by CAD (see Section 4.4.4.2 above).

The native surface of a tantalum film was also studied with the GC-SIMS and with XPS (N. Toyoda, J. Matsuo, T. Aoki, S. Chiba, I. Yamada, D.B. Fenner, and R. Torti, *Mater. Res. Soc. Symp. Proc.*, 647, O5.1, 2001; D.B. Fenner and Y. Shao, *Proc. SIMS Workshop*, Clearwater Beach, Florida, May 2, 2002 (unpublished); D.B. Fenner and Y. Shao, *J. Vac. Sci. Technol. A*, 21, 47–58, 2003). From the XPS analysis that surface is known to have an oxide of Ta that is 3–4 nm thick, i.e., the native oxide. Under exposure to an argon GCIB the oxide is eroded away and the underlying metal exposed. Figure 4.34 shows the emission strength of the TaO^+ ions from

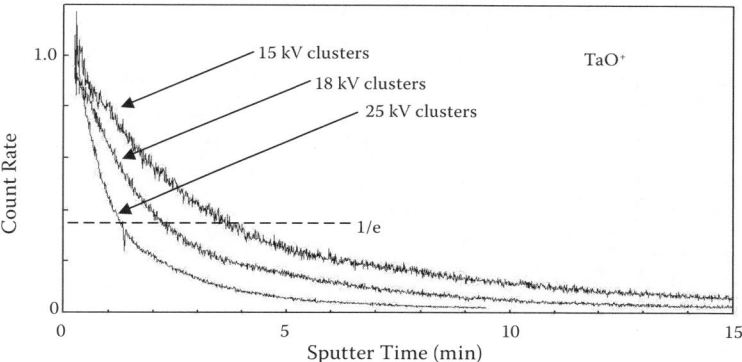

Figure 4.34 Normalized TaO⁺ ion emission measured as the native oxide is sputtered off of a Ta metal surface by an argon GCIB. The emission count rate has arbitrary units.

the native surface as the exposure time increases (D.B. Fenner and Y. Shao, *J. Vac. Sci. Technol. A*, 21, 47–58, 2003). The ion-emission rate is seen to decay in an approximately exponential fashion and to do so more quickly for higher GCIB acceleration. Hence, the higher accelerations etch away the oxide faster, as would be expected. Also, it is found from the measured raw count rate, that the higher acceleration generally results in a stronger yield for secondary ions. Note in Figure 4.34 that at high fluence (long exposure time) the emission of the TaO⁺ ions does not fall to zero but rather to a weak residual amount. That residual is smaller for higher acceleration of the clusters. The residual TaO⁺ ion emission represents a dynamic equilibrium between GCIB etching and cluster-stimulated ambient-gas reoxidation of the Ta (D.B. Fenner and Y. Shao, *J. Vac. Sci. Technol. A*, 21, 47–58, 2003).

Since the GCIB process can reduce the roughness of surfaces to exceedingly low values and additional fluence does not cause the return of roughness (as with well-known stochastic etching and deposition phenomena), the cluster beam would seem to be a superior sputtering tool for analytical depth profiles. Depth profiles accomplished by tracking the secondary-ion emission (SIMS) or the electron emission (XPS or AES) while an ion beam sputters away the surface are important tools in microelectronics. As film layers become thinner and interfaces more abrupt in present and future devices, the need will grow for higher depth resolution with dynamic SIMS, XPS, AES, and similar surface-analytic instruments. Indications are that GC-SIMS will provide superior depth resolution (N. Toyoda, J. Matsuo, T. Aoki, S. Chiba, I. Yamada, D.B. Fenner, and R. Torti, *Mater. Res. Soc. Symp. Proc.,* 647, O5.1, 2001; D.B. Fenner and Y. Shao, *Proc. SIMS Workshop,* Clearwater Beach, Florida, May 2, 2002 (unpublished); D.B. Fenner and Y. Shao, *J. Vac. Sci. Technol. A*, 21, 47–58, 2003).

Chapter 5

Optical Effects of Ion Implantation

P.D. TOWNSEND AND P.J.T. NUNN

CONTENTS

ABSTRACT

Optical effects of ion implantation have historically received far less attention than the electrical properties of semiconductors. This relates to the size of the commercial markets. Developments in photonics are increasing the optical market and the implantation methods offer many potential applications, as well as considerable information about the near-surface structures. In many cases the implantations are the only route to device fabrication in some materials. Consequently the optical literature is advancing rapidly and in particular, the value of ion implantation for producing near-surface optical waveguides has been recognized. In this example the approach is a nearly universal route for guide fabrication, with more than 100 examples so far, as well as excellent examples of waveguide lasers, guides for second harmonic generation, up-conversion, four-wave mixing and electro-optic devices. Equally, the implants can be used in other property changes such as control of luminescence and lasing. Changes in reflectivity and optical absorption, as well as modifications of surface chemistry, have been exploited for optical devices with well-established successes. Many of these topics had been introduced and their potential summarized in an earlier book in 1994 (which included one of the present authors). The present chapter restates some of the principles but also references the many new and successful advances that have developed since that time. There is even more emphasis on the topical aspects of ion-implanted nanoparticle formation for use in photonics. For example, implantation of nanoparticles has generated the fastest electronic optical switching reported by any technique. The chapter underlines basic principles and includes many references to current literature, and indicates that the field is still rapidly expanding.

5.1 AN OVERVIEW OF ION IMPLANTATION IN OPTICAL MATERIALS

The control of surface properties is of paramount importance for a wide range of materials applications, and there is a long history of attempts to combat the problems of corrosion, or to modify such properties as surface hardness, friction or electrical and optical behavior. Quite often the requisite surface properties are incompatible with those of the bulk materials, and so there has always been interest in finding ways to modify surface layers. Processes such as thermal quenching prove effective for hardening steel and glass bottles, but lack the finesse that is required for more sophisticated technology. Somewhat more controllable treatments, including the deposition of surface coatings or diffusion of impurities into the surface layer, have therefore emerged. Ion implantation is particularly attractive since it allows a controlled injection of a well-defined number of ions into the surface, and the choice of ion and the concentration is relatively insensitive to both thermodynamics and the chemistry of the original material. Further, there is the ability to control the depth by variations in ion energy. On the negative side there will be an upper limit to the implant concentration since material will be sputtered from the surface. This will be most apparent with heavy ion implants into light ion targets. Similarly, the ion incidence of energetic ions displaces many of the target ions from their lattice sites and so introduces structural defects. To some extent this damage can often be removed by thermal treatments.

The associated technology requires relatively complex and expensive equipment, so that for usage in commercial applications there must be some economic or technological advantage compared with other surface processing methods. The classic example of the value of ion implantation emerged via the fabrication of semiconductor devices where the spatial precision, dopant control and added value make the ion beam methods an indispensable part of all semiconductor fabrication. Less obvious is that the ion doses for these electrical devices are often very modest, with implants at concentrations as low as 10^{14} ions/cm^2, and involve extremely small regions of the target materials. Further, with high packing density and short penetration depths the requisite ion energies can be down in the 10 to 100 keV range. In contrast, the processing of metals and insulators is often concerned with relatively large areas and, at least initially, the economic advantages are less obvious. For metals, items of small size and high value, such as razor blades that retain their cutting edge, or hip joints and heart valves that do not corrode, wear or fragment, have formed a natural set of initial commodities that are suitable for exploitation by ion implantation. At first sight the optical devices appear to be more challenging in economic terms, since optical effects will frequently require surface modification on the scale of the wavelength of the light, so the penetration depths will require changes over perhaps a few microns into the surface.

Additionally, to make optical changes via the use of dopants the structure may need continuous doping over the entire range, at concentration levels measured in percentage terms. For an implant route this would mean heavy ion MeV accelerators and multi-energy implants at a very high dose. Thus for simple coloration experiments, for example in fabrication of synthetic gem-stone coloration with chromium or titanium doping of aluminum oxide to produce ruby or sapphire, this would be unrealistically expensive. While such optical effects can be demonstrated, the main efforts in terms of applications have developed in other areas. In particular, these include the formation of optical waveguides in many of the most important photonic and electro-optic materials, definition of laser waveguides from bulk laser crystals, addition of impurities to make laser waveguides and development of non-linear responses from implanted nanoparticles. However, some of these achievements cannot be reached by alternative surface processing, which reduces the emphasis on costs. Further, commercialization of the implant routes can greatly reduce the costs compared with those using research laboratory equipment.

Economics of the manufacture of mirrors with specific wavelength responses or sunglasses can also be competitive with coating techniques and have offered some advantages in terms of performance. Examples of these various applications will be presented in this chapter. They are not new in concept and date back over a quarter of a century, but the value of the implantation routes are becoming increasingly popular and to some extent the range of applications has not been exploited except by groups in a limited number of laboratories. An extended review and books that give an overview of the topic of optical effects of ion implantation up to 1994 include Refs. [1–3]. More recent advances are identified in a variety of related reviews [4–6] containing many detailed discussions on theory, ion ranges, limitations of sputtering, methods of characterization, etc., and such material will not be repeated here. Instead the examples will emphasize the key steps and numerous successes so that one can confidently attempt new applications. References to more recent examples of relevant literature are included and one notes that this is steadily increasing, although commercially manufactured items of optical devices using ion implantation are still very limited (excluding semiconductor based items).

It should also be appreciated that all other aspects of implantation that still apply to the wide-energy-gap optical materials and implants can be combined with other processing steps (e.g., implants may be made into or before layer deposition or diffusion). Table 5.1 lists some of the features than can, and have, been modified by implants in insulators.

TABLE 5.1 Properties Influenced by Surface Implants

Mechanical	Chemical	Electrical	Optical
Microhardness	Corrosion	Resistivity	Color
Friction	Passivation	Photoconductivity	Reflectivity
Adhesion	Diffusion	Electron mobility	Transmission
Wear	Reactivity	Semiconductivity	Optoelectronics

5.2 ION-IMPLANTED OPTICAL WAVEGUIDES

Modern photonics is based on methods of optical processing that require signal routing via optical waveguide pathways. The essential step in defining an optical waveguide is to have a higher refractive index than the surrounding material, and alternative routes to this have been to use deposition of high index layers, diffusion doping by introducing new ions of greater polarizability, or out-diffusion of component ions. All the routes have degrees of success in limited systems, but guide formation does not automatically retain the desirable optoelectronic or non-linear properties that existed in the original bulk material. Less obvious problems are that deposited layers may be strained or include columnar growth, and diffusion doping may require a very long development phase. An extremely well-documented example is for waveguide production in lithium niobate ($LiNbO_3$). Here doping with protons or titanium have led to very good quality guides and, despite initial reductions in the electro-optic coefficients, modern devices can now be further processed to regain most of the useful properties. This has not been an easy development since the processing temperatures are high and there are a range of Li_2O/Nb_2O_5 phases and compositions; the material is rarely stoichiometric and can both dissociate at high temperature and be heated above the Curie temperature, thus losing the requisite single domain structure. In real terms the present stage has been achieved after many thousands of research years of effort. Worse, it is not a predictive route for other materials; for example, the related $KNbO_3$ crystal would need a totally different approach since it has a low Curie temperature and low temperature phase transitions. Nevertheless, for $LiNbO_3$ the investment and experience has defined diffusion doping methods as the standard. For many materials of photonic interest (e.g., quartz) no chemical waveguide route has yet been discovered.

Ion implantation offers several alternative ways to make waveguides. The first is that it can be used to implant impurities. Here the advantages are that the process can be conducted at any selected temperature (even, say, 77 K) and by suitable masking, choice of ion energy and dose it can directly define the waveguide patterning. Titanium implants in $LiNbO_3$ do indeed result in waveguide formation, but since the titanium concentration is high it requires a very large implant dose, and a further annealing stage is essential to remove the damage of the amorphized niobate lattice. Titanium is also a relatively heavy ion so that for micron-thick waveguides the implant energy must be several MeV, and this limits the type of accelerator that is capable of this approach.

A more suitable target material is amorphous silica, since it has long been recognized that the synthetic material is not of the highest possible density and implantation can induce relaxation into the higher density state. Hence virtually any ion species will suffice and the glass network will compact into a higher density, higher refractive index, waveguiding zone. The index increase is from 1.46 to 1.48 in most cases. Numerous ions have been used to demonstrate this and optical pathways have been written into both planar and silica fiber structures. A chemical opportunity to reach a higher index host has also been exploited via implants with nitrogen [7]. Here the implant is a fairly light gas ion, so it can be operated with a simple accelerator of modest energy. The resultant composition approaches that of silicon oxynitride and hence, the index can in principle be raised from, say, 1.46 to around 1.5. In waveguide terms this is an enormous index change and gives a sharply defined waveguide. The major advantage of the nitrogen route is that the new compound is a stable material and so annealing, intended to remove any coloration, automatically results in a transparent stable formation of the silicon oxynitride glass. Hence it is a low-loss waveguide. Figure 5.1 describes the refractive index profile with depth for nitrogen-implanted silica at an ion dose of 1.5×10^{17} ions/cm^2 and an energy of 2 MeV [7]. Only a single energy was used in this first example, so the plateau base near the surface is the increase from the compaction damage, but the deeper index increase is from the chemical change.

Other chemically formed implant layers to modify the refractive index could be considered. Increased index is not

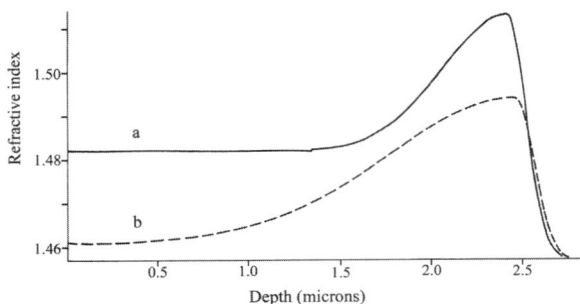

Figure 5.1 Refractive index profile of nitrogen implanted silica. The upper curve (a) shows the index enhancement as a mixture of damage compaction of the silica plus a chemical formation of silicon oxynitride at the end of the ion range. After annealing at 450°C the lower curve (b) has retained the chemical effects but lost most of the damage changes to the index.

the only objective since it may be possible to define an anti-reflective layer by an index reduction, so the scope for invention is enormous. Indeed the only certain fact about implants into optical materials is that the surface index will differ from the original material. In some cases it will increase and in others it decreases, so the problem is to consider how to exploit the change rather than ignore it. Note also that since the antireflective properties are dose, ion, energy and wavelength dependent, a single measurement at one wavelength cannot be immediately be extrapolated to either the magnitude or the sign of the index change elsewhere. For example, in experiments with implanted glass there was an initial conflict between the sign, not just the magnitude, of reflectivity data from different groups [8]. The two groups were using different measurement wavelengths and the conflict was resolved once a full spectral response was measured and monitored continuously during the implantation.

5.2.1 Optical Barrier Waveguides

This view that any significant index change (even a decrease) could be utilized was precisely the basis for the extremely successful route of waveguide formation using light ions of helium or hydrogen proposed by Townsend in the 1970s [9–11]. He capitalized on the reduction of the index caused by ion beam implantation in LiNbO$_3$, since he recognized that for most insulators the implantation of ions resulted in very different processes from energy deposited from electronic and nuclear collisions. As sketched in Figure 5.2 for MeV helium ion implantation into a light mass target such as silica, the bulk of the energy is given up via electronic excitation throughout the ion range. However, the nuclear collision events are limited to the end of the range where the ions have slowed down to the last 50 or so keV. For oxides the ionization events will move charges between ionic sites and form absorbing color centers, but lattice damage, displaced ions, or even total amorphization will be confined to the nuclear stopping processes at the end of the range. In general, amorphization and lattice destruction imply reduced density and consequently a reduction in the refractive index. Thus the low index zone buried beneath the surface acts as an optical barrier that

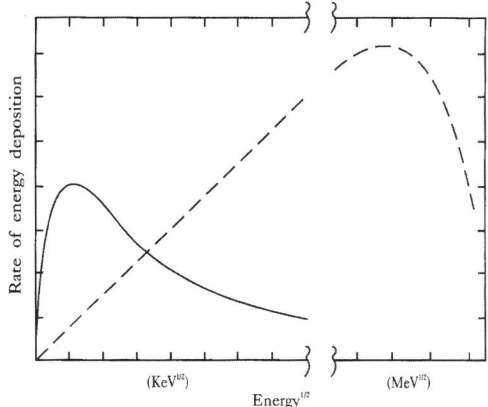

Figure 5.2 Comparison of the rates of energy deposition from electronic and nuclear collisions as a function of energy. Since the ions are slowing down, the lowest energies occur deeper into the material.

confines light to the outer surface layer and has automatically forms an optical waveguide. The process is not target specific and so it should operate in virtually any insulator. The only exceptions are likely to be materials such as the alkali halides where electronic damage processes can exist. In fact, guides can still be formed in many halides although they are of reduced quality.

For good waveguide operation there should be minimal loss and a low index boundary to the guide. The low index could just be a different bulk material but equally it could be a low index barrier zone to isolate the guide from a high index in the bulk crystal. Ideally, the barrier should have a significant index difference from the original material and be wide relative to the effective wavelength. In practice, breadth is not a problem as implants with several energies can broaden the barrier. Figure 5.3 shows the index profile for $KTa_{1-0.08}$ $Nb_{0.08}O_3$ implanted with three helium ion energies. There is an obvious overlap of the barrier zone but three distinct features appear in the computer profile simulation from the experimental refractive index data. For the lower modes the barrier width is greatly increased, leading to better confinement of the waveguide energy. In this example the maximum index change is only ~ 4%. This is typical of the general level of index change exhibited by amorphization (or major lattice disorder) in oxides studied so far. Values often display an index reduction of between 1 and 5%. However, record values are up to ~ 17% for $KTa_{1-x} Nb_xO_3$ and related tanatalate compositions. In most materials the color center formation is initially a problem, but charge trapping at defect sites (i.e., the color centers) can generally be removed by thermal annealing at modest temperature from about 150 to 300°C.

5.2.2 Quartz Waveguides

An early example that offers a "classic" demonstration of the model is for ion-implanted waveguides in crystalline quartz. In this case the heavily damaged region becomes amorphous silica and since this is a very stable material, the annealing can be made to well over 1000°C without any problems of

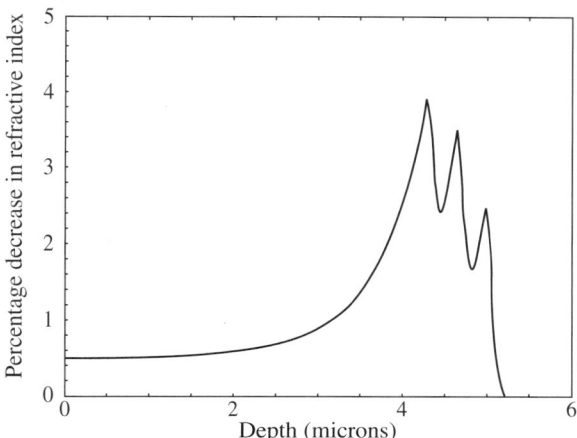

Figure 5.3 Refractive index profile for $KTa_{1-0.08} Nb_{0.08}O_3$ as the result of three He implant energies.

recrystallization [3,12]. Figure 5.4 shows an example of the waveguide profile, plotted as a reduction in index for different ion doses in quartz. This presentation of a decrease in index [3] emphasizes the characteristics of an optical well in which the waveguide mode values are equivalent to the energy states of a simple quantum well. Note that optically, the modes are labeled m = 0, 1, 2…, whereas in quantum mechanics the equivalent uses a notation for the energy levels with numbers n = 1, 2, 3…. The examples of Figure 5.5 also show that on annealing the well takes on an even more square shape since the damage close to the boundary has not fully amorphized the lattice, so high-temperature anneals allow crystalline regrowth. The net index difference in this case is ~ 5%. It should also be noted that chemical routes to waveguide formation in quartz are not available, although there is limited literature on Ge and Ti doping to slightly enhance the index.

The quartz example, Figure 5.4, shows that for MeV helium ions the waveguide dimensions are greater than a few wavelengths, hence the guide zone is multi-mode. The lower modes are totally confined by the barrier but energy in the

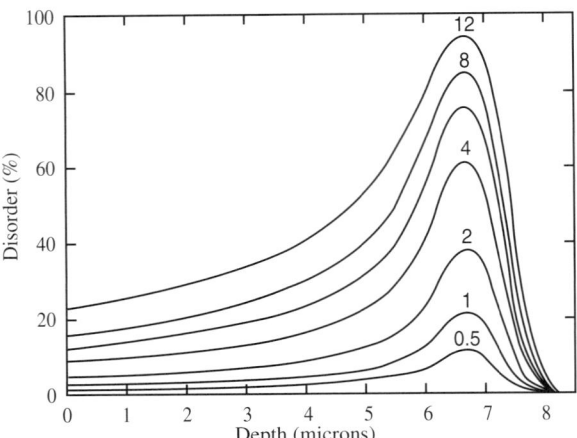

Figure 5.4 Lattice disorder as a function of ion dose in quartz for 2.5 MeV He ions in units of 10^{16} ions/cm^2.

higher modes may tunnel through to the bulk material. Equivalent energies using hydrogen will offer about four times the range, but since the barrier zone is narrower it is often necessary to use several closely spaced energies to achieve the same guide/substrate isolation. For ions from, say, a Van de Graaff accelerator beam, currents of μA/cm^2 will usefully form good quality waveguides in a few hundred seconds. Hence the processing times and conditions are not stringent and are slow enough to be readily controllable.

Ion implanted quartz waveguides have been used in several demonstration devices [3] including a spectrum analyzer in which the guided light was modulated by an acoustic wave [13]; a vertically coupled pair of waveguides in which the two guide boundaries were defined by different energies and control of the barrier width limited the mode coupling between them as a function of wavelength [14]; and a device that showed efficient second harmonic generation (i.e., as expected since the guide confinement maintains a high power density along the laser path). The choice of guide dimensions determined the wavelength at which there was a match between the mode indices of the fundamental and harmonic. Variations

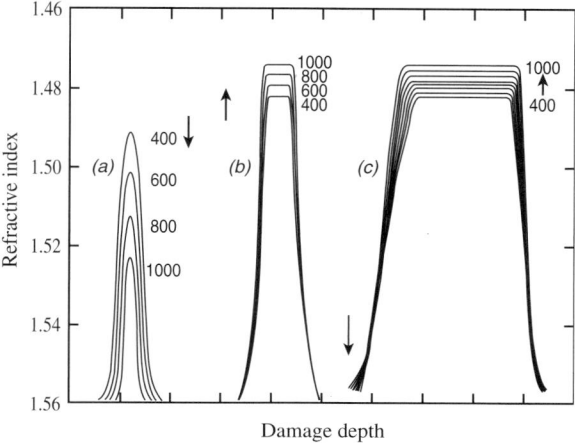

Figure 5.5 Implant profiles of helium-implanted quartz as a function of anneal temperature. Annealing can totally remove the damaged zone for (a) low ion doses, (b) higher doses where saturation damage has been reached and (c) high doses using several ion energies to broaden the low index barrier. The experimental data were obtained at nominally the same depth but are separated here for ease of viewing. Annealing temperatures (°C) are shown.

on these themes have been generated using alternative host lattices and in many cases, the effects are more spectacular since the non-linear or electro-optic properties of $LiNbO_3$, $KNbO_3$, KTP, $BaTiO_3$, etc., far exceed those of quartz [15–18]. For coupled waveguides formed by using several ion energies it is possible to design more efficient mode matching by altering the shape of the guide index profile for the fundamental and harmonic waves [19]. Equally, the guide geometry is well suited for up-conversion of high efficiency, and numerous ion-implanted waveguide examples exist [20–23]. Although not considered by most authors, loss reductions have been achieved by an additional very low energy implant near the surface. This surface index reduction maintains the guide away from contact with the surface and so reduces scattering loss problems.

There are many materials that display ion-implanted waveguide refractive index profiles that resemble the quartz case, although in general the damage tail extends weakly toward the surface since the amorphization product is not a stable entity, as for silica. Thus, very high temperature anneals are prohibited since they could remove the essential damage that defines the optical barrier. In less stable materials, including several glasses, where the color centers and damage barrier are removed at the same rate, it has been possible to preferentially anneal the waveguide, with minor loss of the barrier, by using pulsed laser anneals [24]. In these cases the aim is to choose a laser wavelength that is strongly absorbed within the barrier so that there is a steep temperature gradient that allows guide annealing but minimal change to the barrier.

5.2.3   Complex Waveguide Profiles

Surface waveguide production using the buried damage barrier approach has been recorded in some 100 insulating materials. The most effective and cleanly defined waveguides have been made with light ions (He or H); however, there are examples using heavier ions as well, not for the chemical aspects of the ions but for their damage production. There are numerous examples using ions such as Ni or P (e.g., Refs. [25] or [26]) to form waveguides. The problem in those cases is that the nuclear stopping regime can extend back to the surface so the entire guide region has a reduced index and is inevitably more absorbent as the result of color centers. In addition to a high absorption, the higher displacement damage rates degrade the crystallinity and so undermine useful properties such as electro-optic or non-linear responses. Nevertheless, heavy ion implants may be interesting if the implant itself is crucial for a specific purpose, e.g., implantation of rare earth ions to form a waveguide laser [27–29].

For materials that have low temperature structural changes there is a problem with thermal annealing, and the samples must not rise in temperature during the implantation. For example, two key photonic materials are $KNbO_3$ and

BaTiO$_3$ with phase transitions near 220 and 108°C, respectively. Fortunately, in the case of the KNbO$_3$ the waveguides form very readily and so the ion doses introduce little coloration within the guide while making excellent waveguides, and only require very modest anneals [3,30].

At the other extreme, materials such as sapphire do not readily retain the displacements caused by the passage of the ions and so it is difficult to form a waveguide without using such high helium ion doses that there is a danger of forming a sheet of helium bubbles during the annealing phase. (Nevertheless, even this effect can be exploited, and it has been used to form a thin exfoliated sheet of material). The problem with sapphire is that the rate of the damage retention does not increase significantly until the lattice is already strained. Two methods that have been successfully used to reduce the problem and make waveguide in both undoped and titanium-doped sapphire have been to make the implants at low temperature and use carbon rather than helium ions. The low temperature limits the diffusion of displaced ions and so a higher percentage of the damage is retained while the carbon chemically reacts with the Al$_2$O$_3$. The suggestion is that carbon ions make carbide bonds, which then distort the local lattice structure and destabilize it. Overall, the two factors can reduce the implant dose requirement to a level where waveguides are formed satisfactorily [31]. The sensitivity of aluminum oxide to changes in the implant ion has been studied in some detail by the Oak Ridge group [32], with carbon being the ion most obviously showing effects attributable to chemical interactions.

Not all the refractive index profiles are as simple as those initially predicted and the many detailed studies of LiNbO$_3$ ion-implanted guides. The devices made with them indicate that not only does the damage extend toward the surface, but that there are very different depth profiles for the ordinary and extraordinary polarized light for the two principal refractive indices, as discussed in detail in Ref. [3]. Figure 5.6a and b displays examples of such profiles. The ordinary index, n$_o$, is relatively standard but the extraordinary index, n$_e$, shows enhancements both on the shallower and deeper sides of the

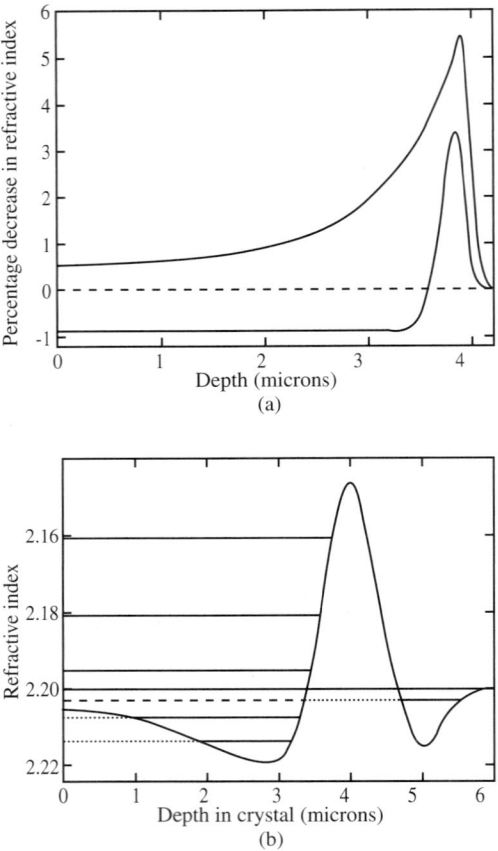

Figure 5.6 (a) Refractive index profiles of 1.75 MeV He implanted to 2×10^{16} ions/cm^2 for LiNbO$_3$ for the n$_o$ and n$_e$ indices. (b) A more complex pattern seen for the n$_e$ index in which there are two regions of increased index, compared with the original bulk value of 2.2 for light at 488 nm. Note that not all of the modes are readily coupled to the surface and may not be detected by normal prism measurements of the indices.

helium ion range. Various models have been proposed that include restructuring of the lattice and/or oxygen or lithium diffusion.

The possibility that a relatively minor encouragement from the ion implantation can lead to quite dramatic structural changes, in terms of not only the ionic lattice but also in terms of bond polarizability, have been discussed for waveguides formed in $Bi_4Ge_3O_{12}$ in which the guide index was raised throughout the implant zone [33,34]. This is not totally surprising, as the Bi/Ge oxide mixture can exist in several metastable phases. Although of a high positive index the initial guide had too much loss to be valuable, and annealing was required. The heat treatments above 400°C resulted in an index profile closer to that of the pattern seen in quartz.

Although not reported in waveguide studies it should also be appreciated that where there are metastable phases of a compound, it may be possible to nucleate or catalyze a phase transition from the strain and disorder zone generated at the end of an ion track. Such a feature has recently been observed during a luminescence study of phase transitions in $SrTiO_3$ [35]. In principle this could occur in other materials if they are used over a temperature range where transitions are feasible, and there are suggestions by various authors that transitions that may only be of second order within the bulk are less restrained in the surface and may appear as first-order events. Certainly the free energy restrictions are reduced for surface relaxations.

5.2.4 Applications

The pattern of behavior described above indicates that for a very large range of insulating materials it is relatively simple to fabricate waveguides. The essential challenge is to design structures, or capitalize on material properties that benefit from optical confinement within a waveguide. Immediately apparent examples are for lasers where the confinement maintains high pumping powers coincident with the laser beam; in non-linear optics for second harmonic generation, again exploiting the confinement and mode matching; in up-

or down-conversion optical pumping; or for four-wave mixing and waveguide parametric amplifiers. The figure of merit for the ion-implanted waveguide geometry of four-wave mixing in $BaTiO_3$ is interesting in that, in addition to the expected gains from a high power density, the implanted material performs more efficiently than the bulk region. It is thus encouraging to recognize that implants can stimulate, rather than degrade, interesting photonic properties. In terms of commercialization of ion-implanted waveguide devices there is still reluctance by industry to embrace the technology, since implantation is rarely an existing in-house facility. However, one suspects that it only requires a few examples of high-performance devices which cannot be fabricated by alternative routes to erode this psychological barrier.

The deep penetration of light ions, coupled with their ability to increase the index by compaction of glasses such as silica, also offers a direct route to waveguide fabrication at several micron depths below the surface of bulk silica, or indeed to modify the refractive index within silica waveguides. There are several examples of this approach and they conform to the expectations described above [36].

5.3 REFLECTIVITY AND REFRACTIVE INDEX CHANGES

The formation of optical waveguides, as described above, is primarily via a buried change in the refractive index. Shallower implants will, however, alter the index in the immediate surface layer. Hence there is the opportunity to make antireflection coatings or other indices that favor a particular application. Indeed there is an extremely early patent example from the 1940s of implantation to reduce surface reflectivity. In principle, the damage and amorphization will lead to lower density materials. In part this is because the distribution of the damage within a cascade is not uniform, and calculations indicate that the core is vacancy rich and interstitials separate to the outer region. Additionally, for near-surface implants there may be loss of component materials and an outward expansion of the lattice driven by stress in the nuclear collision region at the end of the ion track. Such events can be controlled to offer reduced reflectivity. Optical benefits may not always be compatible with other changes; for example, an argon implant into a glass used in TV screens offered us a major reduction of the surface reflections, but the lower density glass became friable and could not withstand mechanical cleaning or polishing [8]. Hence, for the TV anti-reflective role it was not viable, but for optics where there would not be abrasive mechanical contact such a procedure still offers a very simple and rapid formation of an anti-reflective layer.

A further option exists for implants into other types of coating layers, either to define waveguides or change the reflectivity. The implant could thus be equally effective in a diffused region or a surface coating. An interesting possibility for a coating layer arises from the problem of stress and columnar growth being an inherent difficulty in coating deposition. The implants can be adjusted in terms of energy to alter the stress profile and/or aid interface mixing. Equally, they can destroy the columnar structures and offer a more homogeneous layer than might be achieved by conventional coating plus annealing of the layer. Examples in the literature of such opportunities are rather scarce, although research laboratory examples do exist. A further option, although probably of minor interest, is to modify an existing multi-layer coating to correct an error within the initial production stage.

The early examples centered on planar devices and simple waveguides, but with suitable mask patterning there is no reason why much more complex index patterns cannot be created in the surface layer. For example, it would be possible to introduce a phase grating while maintaining a planar surface or have planar lenses within the surface. Such options are simple to demonstrate but require a precise application to justify the effort.

5.4 REFLECTIVITY FROM COLLOIDS AND NANOPARTICLES

A feature of the waveguides was that it was essential to minimize guide losses and consequently, the reflectivity caused by absorption was avoided. In other usage the surface implants may not merely modify the refractive index and adjust the reflectivity, but instead be the mechanism to produce very high absorption and reflection. Examples of this type include the production of mirror-like surfaces (reflection) or filters (transmission). To achieve very high absorption and reflection within the depth restrictions of the implant, many experiments initially used metal ions. Ion sources and implanters that are capable of injecting metal ions tend to be limited to energies of a few hundred keV, and so the overall range of, e.g., silver in a glass is likely to be a few hundred nanometers (i.e., a totally different depth scale from the waveguides). Further, to achieve reflectivity the concentrations of ions must be high and so the doses used are perhaps greater than 10^{16} ions/cm^2. Such concentration levels immediately imply that the ions are well above the saturation limit for a dispersed solution within the target, hence they automatically precipitate out of solution into clusters. In the initial experiments from Sandia Laboratories [37,38] in the 1960s data were discussed in terms of colloids, since the theoretical description of the optical properties of such particulates had been well documented for almost 50 years. Identical experiments conducted from the mid 1990s use a more topical description and discuss these metal clusters as nanoparticles, and have been summarized in numerous reviews [8,39–47].

Small particulates of metal have some properties that resemble those of bulk metal and some that differ. For example, the melting point of metal nanoparticles is size dependent and is greatly reduced for the smaller particles (e.g., for gold [48], iron, tin [49], etc., it can be lowered by several hundred degrees). This is not a feature that seems to have been exploited so far, except unwittingly where annealing has been made to change to size distribution of the particles. During simple furnace annealing the mobile metal that evaporates from the smallest (lower melting temperature) particles can move to the larger ones and hence stabilize them even further. This process is not confined to anneals made subsequent to the implant but may be a key factor in determining the size distribution even while the implant is in progress. In detailed studies of silver-implanted float glass [8] it was found that the very initial beam conditions define size conditions for the particulate growth. This is critically dependent on the ion beam current density and, once established, subsequent current changes are of far less importance. In this example the objective was to make car mirrors that have high reflectivity in the blue and less reflection at longer wavelengths. The glass surface is inevitably heated by the incident beam and a temperature gradient develops across the glass. Surprisingly, it was found to be essential to maintain the monitored temperatures within ~ 5°C in order to have optical properties that are reproducible. These are serious factors that have been ignored by many groups who are studying ion-implanted metal nanoparticles. The temptation is that since the ion dose may be as large as 5×10^{16} or even 10^{17} ions/cm^2, one should use a large ion beam current. This will establish different particle size conditions from those implants using lower beam currents, and certainly will cause severe variations in response if the surface temperature is allowed to rise throughout the implant. An extreme case of ion-beam heating has been used by some groups to obtain rapid anneals of the layer [50,51]. It should also be noted that ion beams of different current density, and/or different energy-density profiles in terms of electronic and nuclear excitation, are effective at altering the colloid size as well as intrinsic defect complexes. In the literature examples exist regarding the color center [3], as well as ion-beam changes to the nanoparticles [52] or indeed ion-beam mixing involving nanoparticles [53].

Production of the blue-biased car mirrors was instructive as it revealed many of these inherent variations [8]. The objective was to make large mirrors of around 10×20 cm that had no apparent visible inhomogeneities. A major problem with high current implants into a large-area insulating surface is that there is surface charging from electron ejection, which alters the ion energy by many keV and defocuses the ion beam. The charging will differ near the edges and the perimeter of the glass and, in the case of a scanned beam, will fluctuate during scanning. Experimental methods to minimize the problems include a coincident injection of an electron beam and/or the use of a very small gas jet, which is ionized by the beam to provide charge neutralization to the surface. In the case of silver the conductivity of the surface layer increases with ion dose, and in a scanning system it can be advantageous to make the first few scans with a very low current beam to raise the conductivity for the remainder of the implant.

5.4.1 Optical Responses with Particle Size

The optical responses of metal nanoparticles are complicated by several factors, since the reflections are size dependent and the size distribution of particles will vary with depth because of the initial ion implant range distribution. Further, it will be specific for each metal as the particles exhibit plasma resonances that are specific to each metallic element. Surface plasmon responses may equally need to be considered. The plasma frequency of the metal depends on the conductivity and the electron velocity within the conduction band, and the knowledge of the metallic plasma resonance offers a first indication of the spectral region where anomalies are likely to occur. Relevant theory has been discussed in many publications [54,55]. However, the fact that the metal is embedded within an insulator introduces a much more complex situation, which has usually been presented in terms of colloid models. An alternative situation for high-dose metal implants can be formation of embedded planes of metal that precipitate from the matrix. While the behavior of planes resembles that of the colloids, the precise wavelengths will differ. Since data for the planes are rarely presented, Figure 5.7 plots the measured extinction coefficient for bulk silver and the changing extinction values for silver/silica mixtures [56]. The isometric plot shows how the peak signals change as a function of the silver-to-silica ratio, and some typical slices from this are also presented. Figure 5.8 describes the reflectance and refractive indices of the effective media. For spherical colloids of silver in glass the resonance appears near 420 nm (and this gives the high blue reflectivity needed for the car mirrors). By adjusting the particle sizes via control of the beam energy, implant temperature, dose and dose-rate reproducible depth

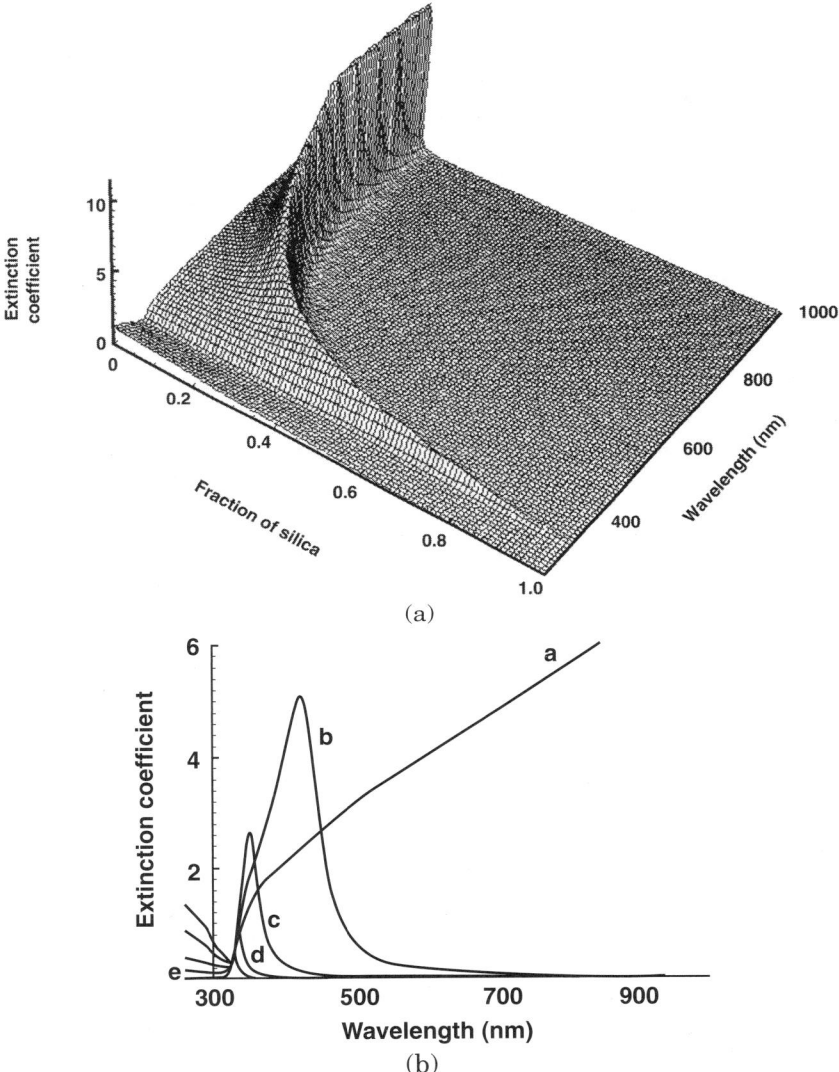

(a)

(b)

Figure 5.7 Extinction coefficient of silver and the computed values of silver/silica planar mixtures. (a) An isometric plot and (b) selected examples for silica concentrations of 0, 25, 50, 75 and 90%, respectively labeled (a) to (e).

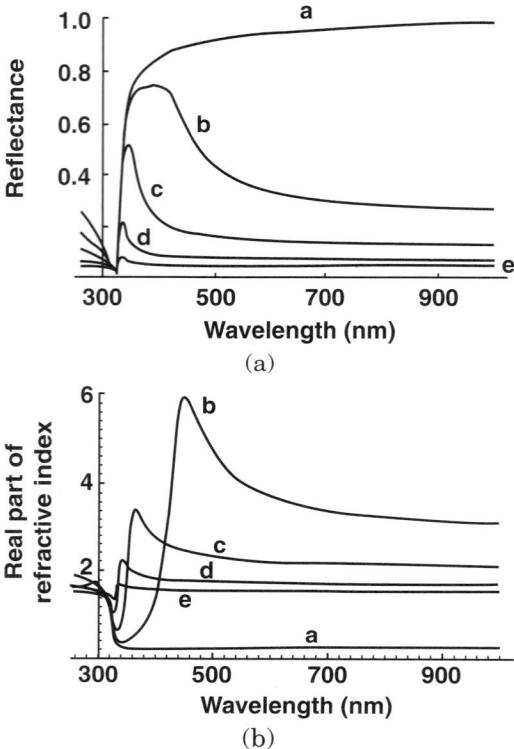

Figure 5.8 (a) Reflectance values and (b) refractive indices for the silver/silica medium with concentrations as for Figure 5.7.

distributions evolved. As noted below, these could be modified by heat treatments to offer control of the relative green-to-red reflectivity. Some examples of these data are shown in Figure 5.9.

The relationship between overall reflection and transmission is complicated by the problems of absorption and scatter that are size- and depth-dependent from the ion range. Equally, the sizes and depth distributions differ as the result of annealing. Figures 5.10 and 5.11 contrast the situations for silver [57] and gallium [58] implants in which the silver has a broad range of nanoparticle sizes at all depths, although the largest particles occur near the surface or peak of the

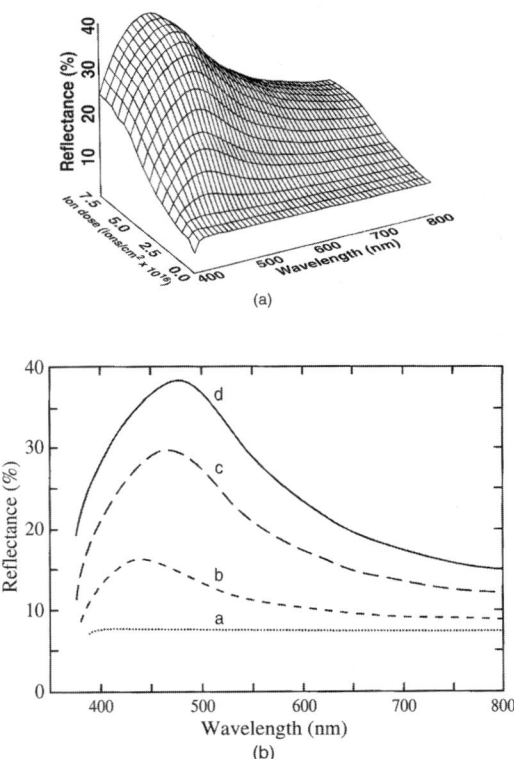

Figure 5.9 Reflectivity of Ag-implanted float glass as a function of ion dose using 60 keV ions. (a) Continuously recorded data measured *in situ* during the implant; (b) dose-dependent wavelength slices.

implant range. In contrast, the gallium data from Rutherford back-scattering (RBS) analyses look perfectly normal in terms of the concentration with implant depth, but the size distribution undergoes a major discontinuity with a few very large particles near the surface and a high density of very small particles beyond a certain critical point. The reasons for the gallium behavior are unclear but relevant factors presumably include the fact that gallium bulk material melts near 29°C and that nanoparticle-sized clusters may differ in their melting temperature. Also, gallium is unusual in that on melting

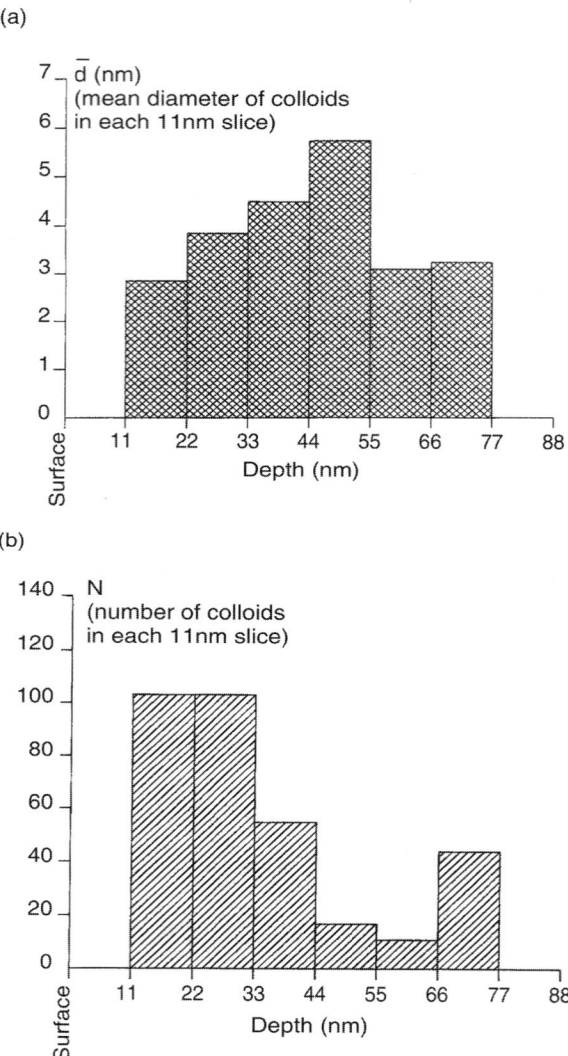

Figure 5.10 Depth and size distributions on Ag nanoparticles in float glass. (a) Variation in average colloid diameter with successive depth sections; (b) number of colloids in successive depth slices.

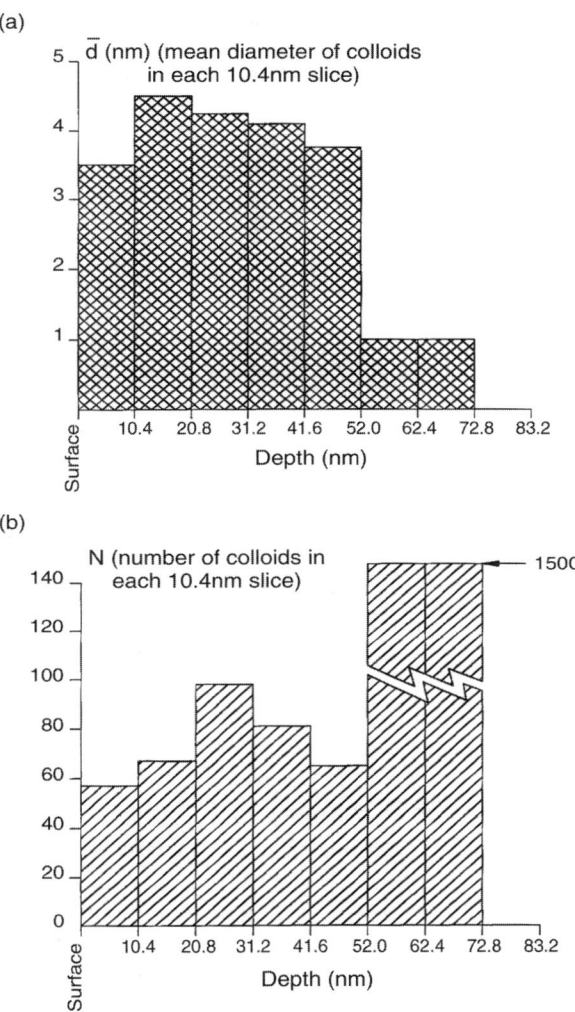

Figure 5.11 Depth and size distributions on Ga nanoparticles in float glass. (a) Variation in average colloid diameter with depth that changes discontinuously at a depth of ~ 52 nm; (b) number of colloids in successive depth slices.

the volume is decreased by ~ 10% (e.g., as for ice to water). Hence the local strains in the glass host can pass a critical change between the two regimes.

5.4.2   Implanted Sunglasses

Note that bulk gallium does not have a reflectivity peak in the visible spectrum and so gives a very "flat" mirror effect, and this has been used in conjunction with a green glass in the manufacture of mirror-front sunglasses that reduce the total light level but offer minimal color distortion [59]. The absorbing green glass played two roles. The first was to define the spectral range with a central maximum transmission that matches the response of the eye, and the second was to minimize back-reflections from the mirror face by light which enters the glasses from the rear. There is thus twice the absorption for light entering from the rear, thus minimizing unwanted reflections. On clear glass sunglasses the front metal would give unacceptably strong rear reflections. In an extreme case of 50% reflection and 50% transmission this would make front- and rear-view images of equal intensity. Increasing transmission and reducing back-reflection is not an ideal solution as this limits the dynamic range of the sunglasses. Therefore, in these glasses, two types of implants were made. The outermost layer was a metal, as before, but beneath this layer a second ion was implanted that gave a strong wavelength-independent absorption [8]. Again this resulted in twice the apparent absorption thickness for the back-view reflections and effective suppression of the rear-view reflections.

5.4.3   Non-Spherical Nanoparticles

Other metals with strong resonances in the visible part of the spectrum include copper and gold, both of which have been studied in great detail. Nanoparticles can also develop from metal liberated from the host lattice and in the float glass example, there was evidence for particles that formed at some ten times the depth of the implant ions; these were thought to be from metal ions displaced from the host glass that had diffused as the result of temperature or stress gradients and internal electric fields. Similarly, in other materials metal nanoparticles have evolved from the host crystal even though the implant was an inert gas such as argon. In crystalline targets the nanoparticles may be accommodated as non-spherical units within the host lattice (e.g., with early examples from Ag implants in Al_2O_3 [60] and $LiNbO_3$ [61]), and possible applications of such structures exist for forming wavelength specific polarizing devices.

5.4.4 Annealing of Nanoparticles

As implanted, the metal implants form nanoparticles with a wide range of sizes as a function of depth. Except for the example of the blue-biased car mirrors where a size distribution was required, many of the predicted uses for the nanoparticles would benefit from a single size distribution. Hence a thermal post treatment is needed. Simple furnace treatments at modest temperatures allow diffusion and growth of the larger units, whereas high-temperature anneals allow evaporation and dissolution of the metal. Hence in an annealing cycle, it is important not only to specify the maximum temperature and duration of the heating but also the speed at which the material is cooled, since during slow cooling a new thermodynamic set of particle sizes can develop. Consequently the trend has been toward rapid thermal anneals. The extreme case of this is to use pulsed laser anneals [62]. Here the advantages are that an extremely high effective peak temperature can be reached (in excess of the melting point of the nanoparticles) and, depending on the pulse duration, this can allow diffusion and/or regrowth of the particles. The choice of wavelength allows some control on the penetration of the deposited power into different depths of the surface. Direct coupling to the host lattice via excimer (ArF, KrF) or infrared lasers (e.g., CO_2) can heat the entire surface region or, at least in principle, for transparent materials the photon energy can be initially absorbed at the metallic particles. All of these options have been used in various situations [62–67]. Figure 5.12 shows some examples of the reflectivity curves for silver in float glass following pulsed laser anneals.

As shown by Figure 5.12, an interesting variation on the method is to address the particles from the rear face of a transparent host so that the laser power develops a temperature gradient in the opposite sense to normal front-face illumination and, importantly, interacts most directly with the particles in the tail of the distribution. The net result is a different size distribution and optical properties compared with the front-face illumination. The changes in reflectivity as a function of directionality will of course exist even without any annealing, and they result from a combination of absorption and reflectivity from a non-symmetric size distribution. To emphasize this, Figure 5.13 shows some early data for front- and rear-face reflectivity of copper-implanted silica [8].

5.4.5   Removal of Nanoparticles

Nanoparticle formation can of course be an unwanted side effect. A classic example is the implantation of rare-earth ions to form a waveguide laser, for example using erbium to give the 1.5 μm emission used in telecommunications. However, in examples with silica or sapphire the high rare-earth concentration, needed to offer sufficient light intensity and gain for a laser, far exceeds the solubility of the rare-earth ions in the host lattices. Also, the implant destroys the host structures and forms color centers. Conventional furnace anneals can remove the color centers but at the temperatures ($\sim 1000°C$) where the host damage is removed the implanted ions fall out of solution to form precipitates (i.e., nanoparticles) [27–29]. The result is that few ions are available for luminescence and the material fails in terms of the objective. To overcome this difficulty pulsed laser anneals were used [63]. In the example shown in Figure 5.14 the ion species is europium, since this was more convenient for the luminescence detection. The laser pulse both removes the lattice damage and dissolves the rare-earth nanoparticles into a thermodynamically metastable arrangement of isolated ions. Since the pulses were only on the nanosecond time scale, the thermal energy is dissipated so quickly that this metastable state is frozen in place. The isolated ions

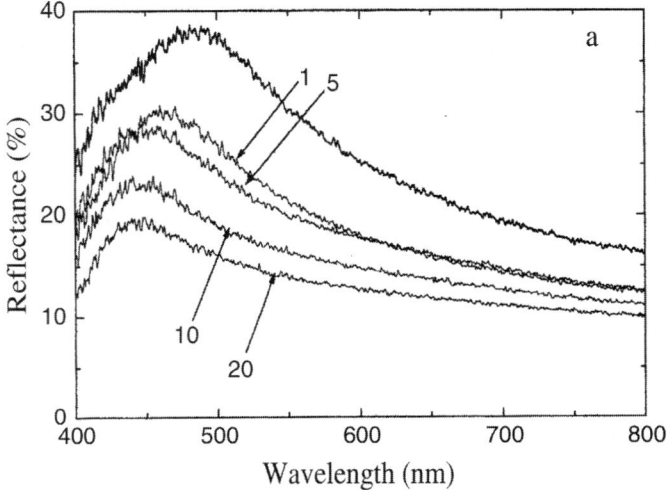

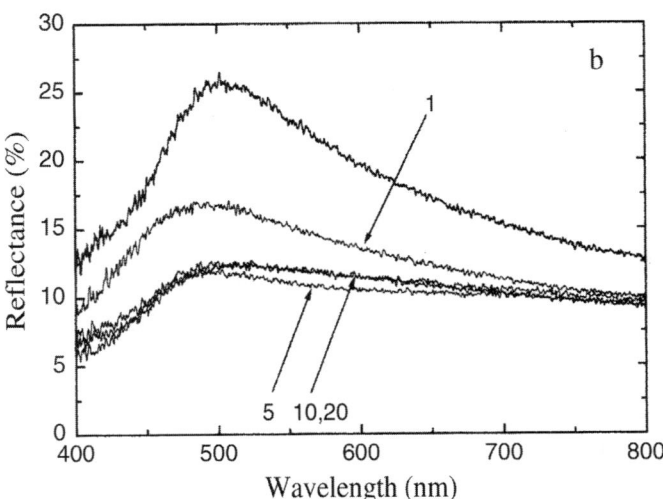

Figure 5.12 Reflectivity data for Ag nanoparticles in glass as the result of pulsed laser anneals. Note that there are differences between light incident on the front and rear face of the particle distribution. The numbers indicate the number of pulses.

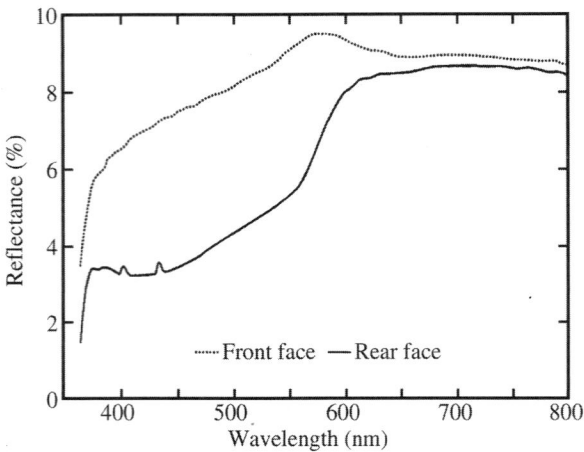

Figure 5.13 Reflectivity of copper-implanted silica. The signals differ between light incident on the front and the rear facs of the implanted sample.

are then efficient luminescent sites. In the example shown the signal intensities after laser treatment are ~ 100 times greater than those obtainable from furnace anneals. Figure 5.14 indicates that both ArF and CO_2 laser pulses were equally effective, and includes an indication of the initial implant intensity plus values after the implant and 1000 and 1200°C furnace anneals. As seen in Figure 5.15, furnace anneals had a significant effect in removing lattice damage (i.e., above about 700°C) but at the highest temperatures the europium became mobile). This pattern is identical to that for erbium implants where transmission electron microscopy data indicated that the dopant ions had precipitated into clusters of rare-earth ions, or a compound that included them.

5.4.6 Optical Non-Linearities of Metal Nanoparticles

One of the driving forces for production of metallic nanoparticles, particularly in the case of copper, is that high pulse

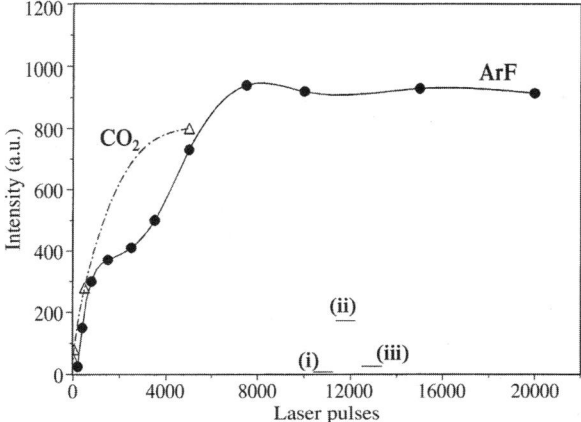

Figure 5.14 A comparison between furnace and pulse laser anneals on the luminescence efficiency of ion-implanted Eu in silica.

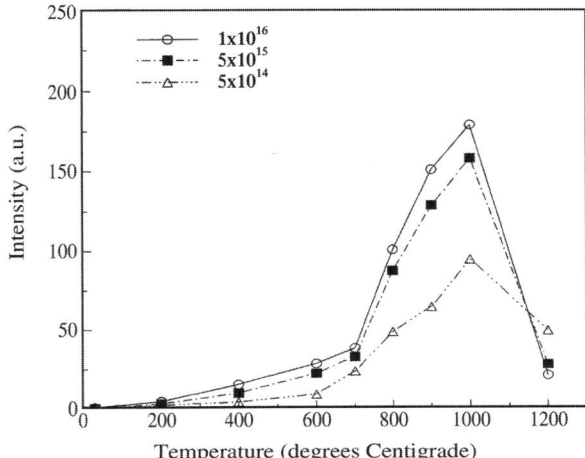

Figure 5.15 Intensity values for Eu luminescence following ion implantation into silica. The pattern is only weakly sensitive to the dopant concentration.

power light near the resonance frequency of the copper particles introduces very strong optical non-linearities via the $\chi^{(3)}$ polarizability [68–70]. This offers a strong change in the refractive index, which in turn can be used as an optical switch. Early data for such changes tended to err in using a very high repetition-rate laser that caused heating and a change from thermal effects, rather than the intended electronic switching. Nevertheless, progress has been made to the point where copper nanoparticles offer very fast responses with large values of $\chi^{(3)}$ that are 100 times greater than those of copper introduced by bulk production methods [69].

5.4.7 Nanoparticles Formed by Multiple Implants

Precipitate phases are not confined to metallic nanoparticles and alternative compounds can form from an implanted metal or other ion reacting with the host lattice. Indeed in many discussions of the metallic nanoparticles, such as copper, there are suggestions that the copper includes particles that are either copper oxide or a metallic particle with an oxide outer layer, or indeed as a hollow structure [71]. Similarly, some authors have suggested that high concentrations of rare-earth impurities in natural minerals (e.g., zircon) are not atomically dispersed in the lattice but precipitate out as rare-earth silicates or oxides [72]. Energetically, such phase inclusions would minimize the lattice strains, even when the ions have charge states that are electrically compatible with the lattice.

To achieve precipitates that have optically interesting features, several groups have explored the options of implanting two impurities with the aim that they form nanoparticle inclusions [41,43,44,73,74]. Nanoparticle inclusions may be aimed at producing non-linear effects, as for the copper example, but materials of the II–VI group are also routinely used in bulk glass material to manufacture optical edge filters. The changing sizes and dopant species offer a series of sharply controlled optical absorption edges from the precipitates within a glass host. Since there is a commercial bulk product, one could envisage making surface layers of similar properties

[75]. No application, however, has yet been cited although combinations of selective optical absorption and transmission could be combined with waveguide and/or other optical structures.

Double implant examples include the use of doped substrates as well as dual implants. Many early examples have been cited by some groups that noted changes by association with the host lattice (e.g., silica) impurities grown in the lattice and/or prior implants and thermal treatments. Precipitate phases can result in, say, the Cu surface plasmon absorption and a shift of the peak to lower energy on addition of F implants. Because there are two implantation stages, the order in which they are made can be important. Each implant introduces many native ion defects and allows mobility from radiation-enhanced diffusion, so the products of A plus B implants can differ from those of B plus A. There have been numerous reviews of the literature including lists of implant pairs that have been used, models on the development of hollow nanoparticles or those with a surface coating, and examples of the types of material properties that are altered (i.e., not just optical effects). The recent examples include the use of energetic heavy inert gas ions to modify the topology of nanoparticles on silicon surfaces, as well as within the surface layers.

Precipitates from the host lattice are of course able to undergo behavior that is characteristic of the precipitate and this includes phase transitions, not just from solid to liquid but also between crystallographic phases. The temperature of the transitions and associated hysteresis effects will be influenced by the host lattice, and examples have been discussed in particular for VO_2 within sapphire. Such a phase transition is that which occurs near 70°C as the VO_2 precipitate goes from a semiconductor to a metal phase [76,77], leading to optical changes in the composite. In passing, one notes that impurity phase inclusions may be more common than is normally considered since in low-temperature cathodoluminescence measurements there are many examples of dramatic intensity changes from inclusions of ice and CO_2. The ice-phase transition at 170 K totally quenches the luminescence signals of the host in many examples [78]. Equally, the sublimation of nanoparticles of CO_2 at 197 K introduces lattice distortions, intensity and wavelength shifts of the Nd lines in Nd:YAG [79]. This phenomenon had not previously been considered but many examples now exist and have been recorded in an initial review [80]. Other authors have similarly considered the role of pressure from trapped gas within implanted materials [81]. Of relevance to the ion implants is that native (impurity) nanoparticles may be playing an important role, but up until now their presence had not even been suspected.

5.5 OPTICAL ABSORPTION, LUMINESCENCE AND LASERS

5.5.1 Optical Absorption

Since the implants are normally confined to the outer layers of a material, within depths as small as a few hundred nanometers (for metals) to a few microns (for helium), the net optical absorption is somewhat limited, even though locally the absorption coefficients can be very high. This does not exclude usage in applications and the example of sunglasses indicated that there was a useful absorption feature, which together with the changes in reflectivity modified the optical transmission. Coloration of transparent materials, such as aluminum oxide with Cr, Ti, or Ni to form ruby, or blue or yellow sapphire, is unrealistic as the depth of color would be minimal. Nevertheless, one could imagine specialist designs where the various gem-stone colors are built into a single implanted surface to produce a picture or design. Surface color modifications of non-transparent materials may have a greater chance of success and, for example, one could envisage implanting the surface of low-cost white jade with iron to make a surface that offers the impression of the higher value green jade. The method is not simple as it needs some additional heat treatments and for such a soft material, it could be detected by removal of the layer by scratching.

In waveguide structures the light is traveling in the plane of the absorbing medium and thus absorption can be significant, and may be an unwanted feature. A more general academic use of absorption introduced by ion implantation is in the study of color centers. Each type of defect within the lattice can have characteristic absorption bands, and the implantation offers routes to test the roles of impurities that influence intrinsic defects. Alternatively, the ion beam can be used to form the defects of the host lattice. This is most apparent with alkali halides where even the electronic stopping region of the ion track will lead to a high density of defects as halide ions are displaced from lattice sites. There have been several such studies on, e.g., implants in LiF with Na, Mg and Ar to consider the relative intensities of colloidal bands from the metal of host lattice and their association with point defect formation (e.g., the simple F, F_2 centers) [3]. In reality "point" defects do not exist and localized distortions of the lattice influence many lattice sites [82]. Consequently there are interactions between defects, and the proximity of formation influences the relative growth of the simple anion vacancy (the F center) and pairs (F_2) or associations into more complex structures. Since an ion track favors a linear generation of defects, compared with the more random isolated events seen by x-ray coloration, the use of an ion beam can be a helpful addition to the routes used in color-center studies. It will also allow very high-density coloration to levels that are not normally achievable by x-rays. A further opportunity exists to vary the rate of defect formation by varying the ion species. This has been done with silica, sapphire and alkali halides. An extreme example comes from the comparisons of normal ion implantation and beams of molecular ions (e.g., H^+, H_2^+, H_3^+). These dissociate at the surface so one has overlapping tracks that are time and spatially coincident, offering an excitation density that cannot be achieved from a conventional ion beam. Adjustments of input energy and current density can maintain equal power dissipation and range so that the alternatives truly reveal dose rate effects. A very early example in which the balance of the F to F_2 and H to V_4 color center could be altered is shown in Figure 5.16.

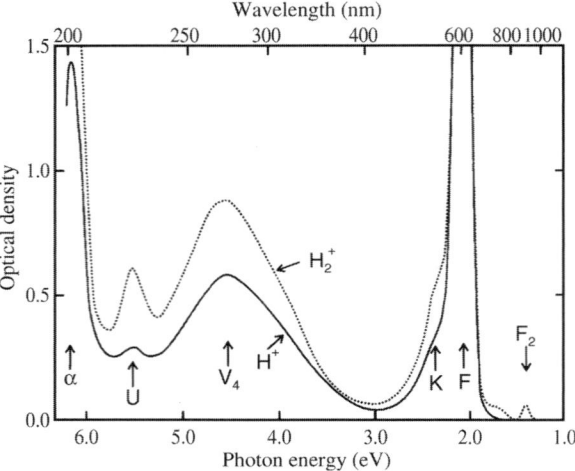

Figure 5.16   Differences in optical absorption produced and measured at 40K in KBr by 400 ke VH⁺ and 800 KE VH₂⁺ ion beams to the same particle dose. Note the changes in the F to F2 (vacancy and di-vacancy) and u and V4 (interstitial and di-interstitial) defect concentrations.

5.5.2   Luminescence During Implantation

The use of silica alignment flags during implantation is standard practice, but very few research teams have monitored emission spectra during implantation, although the signals are often quite intense. The changing signals track the development of defect formation and amorphization as well as indicate the presence of impurity ions. For example, Cr is a frequent trace impurity in Al_2O_3 and is easily detected by the red emission line. The same options are open as for absorption studies, and again comparisons of luminescence spectra from H^+ and H^{2+} in sapphire have separated overlapping emission bands from the primary color centers and have provided evidence for less familiar emission sites. In many respects the ion beam luminescence (IBL) is similar to cathodoluminescence (CL) but for the electron beams the surface penetration is much less than is achieved with the MeV hydrogen ions. This is beneficial, since many insulators have surface states and defects which perturb or quench the luminescence. The IBL thus can offer information more characteristic of the bulk than the surface-dominated signals from cathodoluminescence [83]. Several references are listed for silica [84,85] and sapphire [86–88] to emphasize that IBL can contribute to defect studies in many insulators. The references include an example of probing multi-layer structures, which can be successively addressed by changes in beam energy [89], and the use of luminescence as a diagnostic tool [90,91].

With lithium niobate a study of periodically poled material indicated that there were strong signals from the poled domains, but minimal emission from the distorted inter-domain boundary walls. The data suggest the wall influences almost a micron width, which is far greater than is assumed in many discussions of the material where periodically poled lithium niobate is used in second harmonic generation.

Studies with minerals have also offered better, or different, perspectives on the luminescence properties compared with cathodoluminescence. Both luminescence techniques benefit from the very high sensitivity of photon detection and a dispersive view of the imperfection sites in terms of emission spectra and lifetime resolution. However, an inherent problem of all luminescence studies is the high sensitivity, which results in different responses to impurities and measurable variations between nominally pure and identical samples. While informative, these sample-dependent effects pose problems in detailed defect assignments. Overall, IBL is of value in academic fundamental studies of materials and is vastly under-exploited, and could be a fruitful and productive area for development. In terms of applications IBL has not yet been developed except for beam alignment and simplistic views of sample quality.

5.5.3 Post Implant Luminescence and Lasing

As has already been mentioned, implantation is a convenient route to inject luminescent material into the surface of insulating materials. The simplest examples to consider involve rare-earth implants since the luminescence efficiency is often high and the spectra are clearly identifiable. However, experimentally, rare-earth ions have numerous isotopes and are often cross-contaminated. A particular problem occurs with Er implants with major abundance isotopes of mass 166, 167, 168 and 170. The mono-isotopic elements holmium and thulium occur at 165 and 169, and the mass resolution and/or scanning practices to implant large area samples has resulted in contaminated implants in many samples monitored by us from a variety of laboratories. The problems have not been apparent via RBS analyses, but in spectral terms the impurities can be clearly detected.

For laser applications the implant doses are generally very large and, as mentioned earlier, pulsed laser or rapid thermal anneals are needed to remove implant damage and optimize the dispersion of the rare-earth ions. For high dopant concentration laser materials it may be possible to add ions by implantation to achieve dopant levels that are not compatible with crystal growth. This could have the benefit of avoiding very high dose implants. However, the simpler alternative, particularly for laser waveguides, is to commence with a bulk laser material such as Nd:YAG and use the damage barrier route to define a waveguide geometry. The first example of such a waveguide laser was in 1990 [92] and this was followed by other examples in YAP [93], $LiNbO_3$ [94], GGG [95], Yb:YAG [96], etc. [3]. Many of these show good slope efficiency and low threshold values, and evidence that one can obtain high laser quality from the waveguide structures. The main difficulty in all these experiments is not with the guide definition but in the preparation of the edge quality of the samples where light is coupled in and out through the mirrors. Similar comments apply to second harmonic generation, up- or down-conversion lasers and other non-linear devices.

Although outside the scope of this chapter, the current work on luminescence from surfaces of silicon clearly has similar problems. In many cases the implants are with Er to give the 1.5 μm emission and, since in that range the silicon is transparent, it is equivalent to the visible optical problems of wide-band gap materials. The silicon case is quite complicated as there may be native oxide layers on the semiconductor and there are assumed to be reactions between the Er and the oxygen (e.g., Refs. [97, 98]). Similarly, surface roughness plays a role in the efficiency of the emission process. For CL and IBL surface states, lattice impurities, dislocations and even water from the atmosphere can have dramatic influences on the luminescence efficiency, and all these factors are exacerbated for roughened surfaces. Erbium implants have also been used for subsequent electroluminescence [99].

5.6 INTERACTIONS OF IMPLANTATION WITH OPTICAL AND CHEMICAL FEATURES

Dramatic problems exist for hygroscopic materials such as ZBLAN (ZrBaLaAlNa) fluorides. To avoid the water problem there have been examples of oxygen implants into the surface that cause preferential loss of the light fluorine ions and their replacement with oxygen. The effective result is to clad the material in an aluminum oxide layer, which inhibits the water problems. The oxygen ion implants thus passivate the surface.

In contrast, in many materials such as garnets, niobates or tantalates implant damage can greatly increase the chemical reactivity by factors of 500 to 1000 times, and even quartz, silica and sapphire become more reactive after ion beam damage by factors of 200, 3 or 1.5 times, respectively [3,100,101]. Prior to implantation some problems of surface chemistry are noted even for the nominally hard and chemically inert laser materials such as Nd:YAG, where there is strong evidence to indicate that the surface is not totally passive and is sensitive to the penetration of water from the atmosphere [79]. In this material there are small changes in the CL spectra and efficiency. Nevertheless, other significant problems exist even for well-polished Nd:YAG crystals. The apparently well-prepared surface can disguise a high density of buried polishing grit-induced features. A chemical etch indicates that conventional polishing is responsible both for more than 10^5/cm^2 emergent dislocations and a dense network of "scratch" lines hidden beneath the optically flat surface. For Nd:YAG laser rods this is not critical, but for surface waveguide lasers it is crucial since the light travels in the plane of the surface. The threshold laser power correlates with the dislocation density, and a factor of 10^2 reduction in dislocation density can reduce the lasing threshold power by a factor of two [102]. An ion beam route to reduce the dislocation density at the surface has been to amorphize the optically polished surface. Chemical attack is then some 500 times faster and even residual evidence of dislocations is unimportant. At the end of the amorphized zone, a region emerges that has the same optical flatness as the original surface but dislocation densities as low as $10–100/cm^2$. This is then the basis for waveguide fabrication in a higher quality optical laser material. An immediate monitor of the improvement in sample quality has been a 250% increase in the CL intensity compared with that of the original surface. Further, the removal of dislocations gives significant changes in photoluminescence intensity from subsequent ion-implanted waveguides. As well as variations between transverse electric and transverse magnetic polarizations, the efficiency was raised by up to 40% for the transverse magnetic polarized signals.

5.7 MISCELLANEOUS OPTICAL CHANGES

In studies of laser ablation of insulators there is an inherent problem if the light is not strongly absorbed by the target material. For large band-gap material, such as silica, this means that ablation with an excimer ArF laser at 297 nm is inefficient. Not only must one use high power pulses but also, the charge explosions from the surface induce cracking and fragmentation. Implantation in silica generates a variety of color centers, notably in the ultraviolet, and also has a minor influence on photoconductivity. Consequently we achieved improvements in laser ablation patterning on silica with an ArF laser by a prior implantation. The observations included a threefold reduction in the power threshold and suppression of the surface electrical discharge cracking pattern. Surface conductivity resulting from implants has been noted in other systems as well [103].

5.8 SUMMARY

The use of ion implantation for optical materials has existed for some 30 years and realistically, very few totally new concepts have been proposed recently. However, many more examples of success in surface processing of insulators now exist in the recent literature. For example, there are papers concerned with surface structuring [104,105], LiF colloids [106,107], photorefractive effects [108], light emission [109,110] SHG laser [111,112], and waveguides and waveguide lasers [113,114]. In addition there are numerous Chinese examples of waveguide formation using relatively heavy ions (e.g. Refs. [115–117]). It is clear that acceptance of the concepts and wider availability of ion implanters and accelerators to industrial as well as academic laboratories means that many of the ideas and demonstrations are likely to emerge within commercial devices. The overall view of the topic is thus extremely positive in terms of an expanding future.

REFERENCES

1. Townsend, P.D. (1987) *Rep. Prog. Phys.* 50, 501.

2. Mazzoldi, P. and Arnold, G.W. (Eds.) (1987) *Ion Beam Modification of Insulators,* Elsevier, Amsterdam.

3. Townsend, P.D., Chandler, P.J. and Zhang, L. (1994) *Optical Effects of Ion Implantation*, Cambridge University Press, Cambridge.

4. Weeks, R.A. (1991) *Mater. Sci. Technol.* 9, 331.

5. Buchal, Ch., Withrow, S.P., White, C.W. and Poker, D.B. (1994) *Annu Rev. Mater. Sci.* 24, 125.

6. Buchal, C., Fluck, D. and Gunter, P. (1999) *J. Electroceramics* 3, 179.

7. Faik, A.B., Chandler, P.J., Townsend, P.D. and Webb, R. (1986) *Radiat. Eff.* 98, 233.

8. Hole, D.E. (1994) *Optical Effects of Ion Implantation into Glass*, D.Phil. thesis, Sussex.

9. Townsend, P.D. (1976) *Ion Implantation in Optical Materials*. Inst. Phys. Conf. series 28, chap. 3.

10. Destefanis, G.L., Townsend, P.D. and Gailliard, J.P. (1978) *Appl. Phys. Lett.* 32, 293.

11. Destefanis, G.L., Gailliard, J.P., Ligeon, E., Townsend, P.D., Valette, S., Perez, A. and Farmery, B.W. (1979) *J. Appl. Phys.* 50, 7898.

12. Chandler, P.J., Zhang, L. and Townsend, P.D. (1990) *Nucl. Instrum. Methods B* 46, 69.

13. Pitt, C.W., Skinner, J.D. and Townsend,P.D. (1984) *Electron. Lett.* 20, 4.

14. Chandler, P.J., Lama, F.L., Townsend, P.D. and Zhang, L. (1988) *Appl. Phys. Lett.* 53, 89.

15. Zhang, L., Chandler, P.J., Townsend, P.D., Alwahabi, Z.T., Pityana, S.L. and McCaffrey, A.J. (1993) *J. Appl. Phys.* 73, 2695.

16. Hamelin, N., Lifante, G., Chandler, P.J., Townsend, P.D., Pityana, S. and McCaffrey, A.J. (1994) *J. Mod. Optics*, 41, 1339.

17. Rams, J., Olivares, J., Chandler, P.J. and Townsend, P.D. (1998) *J. Appl. Phys.* 84, 5180.

18. Jazmati, A.K., Vazquez, G. and Townsend, P.D. (2000) *Nucl. Instrum. Methods B* 166/167, 592.

19. Lifante, G. and Townsend, P.D. (1992) *J. Modern Optics* 39, 1353.

20. Zhang, L., Townsend, P.D., Chandler, P.J. and Silversmith, A. (1994) *Electron. Lett.* 30, 1063.

21. Van den Hoven, G.N., Snoeks, E., Polman, A., van Dam, C., van Uffelen, J.W.M. and Smit, M.K. (1996) *J. Appl. Phys.* 79, 1258.

22. Herreros, B., Lifante, G., Cusso, F., Townsend, P.D. and Chandler, P.J. (1997) *J. Lumin.* 72/4, 198.

23. Huang, X., Cuthina, N., Alcazar de Velasco, A., Chandler, P.J. and Townsend, P.D. (1998) *Nucl. Instrum. Methods B* 142, 50.

24. Vazquez, G.V. and Townsend, P.D. (2002) *Nucl. Instrum. Methods B* 110.

25. Chen, F., Wang, X.L., Lu, Q.M., Fu, G., Li, S.L., Lu, F., Wang, K.M. and Shen, D.Y. (2002) *Nucl. Instrum. Methods B* 194, 355.

26. Chen, F., Hu, H., Wang, K.M., Teng, B., Wang, J.Y., Lu, Q.M. and Shen, D.Y. (2001) *Optics Lett.* 26, 1993.

27. Polman, A., Jacobson, D.C., Eaglesham, D.J., Kistler, R.C. and Poate, J.M. (1991) *J. Appl. Phys.* 70, 3778.

28. Polman, A. and Poate, J.M. (1993) *J. Appl. Phys. 73,* 1669.

29. Polman, A. (1997) *J. Appl. Phys.* 82, 1.

30. Fleuster, M., Buchal, Ch., Fluck, D. and Gunter, P. (1993) *Nucl. Instrum. Methods B* 80/81, 1150.

31. Townsend, P.D., Chandler, P.J., Wood, R.A., Zhang, L., McCallum, J.C. and McHargue, C.W. (1990) *Electron. Lett.*, 26, 1193.

32. White, C.W., McHargue, C.J., Sklad, P.S., Boatner, L.A. and Farlow, G.C. (1989) *Mater. Sci. Rep.* 4, 41.

33. Mahdavi, M. and Townsend, P.D. (1990) *Trans Faraday Soc.*, 86, 1287.

34. Mahdavi, M., Chandler, P. and Townsend, P.D. (1989) *J. Phys. Appl. Phys.* 22, 1354.

35. Yang, B., Townsend, P.D. and Fromknecht, R., *Nucl. Instrum. Methods B* 217, 60; ibid *J. Phys: Condens. Matter* 16, 8377.

36. Von Bibra, A.U., Roberts, A. and Dods, S.D. (2000) *Nucl. Instrum. Methods B*, 168, 47.

37. Arnold, G.W. and Borders, J.A. (1977) *J. Appl. Phys.* 48, 1488.

38. Arnold, G.W., Brow, R.K. and Myers, D.R. (1990) *J. Non-Cryst. Solids* 120, 234.

39. Magruder, R.H., Morgan, S.H., Weeks, R.A. and Zuhr, R.A. (1990) *J. Non-Cryst. Solids* 120, 241.

40. Magruder, R.H., Zuhr, R.A. and Weeks, R.A. (1991) *Nucl. Instrum. Methods B* 59/60, 1308.

41. White, C.W., Budai, J.D., Withrow, S.P., Zhu, J.G., Sonder, E., Zuhr, R.A., Meldrum, A., Hembree, D.M., Henderson, D.O. and Prawer, S. (1998) *Nucl. Instrum. Methods B* 141, 228.

42. Gonella, F. (2000) *Nucl. Instrum. Methods B* 166/167, 831.

43. Meldrum, A., Haglund, R.F., Boatner, L.A. and White, C.W. (2001) *Adv. Mater.* 13, 1431.

44. Meldrum, A., Boatner, L.A., White, C.W. and Ewing, R.C. (2000) *Mater. Res. Innovat.* 3, 190.

45. Stepanov, A.L., Hole, D.E. and Townsend, P.D. (1999) *J. Non-Cryst. Solids* 260, 65.

47. Stepanov, A.L., Hole, D.E. and Townsend, P.D. (2000) *Nucl. Instrum. Methods B* 161–163, 913.

48. Buffat, P. and Borel, J.-P. (1976) *Phys. Rev. A* 13, 2287.

49. Bachels, T., Güntherodt, H.-J. and Schäfer, R. (2000) *Phys. Rev. Lett.* 85, 1250.

50. Bayazitov, R.M., Zakirzyanova, L.Kh., Khaibullin, I.B. and Remnev, G.E. (1997) *Nucl. Instrum. Methods B* 122, 35.

51. Stepanov, A.L., Bayazitov, R.M. Hole, D.E., and Khaibullin, I.B. (2001) *Phil. Mag. Lett.* 81, 29.

52. Snoeks, E., van Blaaderen, A., van Dillen, T., van Kats, C.M., Velikov, K., Brongersma, M.L. and Polman, A. (2001) *Nucl. Instrum. Methods B* 178, 62.

53. Thome, L. and Garrido, F. (1997) *Nucl. Instrum. Methods B* 121, 237.

54. Kriebeig, U. and Vollmer, M. (1995) *Optical Properties of Metal Clusters*, Springer-Verlag, Berlin/Heidelberg.

55. Träger, F. (2001) (Editor of Special Issue)*Appl. Phys. B*, 73 No. 4.

56. Harmer, S.W. (2002) private communication.

57. Nistor, L.C., van Landuyt, J., Barton, J.D., Hole, D.E., Skelland, N.D. and Townsend, P.D. (1993) *J. Non-Cryst. Solids* 162, 217.

58. Hole, D.E., Townsend, P.D., Barton, J.D., Nistor, L.C. and Van Landuyt, J. (1995) *J. Non Cryst. Solids* 180, 266.

59. Townsend, P.D. and Menaguale, D. (1994) *Filter for solar radiation, particularly suitable for sunglass lenses*, European Patent 94830012.4.

60. Rahmani, M., Abu-Hassan, L.H., Townsend, P.D., Wilson, I.H. and Destefanis, G.L. (1988) *Nucl. Instrum. Methods B* 32, 56.

61. Rahmani, M. and Townsend, P.D. (1989) *Vacuum* 39, 1157.

62. Wood, R.A., Townsend, P.D., Skelland, N.D., Hole, D.E., Barton, J. and Afonso, C.N. (1993) *J. Appl. Phys.* 74, 5754.

63. Can, N., Townsend, P.D., Hole, D.E., Snelling, H.V., Ballesteros, J.M. and Afonso, C.N. (1995) *J. Appl. Phys.* 78, 6737.

64. Gonella, F., Mattei, G., Mazzoldi, P., Cattaruzza, E., Arnold, G.W., Battaglin, G., Calvelli, P., Polloni, R., Bertoncello, R. and Haglund, R.F. (1996) *Appl. Phys. Lett.* 69, 3101.

65. Townsend, P.D. and Olivares, J. (1997) *Proc. E-MRS Appl. Surf. Sci.* 110, 275.

66. Stepanov, A.L., Hole, D.E., Bukharaev, A.A., Townsend, P.D. and Nurgazizov, N.I. (1998) *Appl. Surf. Sci.* 136, 298.

67. Stepanov, A.L., Hole, D.E. and Townsend, P.D. (2000) *Nucl. Instrum. Methods B* 166/167, 882.

68. Haglund, R.F. (1998) *Mater. Sci. Eng. A* 253, 275.

69. Olivares, J., Requejo-Isidro, J.M., del Coso, R., de Nalda, R., Solis, J., Afonso, C.N., Stepanov, A.L., Hole, D., Townsend, P.D. and Naudon, A. (2001) *J. Appl. Phys.* 90, 1064.

70. Takeda, Y., Lee, C.G. and Kishimoto, N. (2002) *Nucl. Instrum. Methods B* 191, 422.

71. Meldrum, A., Honda, S., White, C.W., Zuhr, R.A. and Boatner, L.A. (2001) *J. Mater. Res.* 16, 2670.

72. Karali, T., Can, N., Townsend, P.D., Rowlands, A.P. and Hanchar, J. (2000) *Am. Mineralogist* 85, 668.

73. Hosono, H. (1995) *Phys. Rev. Lett.* 74, 110.

74. Hosono, H. (1995) *J. Non-Cryst Solids* 187, 457.

75. Okur, I. and Townsend, P.D. (1997) *Nucl. Instrum. Methods B* 124, 76.

76. Gea, L.A., Budai, J.D. and Boatner, L.A. (1999) *J. Mater. Res.* 14, 2602.

77. Gea, L.A., Boatner, L.A., Rankin, J. and Budai, J.D., (1995) *Mater. Res. Soc. Symp. Proc.* 382, 107.

78. Kurt, K., Ramachandran, V., Maghrabi, M., Townsend, P.D. and Yang, B. (2002) *J. Phys. Condensed Matter* 14, 4319.

79. Maghrabi, M., Townsend, P.D. and Vazquez, G. (2001) *J. Phys: Condensed Matter* 13, 2497.

80. Townsend, P.D., Maghrabi, M. and Yang, B. (2002) *Nucl. Instrum. Methods B* 191, 767.

81. Fleischer, E.L. and Norton, M.G. (1996) *Heterodyne Chem. Rev.* 3, 171.

82. Townsend, P.D., Jazmati, A.K., Karali, T., Maghrabi, M., Raymond, S.G. and Yang, B. (2001) *J. Phys. Condensed Matter* 13, 2211.

83. Brooks, R.J., Hole, D.E. and Townsend P.D. (2002) *Nucl. Instrum. Methods B* 190, 136.

84. Jaque, F. and Townsend, P.D. (1981) *Nucl. Instrum. Methods B* 182/183, 781.

85. Abu-Hassan, L.H. and Townsend, P.D. (1998) *Nucl. Instrum. Methods B* 32, 293.

86. Al Ghamdi, A. and Townsend, P.D. (1990) *Nucl. Instrum. Methods B* 46, 133.

87. Al Ghamdi, A. and Townsend, P.D. (1990) *Radiat. Eff. Def. Solids* 115, 73.

88. Brooks, R.J., Ramachandran, V., Hole, D.E. and Townsend, P.D. (2001) *Radiat. Eff. Def. Solids* 155, 177.

89. Brooks, R.J., Hole, D.E., Townsend, P.D., Wu, Z., Gonzalo, J., Suarez-Garcia, A. and Knott, P. (2002) *Nucl. Instrum. Methods B* 190, 709.

90. Buchal, C., Wang, S., Lu, F., Carius, R. and Coffa (2002) *Nucl. Instrum. Methods B* 190, 40.

91. Kishimoto, N., Bandourko, V.V., Takeda, Y., Umeda, N. and Lee, C.G. (2002) *Nucl. Instrum. Methods B* 207, 1990.

92. Chandler, P.J., Field, S.J., Hanna, D.C., Shepherd, D.P., Townsend, P.D., Tropper, A.C. and Zhang, L. (1990) *Electron. Lett.* 25, 985.

93. Field, S.J., Hanna, D.C., Shepherd, D.P., Tropper, A.C., Chandler, P.J., Townsend, P.D. and Zhang, L. (1990) *Electron. Lett.* 26, 1826.

94. Field, S.J., Hanna, D.C., Shepherd, D.P., Tropper, A.C., Chandler, P.J., Townsend, P.D. and Zhang, L. (1991) *Optics Lett.* 16, 481.

95. Field, S.J., Hanna, D.C., Large, A.C., Shepherd, D.P. Tropper, A.C., Chandler, P.J., Townsend, P.D. and Zhang, L. (1992) *Optics Lett.* 17, 52.

96. Hanna, D.C., Jones, J.K., Large, A.C., Shepherd, D.P., Tropper, A.C., Chandler, P.J., Rodman, M.J., Townsend, P.D. and Zhang, L. (1993) *Optics Commun.* 99, 211.

97. Kenyon, A.J., Chryssou, C.E., Pitt, C.W., Shimizu-Iwayama, T., Hole, D.E., Sharma, N. and Humphreys, C.J. (2002) *J Appl. Phys.* 91, 367.

98. Kik, P.G. and Polman, A. (2000) *J Appl. Phys.* 88, 1992.

99. Wang, S., Amekura, H., Eckau, A., Carius, R., and Buchal, C., (1999) *Nucl. Instrum. Methods B.* 148, 481.

100. Nunn, P.J.T., Olivares, J., Spadoni, L., Townsend, P.D., Hole, D.E. and Luff, B.J. (1997) *Nucl. Instrum. Methods B* 127/128, 507.

101. Som, T., Navati, M.S. and Kulkarni, V.N. (2001) *Nucl. Instrum. Methods B* 179, 551.

102. Fuxi Gan (1995) *Laser Materials*, World Scientific, Singapore.

103. Liu, F., Dickinson, M.R., MacGill, R.A., Anders, A., Monteiro, O.R., Brown, I.G., Phillips, L., Biallas, G. and Siggins, T. (1998) *Surf. Coat. Technol.* 104, 46.

104. Crunteanu, A., Janchen. G., Hoffmann, P., Pollnau, M., Buchal, C., Petraru, A., Eason, R.W. and Shepherd, D.P. (2003) *Appl. Phys. A* 76,1109.

105. Crunteanu, A., Hoffmann, P., Pollnau, M. and Buchal, C. (2003) *Appl. Surf. Sci.* 208, 322.

106. Mussi, V., Somma, F., Moretti, P., Mugnier, J., Jacquier, B., Montrereali, R.M. and Nichelatti, E. (2003) *Appl. Phys. Lett.* 82, 3886.

107. Mussi, V., Cricenti, A., Jacquier, B., Marolo, T., Montrereali, R.M., Moretti, P., Nichelatti, E. and Somma, F. (2003) *Rad. Eff. Def. Solids* 158, 181.

108. Kostritskii, S.M. and Moretti, P. (2002) *Infrared Holography for Optical Communications* 86, 59.

109. Buchal, C., Wang, S., Lu, F., Carius, R. and Coffa, S. (2002) *Nucl. Instrum. Methods B* 190, 40.

110. Lu, F., Carius, R., Alam, A., Heuken, M., Risi, A. and Buchal, C. (2003) *Thin Solid Films* 425, 171.

111. Vincent, B., Boudrioua, A., Loulergue, J.C., Moretti, P., Tascu, S., Jacquier, B., Aka, G. and Vivien, D. (2003) *Optics Lett.*, 28, 1025.

112. Kremer, R., Boudrioua, A., Moretti, P. and Loulergue, J.C. (2003) *Optics Commun.* 219, 389.

113. Vazquez, G.V., Rickards, J., Lifante, G., Domenech, M. and Cantelar, E. (2003) *Optics Express* 11, 1291.

114. Vazquez, G.V., Rickards, J., Marquez, H., Lifante, G., Cantelar, E. and Domenech, M. (2003) *Optics Commun.* 218, 141.

115. Chen, F., Wang, X.L., Li, S.L., Fu, G., Wang, K.M., Lu, Q.M., Shen, D.Y., Nie, R. and Ma, H.J. (2003) *J. Appl. Phys.* 94, 4708.

116. Chen, F., Wang, X.L., Wang, K.M., Shi, B.R., Lu, Q.M., Shen, D.Y. and Rui, N. (2003) *Mater. Lett.* 57, 1197.

117. Chen, F., Wang, X.L., Lu, Q.M., Fu, G., Li, S.L., Lu, F.,Wang, K.M., Chen, H.C., Shen, D.Y. and Nie, R. (2002) *Appl. Surf. Sci.* 202, 86.

Chapter 6

Metal Alloy Nanoclusters by Ion Implantation in Silica

P. MAZZOLDI, G. MATTEI, C. MAURIZIO,
E. CATTARUZZA, AND F. GONELLA

CONTENTS

ABSTRACT

In the last decade, metal nanoclusters (NCs) embedded in insulating matrices have received increasing interest due to their peculiar optical, magnetic and catalytic properties when the size becomes comparable to or less than the electronic mean free path. Glass-based composites are, in general, expected to play an important role as materials for various nanotechnology application, due to the low cost, ease of processing, high durability, resistance and high transparency, as well as the possibility of tailoring the behavior of the glass-based structures. Metallic NCs embedded in glass can increase the optical third-order susceptibility $\chi^{(3)}$ of the matrix by several orders of magnitude, making such systems interesting candidates to be used as optical switches. Among different possible synthesis processing, ion-beam-based techniques proved to be very suitable in synthesizing NC-containing glasses. Moreover, the composition of the clusters can be varied easily by sequential ion implantation in the matrix of two different elements whose energy and dose can be tailored so as to maximize the overlap between the implanted species and to control their local relative concentration. Nevertheless, for achieving tunability of the NCs properties for actual devices, a careful control over alloy clusters synthesis and stability has to be achieved in order to clarify which are the parameters (i.e., implantation conditions, subsequent thermal or laser annealings, ion irradiation, etc.) that can promote separation (via oxidation, for instance) instead of alloying of the implanted species.

This chapter deals with metal nanocluster composites synthesized by ion implantation and formed by clusters of binary transition metals alloy embedded in silicate glasses, in which the cluster concentration is below the percolation limit (dispersed clusters). Some case studies will be presented in which ion-beam-based techniques are exploited either for the synthesis or for the structural or compositional modification of composites. In particular, the analysis will be focused on the formation of metal alloy nanoclusters in SiO_2 matrix obtained by sequential ion implantation of Au+Cu, Au+Ag,

Co+Ni, Cu+Ni and Co+Cu. The most relevant technological parameters for ion implantation (dose, energy) and for the subsequent treatments (thermal annealing, ion-irradiation with light ions or laser annealing) are investigated to control cluster composition, size and stability. As far as the technological properties of the studied systems are concerned, some results will be presented about the nonlinear optical properties of Au-based alloy NCs composites.

6.1 INTRODUCTION

Owing to their peculiar properties, which are suitable for application in several fields, metal NCs embedded in glass matrices give rise to composite materials that have been the object of numerous studies for several decades. The composites formed by nanometer-sized metal clusters embedded in glass matrices exhibit striking nonlinear optical effects, particularly interesting for application in the field of nonlinear integrated optics as part of all-optical devices. Moreover, NCs dispersed in dielectric matrices may exhibit superparamagnetic properties, enhanced coercivity, shift of the hysteresis loop and large magnetotransport properties [1]. Nanostructured materials are also used in catalysis, where high surface-to-volume ratios are required [2].

Glass-based composites are in general expected to play an important role as materials for various nanotechnology application, due to the low cost, ease of processing, high durability, resistance and high transparency, as well as the possibility of tailoring the behavior of the glass-based structures. Since the first attempt by Faraday [3] to explain the nature of the color induced in glasses by small metallic precipitates, many studies have been dedicated to the properties of metal nanocluster composite glass (MNCG). In general, the physical properties of these systems change dramatically in the transition from atom to molecule to cluster to solid, where the cluster regime is characterized by the confinement effects that make MNCGs peculiarly interesting. For example, experiments that showed the transition from atomic over molecular to bulk plasmon absorption features have been performed with Ag clusters in photosensitive glasses during the last four decades. Updated review articles dealing with MNGCs are currently published, each one covering one or more particular aspect ranging from preparation techniques to properties and characterization. A general treatment of quantum dot materials including MNCGs is presented in Ref. [4], while glasses for optoelectronic devices, particularly MNCGs, are treated in Refs. [5–7]. The theoretical aspects of MNCG properties are approached in the literature from a great variety of viewpoints, for example, by treating the electronic properties of metal clusters [8], or focusing on the quantum size effect [9]. Nonlinear optical properties of small metal particles are treated in detail in Refs. [10, 11], while Refs. [12, 13] present extended reviews on theoretical and experimental aspects of the optical response of metal clusters. Recent aspects concerning the interface properties of MNCGs are in Ref. [14], and Ref. [15] presents a review of all-optical switching via nonlinear optical materials. This chapter deals with metal nanocluster composites formed by clusters of binary transition-metal alloys embedded in silicate glasses, in which cluster concentration is below the percolation limit (dispersed clusters). The scale regime has characteristic lengths, such as correlation lengths among clusters or electron mean free paths of the bulk, that are larger than the cluster size. On the other hand, the clusters are considered to be large enough to exhibit electronic behavior features of the metal (usually, larger than 1 nm diameter), being small enough even to exhibit strong confinement effects (up to tens of nm diameter).

Ion implantation of metals into glass has been explored since the last decade as a useful technique to produce nanocomposite materials in which nanometer-sized metal or semiconductor particles are embedded in dielectric matrices [5, 16–22]. Furthermore, ion implantation has been used as the first step of combined methodologies that involve other treatments such as thermal annealing in controlled atmosphere, laser or ion irradiation [23, 24]. In general, ion implantation is well established as a suitable technique for improving mechanical, optical and structural near-surface properties of glasses. It has several advantages, such as low-temperature processing, control of distribution and concentration of dopants, and availability of chemical states that cannot be realized via conventional techniques, thus overcoming solubility limits. Moreover, ion implantation can be exploited for designing waveguiding structures along prescribed patterns. The implantation of ions (of typical energies in the range from keV to MeV) into materials results in various modifications that depend on the glass composition, the ion species, the fluence, the implantation energy, and, in some cases, the interaction with the external atmosphere when implanted glasses are removed from the implantation chamber. These modifications are here treated in the context of silicate glass matrices, mostly fused silica and soda-lime glass, both of which are used for the synthesis of quantum dot materials. Energetic ions lose their energy in implanted dielectrics by either interaction with the electronic system or by direct hard-sphere collisions with the nuclei. The partition of the ion energy into electronic and collisional processes, as well as the projected ion range and the straggling, are usually obtained by the TRIM code [25]. Both nuclear and electronic stopping give rise to structural changes in materials, and in several cases the process of ion implantation gives rise to the clusterization of implanted metal ions (for comprehensive reviews, see Refs. [5, 16, 18]).

The physical mechanisms governing cluster formation are presently under debate, as are the relative roles of electronic and nuclear energy release [26]. Perez et al. [27, 28] first studied the state of implanted atoms using a simple statistical model and describing the role of a crystalline host matrix structure in the determination of the final compounds upon ion irradiation. Still useful as a rough picture, Perez's statistical model is actually reliable only (and not always [29]) for ion implantation in crystalline substrates. Hosono [30] proposed a criterion to predict the formation of small clusters by ion implantation in SiO_2 glass, based on physical and chemical considerations. The main defects produced during ion implantation in silica are oxygen-deficient centers, namely, Si-Si bonds and neutral oxygen monovacancies. The concentration of Si-Si bonds results highly ion-specific, i.e., depending critically on the chemical interaction among the implanted element and the silicon and oxygen atoms of the glass matrix. In the case of strong chemical interaction, implanted ions (M) tend to form M-O bonds, leaving Si-Si bonds, whereas for weak chemical interaction, a large part of the implanted atoms do not react with oxygen atoms, with a very low resulting concentration of Si-Si bonds. Implanted (M) and silicon ions compete for bonding oxygen, and cluster formation will occur when the chemical affinity of M ions for the oxygen is smaller than that of Si^{4+}. The Hosono model takes the free energy for the oxide formation as the quantity giving the measure of the chemical interaction. Clusterization is expected to occur whenever the Gibbs free energy for an oxide formation with the implanted element M is greater than that for SiO_2. The chemistry considered here is the formation of the M_xO_2 oxide starting from the element M in the metallic form and the molecular gaseous oxygen, at the estimated effective temperature of T = 3000 K, as a consequence of the thermal spike phenomenon. A more general approach accounting for the different compounds observed to form in silica upon ion implantation is found in Refs. [31, 32]. This approach starts from considering the concentration of the main implantation-induced defects in silica, pointing out the importance of the chemical interaction among implanted and host matrix atoms as a determining factor that gives rise to the observed defects. In particular, it is assumed that the primary factor controlling the chemical interaction of the implanted ions M with the matrix structure is the electronegativity (EN) of the implanted element. Three different categories are so defined, depending on the behavior of the element when implanted in fused silica. For elements with EN < 2.5, the extraction of oxygen atoms from the silica network and the formation of M-O bond take place, while implants of elements for which EN > 3.5 give rise to knock-on of oxygen atoms from the silica network, thus forming Si-M bonds. The third kind of implant is that of noble metals, which do not exhibit significant chemical reactions.

Research groups from the Universities of Padova and Venezia have proposed a model [16] originating from the investigation of Kelly [33] on the compositional modification induced by ion implantation on alloys, oxides and other substances. Kelly first pointed out the important role of the chemistry in the ion-beam mixing process, even though, during ion implantation, both the incoming ion energy (several keV) and the recoil energy of the atoms are very large compared with the chemical energies coming into play. He introduced a new term in the ion-beam mixing diffusion equations, describing the effects of chemically guided steps along preferential trajectories. Some sort of chemical driving force taking place at the end of the collisional cascades could therefore significantly affect the ion-mixing process despite the energy disparity, with no need for a thermal spike picture. After the first few tenths of a ps in the cascade (during which the temperature concept is meaningless), the effective local temperature is of the order of a few thousand K for a time interval of 10–20 ps, and at this level, the self-diffusion coefficient is typical of a liquid state. On the other hand, after few tenths of a ns the local temperature becomes that of the substrate. Indeed, molecular dynamics simulations suggest that the described chemical (thermodynamical) processes should take place at the substrate temperature. So the proposed model distinguishes two different steps in the ion implantation process. First, a high-energy ballistic regime takes place, giving rise to the substrate damage (defects, knock-on events, etc.). Second, low-energy, chemically guided processes induce the formation of compounds that can be determined on the basis of thermodynamical considerations by calculating the Gibbs energy variation for a chemical reaction between the implanted element (in the gaseous form) and the silica molecule at the matrix temperature. The more negative this value, the more likely is the formation of the particular compound. The two-step model correctly predicts the formation of various compounds and the formation of metallic clusters at room temperature, yet failing in the prediction of silicide formation in the case of Fe, Co and Ni implants, which have significant magnetic behavior [16]. Generally speaking, exact calculations are in any case possible only for stoichiometric silica, even if qualitative considerations can be put forward as well. Indeed, thermodynamical concepts are limited by the uncertainty in the evaluation of the local temperature, as well as on the assumption that chemical interactions take place between the implanted atoms and the substrate compounds (not the individual atomic species). The possibility of such modelization for cluster formation upon ion implantation is further complicated when dealing with sequential double implantation of two different elements.

The preparation of "mixed" colloidal structures, containing clusters of either different metals or formed by metal alloys, has recently attracted much attention. With respect to the single-element case, binary metal nanoclusters offer further degrees of freedom for the control of the material features, namely, cluster composition and crystal structure. On the other hand, the suitability of MNCGs containing binary NCs for actual devices requires a more detailed knowledge of the conditions for clusterization to occur, in order to control the parameters that may induce separation instead of alloying of the implanted species.

Sequential ion implantation of two different metal species may give rise to various different nanocluster structures, with the presence of separated families of pure metal clusters, crystalline alloy clusters, or core-shell structures (see Figure 6.1). Composition of the clusters can be varied by sequentially implanting in the matrix the two different elements at energies and doses such as to maximize the overlap between the implanted species to control their local relative concentration. The formation of clusters of a certain nature depends critically on the implantation parameters such as ion fluence and energy, implant sequence and the temperature at which the process takes place. Moreover, post-implantation treatments such as annealing in controlled atmosphere or ion or laser irradiation have been demonstrated to be effective in driving the system toward different stable clusters structures [23,24]. An effective phenomenological model describing the formation of binary clusters is, in any case, still lacking. Miscibility of the implanted elements comes into play together with the

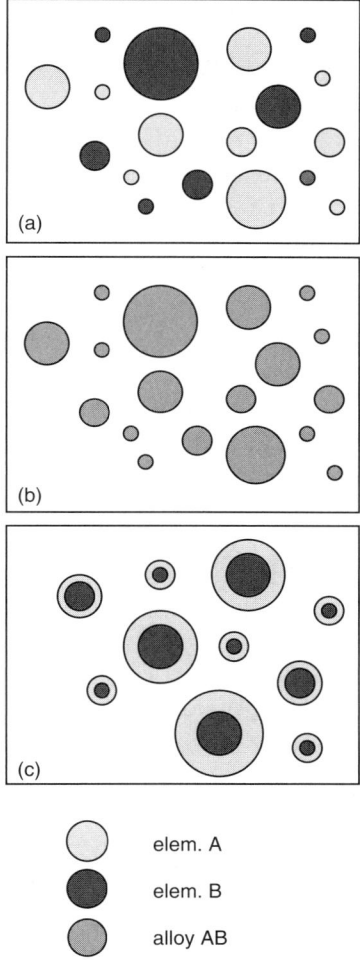

Figure 6.1 Sketch of possible families of clusters formed by sequential ion implantation of two different species A and B: (a) separate A and B clusters, (b) A-B alloy, and (c) core-shell clusters.

chemical reactivity with the matrix and the presence of radiation-induced defects in the definition of the resulting structure. Systematic experimental data are presently available for only a few cases, as discussed in the following.

Multicomponent nanoclusters of Ag-Cu, In-Cu, In-Ag, Ag-Sb and Cd-Ag were first claimed to be formed by sequential implantation in silica [34–37]. More recently, Cu-Ni, Ag-Ni, Au-Cu, Au-Ag, Cu-Ag systems have been studied for their optical properties, and the formation of alloy nanoclusters was unambiguously observed for the Cu-Ni, Au-Cu and Au-Ag systems [23, 38, 39]. In these cases, the interest is focused on both the third-order optical nonlinearity of the composites [40, 41] and the optical absorption response in terms of tunability of the plasma resonance [23]. Furthermore, Cu+Co, Co+Ni, Co+Fe, Pt+Fe, Fe+Al implanted glass nanocomposites were studied for their peculiar magnetic features [42–44]. The possibility to synthesize alloy-based composites permits tailoring of the magnetic properties of these nanostructured materials. For example, preliminary results have been recently obtained [45] for sequential ion implantation of Cu^+ and Co^+ in silica at different dose values. Indeed, this pair of elements is particularly interesting for the possibility of obtaining Co-Cu alloy nanoclusters, despite their immiscibility in the bulk phase. Pd-Ag and Pd-Cu have also been studied for their properties in the field of catalysis [46]. It is worth remarking that sequential ion implantation is also currently studied for the formation of semiconductor binary systems in dielectrics. For example, the bandgap tunability of semiconductor nanoclusters draw the study of Ga-N, As-Ga, In-P, Cd-S, Cd-Se, Cd-Te, Ga-P, In-As, Pb-S, Zn-Te, Zn-Se, Ag-S, Zn-S couples [22, 47–49]. In the following, we will focus the discussion on the systems formed by sequential ion implantation of transition metal elements, presenting some case studies for which an extended experimental body has been already obtained.

6.2 EXPERIMENTAL

Sequential ion implantations with different elements were performed on silica substrates at INFN-INFM Ion Implantation Laboratories (Italy) of INFN-Legnaro National Laboratories. The current densities were maintained lower than $2\mu A/cm^2$ in order to avoid sample heating during the implantation, and ion beam energies were chosen to get the same projected range for the implanted species (typical energies were up to 200 keV to modify a sub-surface layer of 100 to 200 nm of the host matrix), and ion doses ranged from 1×10^{16} ions/cm^2 up to 4×10^{17} ions/cm^2. Ion-implanted slides were then heat-treated in a conventional furnace at different temperatures and time intervals in air, N_2 or in an H_2 4%– N_2 gas mixture. The samples were studied by several complementary techniques, in particular transmission electron microscopy (TEM), extended x-ray absorption fine structure (EXAFS), grazing-incidence x-ray diffraction (GIXRD), x-ray photoelectron spectroscopy (XPS), Rutherford backscattering spectrometry (RBS), secondary ion mass spectroscopy (SIMS). EXAFS and GIXRD experiments [50] were performed at the European Synchrotron Radiation Facility (ESRF) in Grenoble (France), on the Italian beamline GILDA and the ID09 beamline, respectively. Optical characterization has been performed by linear absorption in the UV-Vis spectrum (OD). In particular, it is worth stressing the strict interplay between optical and structural techniques for effective alloy detection: in the case of noble metals alloy, the surface plasmon resonance (SPR) band is located between those of the pure elements, and is triggered by the complex interplay between the modified free electron density and interband absorptions [12, 24, 51]. Linear absorption (OD) measurements can therefore give the first indication of possible alloy formation. Nevertheless, in systems containing transition metals (Pd-Ag, Co-Ni, ...) such a simple technique is no longer effective, as interband transitions completely mask the SPR peak, resulting in a structureless absorption that hinders any unambiguous identification of the alloy. In such cases, one has to rely on structural techniques like TEM (coupled to selected area electron diffraction-SAED and energy dispersive x-ray spectrometry-EDS), GIXRD, or EXAFS to establish alloy formation.

6.3 Au-BASED ALLOYS

Generally, the criterion (valid for bulk systems) of miscibility of the two elements as a requirement for alloying is not so stringent in the case of nanoclusters, due to the incomplete onset of the bulk properties linked to the large number of atoms at the surface, making a cluster more similar to a molecular than to a massive system [52]. This leads to new possible alloy phases, which may not be thermodynamically favored in the bulk. In the case of noble metal-based systems (Au-Cu, Au-Ag, Pd-Ag and Pd-Cu) perfect miscibility is expected from the bulk phase diagrams and in fact sequentially as-implanted samples exhibit direct alloying [23, 24, 53]. We now briefly review some of the most relevant results obtained by the Padova and Venezia groups on these systems, whose structural and compositional features are summarized in Table 6.1.

6.3.1 Au-Cu Nanocluster Formation

The samples were labeled according to the following rules: the first (second) element symbol in the name designates the first (second) implanted ion species; the number after each element symbol the implantation dose in 10^{16} ions/cm^2 units; the last letter specifies the annealing atmosphere, oxidizing or reducing (A = air, H = $H_2(4\%) - N_2$). For example Au3Cu3H

TABLE 6.1 Summary of the TEM and OD Results on Au_xCu_{1-x} Alloy Systems

Sample	Annealing Conditions	Size^a (nm)	Structure (SAED)	x alloy	lattice const. (nm)	SPR (nm)
Au3Cu3	—	3.8 ± 1.6	Au_xCu_{1-x} (fcc)	0.67(3)	a = 0.3958(15)	550
Au3Cu3H	H_2, 900°C, 1h	8.7±2.5	Au_xCu_{1-x} (tetrag.)	0.50(3)	a = 0.3960(12), c = 0.3670(12)	550
Au3Cu3A	air, 900°C, 1h	33 ± 20	Au_xCu_{1-x} (fcc) + CuO (tenorite)	0.97(3)	a = 0.4060(12)	532
Au3Cu3HA	H_2, 900°C, 1h + air, 900°C, 15'	28 ± 17	Au_xCu_{1-x} (fcc) + Cu_2O (cuprite)	0.93(3)	a = 0.4051(12)	535 (620)

Note: H_2 refers to $H_2(4\%)N_2$ atmosphere.
^{a}average diameter ± standard deviation of the TEM experimental size distribution.

indicates the sequential implantation of 3×10^{16} Au$^+$/cm^2 + 3 $\times$ 10^{16} Cu$^+$/cm^2, followed by an annealing in H$_2$(4%) – N$_2$ atmosphere.

Figure 6.2 shows the optical absorption spectra of the sample Au3Cu3 before (as-implanted) and after annealing in air or in H$_2$– N$_2$ at 900°C for 1 hour. In comparison with the as-implanted case (which shows a faint absorption band at about 560 nm), the Au3Cu3A sample exhibits a well defined absorption peak at 530 nm that is consistent with the SPR of gold metal clusters with size in the nanometer range [12]. The optical absorption of the sample Au3Cu3H annealed in H$_2$ – N$_2$ is similar to the as-implanted one, even if the absorption band at 560 nm is now sharper and more intense, suggesting that the average cluster size is increased. It is worth noting that the SPR peak position of Au3Cu3 and Au3Cu3H is located between that of pure Au and Cu nanoclusters (which are at about 530 nm and 570 nm[54], respectively). Therefore, the absorption band observed in Figure 6.2 on the Au3Cu3 and Au3Cu3H samples suggests that a possible intermetallic Au-Cu alloy could have been formed instead of two separate metallic systems, which on the contrary would give rise to double-peaked spectra.

Figure 6.3 shows the bright-field TEM (BFTEM) planar views of: Au3Cu3 (Figure 6.3(a)), Au3Cu3H (Figure 6.3(b)), and Au3Cu3A (Figure 6.3(c)) samples. In the as-implanted sample, rather densely packed spherical clusters are visible with an average diameter $\overline{D} = (3.8 \pm 1.6)$ nm. SAED analysis shows Debye-Scherrer rings, which are typical of crystalline clusters with mutual random orientation. Even if EDS microanalysis shows the presence of both gold and copper in the examined area, the SAED pattern can be indexed according to a single face-centered cubic (fcc) phase with lattice constant $a = (0.3958 \pm 0.0015)$ nm. This value is not consistent with the experimental values of the pure bulk phases of either gold ($a_{Au} = 0.40786$ nm) or copper ($a_{Cu} = 0.36150$ nm), thus showing that an $Au_x Cu_{1-x}$ intermetallic alloy has been formed. It is known [55] that the lattice parameters of the continuous Au-Cu solid solution phase exhibits a positive deviation from an assumed Vegard's law, as in Equation (6.1).

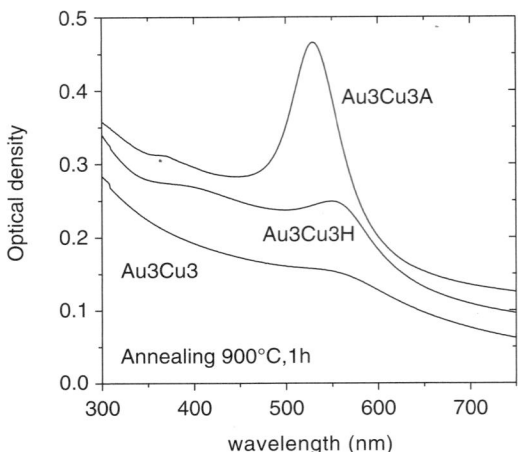

Figure 6.2 Comparison of the optical density spectra of the Au+Cu implanted samples: as-implanted (Au3Cu3), annealed in $H_2(4\%)$-N_2 at 900°C for 1 h (Au3Cu3H), and annealed in air at 900°C for 1 h (Au3Cu3A).

$$a_{alloy} = x \cdot a_{Au} + (1-x) \cdot a_{Cu} + 0.01198 \cdot x \cdot (1-x) \qquad (6.1)$$

Inserting our measured lattice parameter in Eq.(6.1), a value of $x = 0.68$ was obtained, which corresponds to a ratio $\alpha = [Au]/[Cu] = 2.1$. It is worth noting that EDS compositional analysis obtained at the *AuM* and *CuK* peaks in the spectra gives an average value $\alpha_{EDS} = 1.5$, which is slightly lower than the SAED results. This could imply that part of the copper atoms do not form the alloy but remain, after implantation, in a different (atomic or oxidized) state in the matrix.

Figure 6.3(b) shows the BFTEM micrograph of the sample annealed in reducing atmosphere: the clusters are spherical and their average diameter ($\overline{D} = 8.7 \pm 2.5$ nm) is increased with respect to the as-implanted sample. In this case, the SAED analysis exhibits the formation of an ordered Au-Cu phase, which is consistent with the tetra-auricupride phase (tetragonal, $a = 0.3960$ nm, $c = 0.3670$ nm) with equal Au-Cu atomic concentration. Sample Au3Cu3A shows near the surface large and irregularly shaped crystallites (not shown),

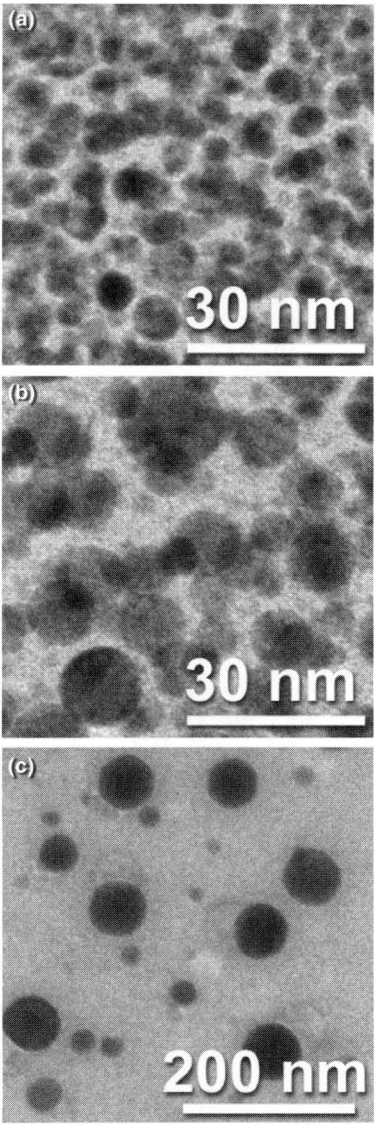

Figure 6.3 Bright-field TEM planar views of the Au+Cu implanted samples: (a) as-implanted (sample Au3Cu3), (b) annealed in $H_2(4\%) - N_2$ (sample Au3Cu3H), and (c) annealed in air (sample Au3Cu3A).

whose dimensions are even in the µm range, which, in EDS analysis, gives a strong Cu signal with almost no Au present. SAED patterns show that these structures are composed of crystalline domains of CuO in the tenorite phase, with very small amount of crystalline Cu_2O. This indicates that during annealing in air, Cu migrates toward the surface, where it is oxidized. Such a picture has been confirmed by RBS measurements, which evidenced in the Au3Cu3A sample a significant rearrangement of copper ions, which pile up at the surface, whereas gold ions diffuse deeper into the samples. A loss of about 25% of the copper has been detected for this sample, while no loss occurs for the Au3Cu3H one. Therefore, in comparison with Au3Cu3 and Au3Cu3H samples, which exhibit almost coincident gaussian in-depth distribution of the implanted species, Au3Cu3A samples show a splitting of the two distributions with inward and outward migration of gold and copper, respectively. Moreover, XPS measurements clearly indicate the presence of oxidized copper (CuO) at the surface. Below these copper oxide crystallites, the BFTEM micrograph of Figure 6.3(c) shows almost spherical clusters with mean diameter $\overline{D}$ = 33 ± 15 nm. The SAED pattern in this region shows the presence of Debye-Scherrer rings of an fcc phase, together with some spots coming from the copper oxide at the surface. From the SAED results, an estimation of the lattice constant a = 0.4067 ±0.0015 nm for the fcc phase was obtained, which is consistent with (even if slightly shorter than) that of the gold bulk lattice parameter. It has to be pointed out that, along with these strong rings, some weaker spots can be detected in the SAED pattern that correspond to an fcc phase with a smaller a, indicating that very few alloy clusters are still present even after annealing, as confirmed by EDS measurements on single clusters.

The picture that emerges from TEM and optical results (summarized in Table 6.1) is that sequential ion implantation is able to give directly alloyed Au-Cu colloids. Subsequent annealing in reducing atmosphere can be used to increase the average size of the clusters with a change in the structural ordering of the alloy. On the contrary, if the as-implanted sample is annealed in an oxidizing environment, a separation

of the two species takes place, assisted by a chemical interaction between copper and oxygen, as will be described in
more detail in the next paragraph. It is interesting to note
that this can be done in a reversible way: preliminary results
on Au3Cu6 system annealed first in air and then in hydrogen
again show the formation of Au-Cu alloy clusters. A possible
mechanism for the alloy formation is an enhanced diffusion
of copper in small gold clusters (as reported in Refs.[52, 56])
during the implantation. The subsequent annealing in reducing atmosphere at 900°C drives copper atoms that were dispersed in the matrix toward the already formed clusters,
increasing their size and shifting the Au/Cu concentration
toward the nominal 1:1 ratio, and therefore inducing a structural disorder-order phase change. The question is whether
this transition follows a solid–solid path or is assisted by an
intermediate solid–liquid transition of the alloyed clusters.
Indeed, from the phase diagrams of the Au-Cu system, the
melting point of the 1:1 Au:Cu alloy is just at 900°C: moreover,
due to the nanometric size of the clusters, a depression of this
melting temperature could take place according to the thermodynamic size effect [57]. The fact that annealing in air at
the same temperature is able to separate the two species may
indicate that the latter mechanism could play a significant
role.

6.3.2 Au-Ag Nanocluster Formation

The samples were labeled according to the same rules introduced in section 6.3.1.

Silver nanoclusters glasses are among the first nanocomposites realized by ion implantation [58–60]. Nanoclusters of
silver-gold alloy in silica matrix are, in many respects, a good
system for investigating the physical processes involved in
alloy formation. In fact, because of the complete mixing of the
two components and their reactivity with other atomic species, the alloy formation and clustering process is expected to
depend mainly on the diffusivity of the two metals in silica:
this should have in principle a relevant role especially during
a post-implantation thermal annealing. The drawback of this

system regards the difficulty of its characterization: the absence of a mismatch in the lattice parameter of the two single metal makes diffraction techniques unsuitable for detecting alloy formation, especially for nanoclusters, whose reduced size results in a broadening of the diffraction peak. On the other hand, by means of EXAFS spectroscopy, performed for the absorption edge of both Au and Ag atoms, alloy formation can be unambiguously detected, since different atom species present in the local environment around the implanted ones can be distinguished.

The investigated systems were prepared by sequential Au+Ag ion implantation in silica slides: the implantation energy was 190 keV for Au, and 130 keV for Ag ions. The implantation dose was $3 \times 10^{16} Au^+/cm^2$ and $3 - 6 \times 10^{16} Ag^+$ /cm². In Figure 6.4 the optical absorption spectrum of the Au+Ag implanted sample (Au3Ag6) is shown; it exhibits an absorption band that peaks at 490 nm. Upon annealing at

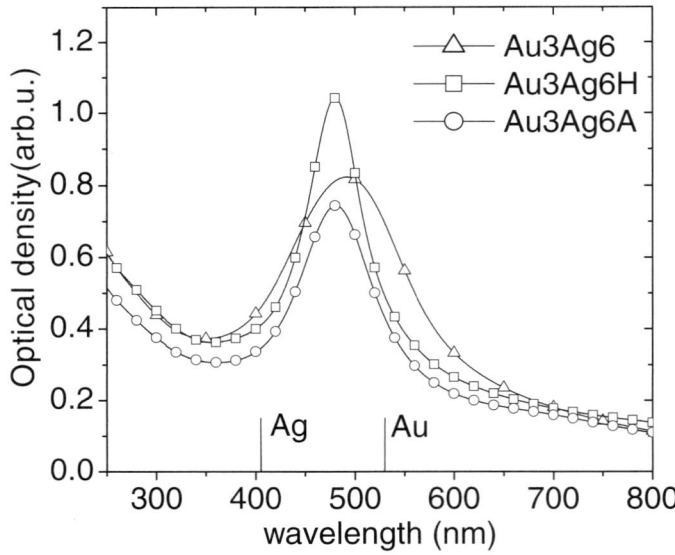

Figure 6.4 Optical density of Au+Ag-implanted silica before (Au3Ag6 sample) and after thermal annealing in reducing (Au3Ag6H sample) or oxidizing (Au3Ag6A) atmosphere.

800°C for 1 hour, this band shifted to a lower peak value (480 nm), irrespective of whether the annealing was performed in a reducing (sample Au3Ag6H) or oxidizing (sample Au3Ag6A) atmosphere. The peak height is higher in the air-annealed sample than in the $H_2(4\%)$–N_2-annealed one. In all cases, for the double-implanted samples, the absorption peak is located between those of the single Au or Ag SPR bands for nanometer-size clusters (530 nm for Au and 410 nm for Ag, respectively), indicating that an intermetallic gold-silver alloy was possibly formed.

In Figure 6.5(a) the 2D XRD pattern recorded in grazing-incidence geometry for the sample Au3Ag3 is reported. The

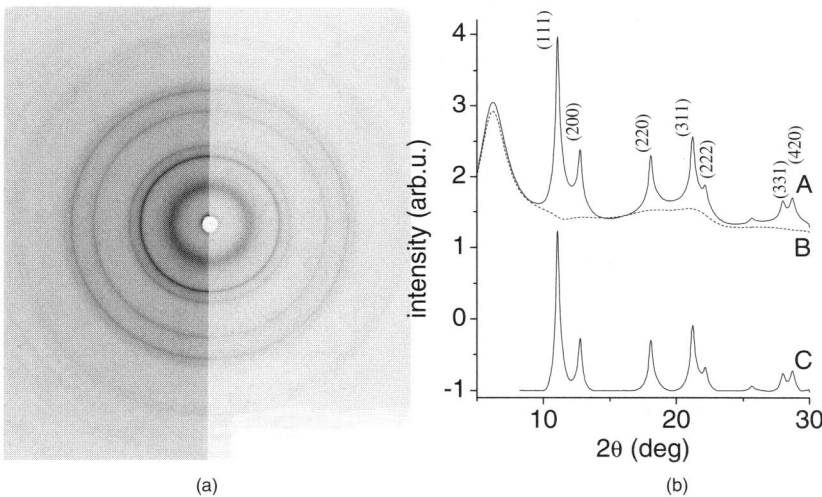

(a)   (b)

Figure 6.5 (a) GIXRD diffraction pattern from silica containing Au-Ag alloy nanoclusters (sample Au3Ag3), recorded with an imaging plate (x-ray energy = 27 keV): the diffraction rings from Au-Ag alloy nanocrystals are visible. (b) The signal obtained from the diffraction pattern, radially integrated, is shown as a function of the scattering angle 2θ (curve A): the first broad peak at $2\theta \sim 7$ deg is the scattering contribution of the silica matrix. For comparison, curve B represents the radially integrated scattering profile from the backside of the sample. After baseline subtraction the scattering contribution of the alloy nanocrystals is obtained (curve C); the reflections from fcc nanocrystals are labeled.

Debye-Scherrer rings from randomly oriented crystalline nanostructures are clearly visible. After radial integration the scattering signal is reported in Figure 6.5(b). The reflections are labeled according to the fcc phase, and the lattice parameter is in agreement with those of gold and silver bulk phases. Figure 6.6 shows the Fourier transform moduli of EXAFS spectra for the Au L_{III} and Ag K absorption edges of Au+Ag implanted samples. The peak at about 2.5 Å (double in the case of sample Au3Ag3) indicates an intermetallic first-shell correlation; no evidence of metal–oxygen correlation is found. Details on the fitting procedure are found in Refs. [61, 62], while fit quality is shown in Figure 6.6. It was found that, for

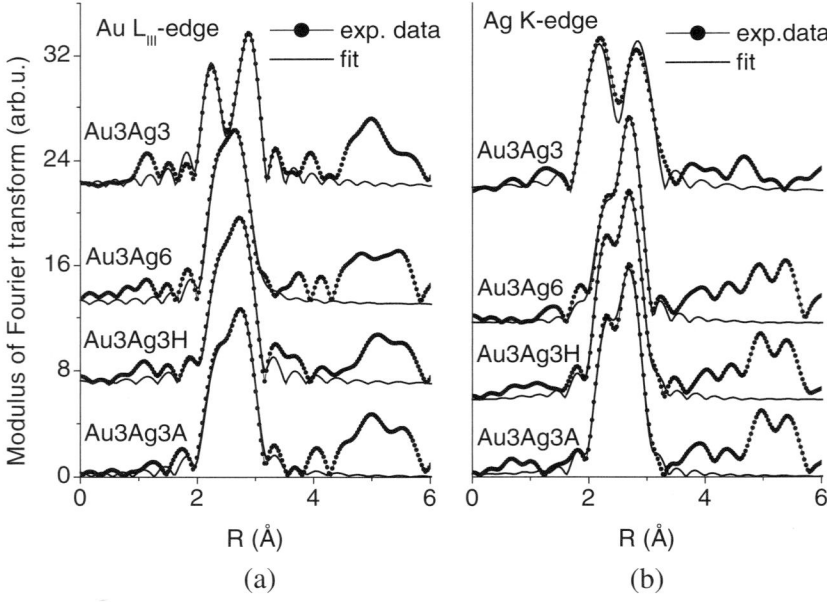

Figure 6.6 k^3-weighted Fourier transform moduli and relative first shell fits of EXAFS spectra of the Au+Ag-doped samples measured at (a) Au L_{III}-edge (in the range k=3.9-12.8 Å^{-1}), and (b) Ag K-edge (in the range k=2.7-12.6 Å^{-1}, except for Au+Ag sample whose spectrum was truncated at k = 9.8 Å^{-1}). All the spectra were measured cooling the sample at 77 K.

both the as-implanted samples, on the average, either silver and gold atoms are surrounded in the first shell by silver and gold atoms. This indicates that gold-silver binary alloy nanoclusters are formed after sequential Au+Ag implantation, independently of the dose of Ag ions. Moreover, with respect to the situation before silver implantation, the total number of atoms in the first shell increases slightly, suggesting a growth of gold-containing NCs, probably supported by silver atoms that fed preexisting gold clusters. For the local environment around silver atoms, as already noticed, two coordinations (Ag-Ag and Ag-Au) were found. The corresponding interatomic distances are compatible with those of the metallic silver and gold bulk phases in the case of the Au3Ag6 sample (high Ag dose), whereas they shrank (especially the Ag-Ag one) for Au3Ag3 sample (equal doses of Au and Ag implants). This suggests that in the Au3Ag3 sample the cluster size is small (around 1–2 nm), and smaller than in the Au3Ag6 sample. It is known that the crystallization of the alloy occurs, at least in the bulk phase, as a solid solution [63]. Thus, from the relative percentage of the two kinds of atoms in the first shell, the NC alloy composition can be estimated. This can be done starting from some hypotheses. The first is that the alloy takes place in the cluster as a solid solution, at least for the first shell ordering. The second is that alloy clusters are the dominant phase, i.e., a small percentage of single-metal clusters are present. Based on these assumptions, for the Au3Ag6 sample (high Ag dose), the ratio of silver/gold atoms in the first shell (and so in the alloy) yields approximately [Ag]/[Au] ~ 2 (2.0 when calculated from the first shell around silver, 1.7 from the first shell around gold): this value is consistent with the implantation doses. For the sample with equal Au and Ag doses (sample Au3Ag3), the same considerations lead to a silver/gold ratio [Ag]/[Au] ~ 0.5.

So, on the hypothesis that most clusters consist of Au-Ag alloy, it is clear that different silver implantation doses determine different alloy compositions. In particular, for equal Au and Ag implantation doses, the alloy is rich in gold, probably because a part of the silver atoms feeds the preexisting gold clusters, thus leading to Au-Ag clusters, and a part of the silver atoms remain dispersed in the alloy. In fact, from EXAFS results, the total number of atoms in the first shell around silver atoms is slightly lower than the total number of atoms around gold atoms. This indicates that the percentage of silver atoms dispersed in the matrix is higher than that of gold atoms, so that the alloy clusters are rich in gold, even if the implant occurred with equal doses for the two components. Concerning the Au3Ag6 sample (high Ag dose), the alloy composition is consistent with the implantation doses, indicating that in this case the same relative fraction of Au and Ag atoms remains dispersed in the host.

Concluding, double implantation of Au+Ag in silica determines the formation of silver-gold fcc NCs; the mean composition of the alloy clusters depends on the relative implantation doses of the two dopants. The alloy is poor in silver for equal Au and Ag implantation doses, and is in agreement with the total doses of dopants present in the matrix (about 2Ag:1Au), for an Ag dose double that of the Au one. Part of the silver and gold atoms remained dispersed in the matrix. No evidence of an oxide phase was found.

Considering the effect of the thermal treatments performed for 1 h at T=800°C on the Au+Ag implanted system, RBS analysis did not detect significant changes in the dopant concentrations. This is found no matter what kind of atmosphere (reducing or oxidizing) is used for the thermal treatments. The TEM analysis on the Au3Ag6A sample annealed at 800°C in air evidenced clusters with size 14.3 ± 7.9 nm, as shown in Figure 6.7. The SAED analysis found randomly oriented fcc NCs, whose lattice parameter is in agreement with those of bulk Au and Ag. The compositional analysis, performed with a field-emission gun (FG) scanning TEM (STEM) microscope by focusing a nanometric electron probe on single clusters [53], indicated the presence of both Au and Ag atoms in the same NC, thereby confirming the alloy formation. It is worth stressing that the compositional analysis on a single cluster is mandatory in order to avoid possible misinterpretations about the alloy formation, due to the presence of EDS signals from stacked clusters along the beam path. EXAFS analysis was performed for both Au and Ag

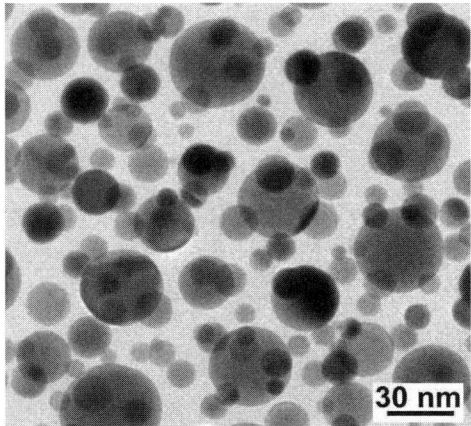

Figure 6.7 Planar bright-field TEM micrograph of the Au3Ag6A sample.

absorption edges on two samples with equal Au and Ag concentrations (corresponding to an effective implantation dose of 3×10^{16} ions/cm^2 for each species), annealed in either reducing or oxidizing atmosphere. Results of the fitting procedure in both cases revealed the presence of Au-Ag alloy NCs. The main difference with respect to the as-implanted case is in the alloy composition. In the case of annealed samples the alloy composition is balanced (about 1:1). This is consistent with the local relative concentration of the two dopants before the annealing treatment. So the effect of annealing, starting from gold-rich alloy clusters, is mainly to promote the aggregation of silver atoms in the alloy NC, that in the as-implanted sample were dispersed in the matrix, thus shifting the alloy composition toward the retained doses, as was seen in the Au-Cu system.

Silver out-diffusion upon thermal annealing in air was clearly observed in silica samples containing silver NCs, prepared by sol-gel method [64]; moreover, it has been shown that, upon annealing in air (at 900°C) silica containing $Pd_{0.5}Ag_{0.5}$ NCs (Pd-Ag bulk alloy is thermodynamically possible in the whole composition range, as in the case of Au-Ag), the alloy composition became $Pd_{0.7}Ag_{0.3}$, because part of the silver left the sample [46]. At the annealing temperature used here (T = 800°C), the bulk phase is still solid (the melting temperature for $Au_{0.5}Ag_{0.5}$ is about 1030°C). On the other hand, it is known that a decrease of the melting temperature (as large as 15% for 5 nm gold particles) could take place in small clusters because of the presence of cluster surface tension that comes into play in the expression for the chemical potential [57]; although this should be experimentally checked, the heating temperature is probably low enough to preserve NCs in the solid solution; in any case, the silver out-diffusion did not take place in this instance.

Thus, the effect of thermal treatment at 800°C for 1h of Au+Ag implanted silica is to promote further aggregation of dopant atoms in clusters that are mainly formed by silver-gold alloy; their composition, starting from Au-rich alloy (in the Au3Ag3 sample before annealing) respects the implantation doses. This indicates that the effect of annealing is to promote silver diffusion into the clusters. The selective influence on the cluster stability of the annealing atmosphere, observed for the Au/Cu system (see next section), is less effective for the Au/Ag one, considering the lower chemical interaction between gold or silver and oxygen.

6.4 THERMAL AND ION-INDUCED DE-ALLOYING

6.4.1 Au-Cu

In Figure 6.8(a) the bright-field TEM image of the reference Au3Cu3H sample is reported. The size distribution of the clusters exhibits an average diameter $\overline{D}$ = 13.8 nm and a standard deviation of the experimental bimodal distribution σ = 11.7 nm. After thermal annealing in air at 900°C for 15 min (Figure 6.8(b)) Cu is extracted from the alloy as Cu_2O (cuprite, cubic, a = 0.4267 nm). The oxide and the alloy cluster form a double interconnected structure in which the Cu_2O is stabilized by the crystallographic cubic template of the Au_xCu_{1-x} alloy (zone B and A in Figure 6.8(b), respectively).

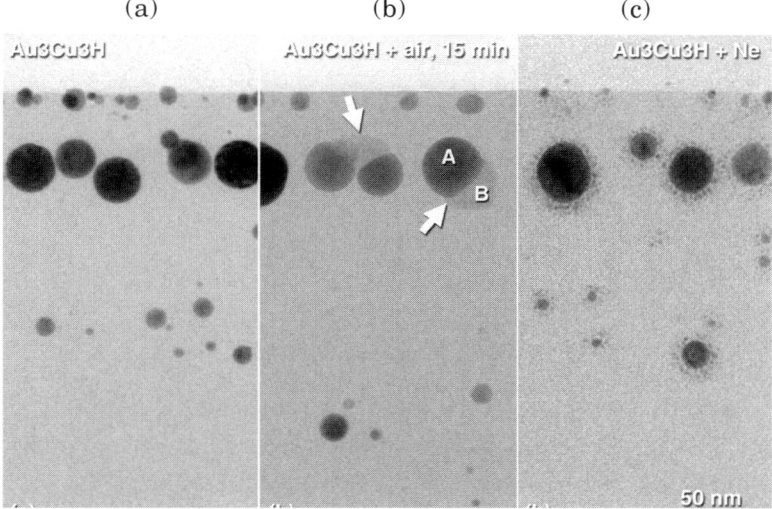

(a)   (b)   (c)

Figure 6.8 Bright-field TEM cross-sectional micrographs of the sample Au3Cu3H (annealing in $H_2(4\%)$–N_2 atmosphere at 900°C for 1h) before (a) and (b) after a thermal annealing in air at 900°C for 15 min, and (c) after irradiation at room temperature with 190 keV Ne ions, at a dose of 1×10^{17} ions/cm^2.

It is interesting to note that if the annealing time interval is increased up to 5 h, an asymptotic configuration is reached in which all the Cu atoms are extracted from the alloy and are driven to the sample surface where they are oxidized in large polycrystalline clusters of CuO [65] (tenorite, monoclinic, as previously discussed).

This can be seen in the optical spectra of Figure 6.9 after different annealing intervals. Upon 5 h air annealing, only a single SPR band of pure Au clusters is detected, while the band at 620 nm is gradually depressed, indicating a complete decoupling of the Au-Cu systems.

The effect of irradiating the Au3Cu3 sample with Ne ions at 190 keV is shown in Figure 6.8(c): around each original cluster a set of satellite clusters of about 1 to 2 nm are present with an average distance of about 3 nm from

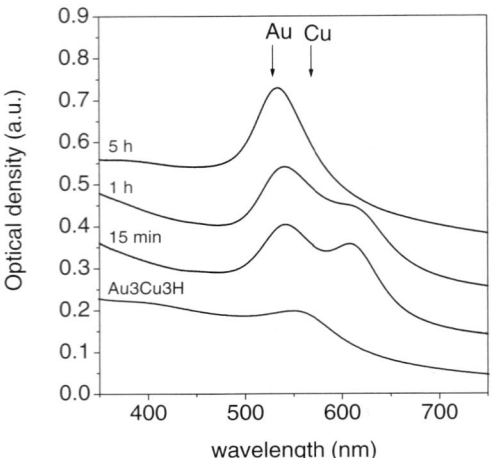

Figure 6.9 Optical density of the sample Au3Cu3H (annealing in $H_2(4\%)$–N_2 atmosphere at 900°C for 1h) before and after thermal annealings in air at 900°C for 15 min, 1h, 3h and 5h. The spectra were vertically shifted for clarity.

the cluster surface, similar to what was reported in an ion-beam mixing experiment of Au islands irradiated by Au MeV-ions [66, 67]. RBS analysis indicated only a small broadening of the Au and Cu concentration profiles, suggesting that only a local rearrangement of the metallic atoms took place, in agreement with the cluster topology evidenced by TEM. The EDS compositional analysis with a focused 2 nm electron beam of the FEG-TEM on the original cluster and on the satellites gives an Au/Cu atomic ratio (measured at AuL and CuK) of 1.3 ± 0.1 and 3.2 ± 0.3, respectively, indicating a preferential extraction of Au from the original clusters during ion irradiation. It is interesting to note the effect of the different treatments on the optical spectra in the UV-Vis range, reported in Figure 6.10. The reference sample exhibits an SPR band centered at 550 nm, which is between those of pure Au clusters (SPR at 530 nm) and of pure Cu clusters (SPR at 570 nm) in agreement with the presence of an Au_xCu_{1-x} alloy [17]. Upon thermal annealing in air for 15

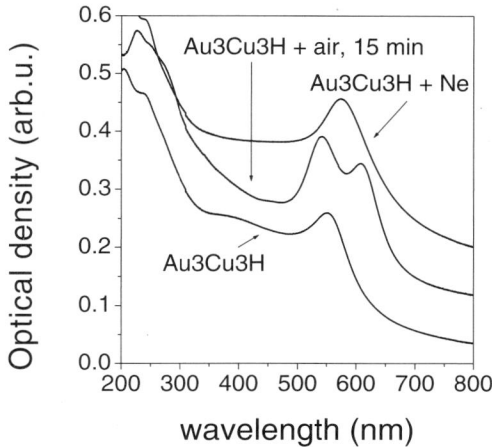

Figure 6.10 Optical density of the sample Au3Cu3H (annealing in $H_2(4\%)$–N_2 atmosphere at 900°C for 1h) before (a), and (b) after a thermal annealing in air at 900°C for 15 min, and (c) after irradiation at room temperature with 190 keV Ne ions, at a dose of 1 $\times 10^{17}$ ions/cm^2. The spectra were vertically shifted for clarity.

min, two absorption bands develop, centered at 540 nm and 610 nm, respectively. This spectrum can be simulated within the framework of Mie theory with the presence of Au-rich alloy clusters (partially) surrounded by copper oxide, in agreement with TEM findings. More complicated is the understanding of the spectrum of the sample irradiated with Ne, which exhibits a single band at 575 nm. In the framework of Mie theory, this band could be ascribed to pure metallic Cu clusters in SiO_2, therefore implying a strong compositional change in the original alloy clusters (practically a complete depletion of Au), which is not the present case, according to the TEM and EDS analyses. A possible explanation is that the satellites around each cluster modify the surrounding effective dielectric function with a global increase of the refractive index just near the surface of the alloy clusters, therefore red-shifting their SPR band. A slight asymmetry in the satellite cluster density (more pronounced in the lower part of the clusters, i.e., in the opposite side with respect to the irradiating beam) supports the hypothesis of a relevant contribution of the nuclear components of the energy loss in the creation of an elemental-selective vacancy formation in the NCs, which could be responsible for the preferential Au out-coming from the original alloy, further influenced by damage cascades or radiation-enhanced diffusion processes [68]. Additional effects of preferential (Au) elemental segregation [69] (mostly in medium-large-size clusters) can give further driving force to the de-alloying mechanism via Au extraction. The satellite clusters are then formed by precipitation of Au atoms when the local concentration overcomes the solubility limit [70].

6.4.2 Au-Ag

The temperature and time dependence of the alloy structure after air annealing is also interesting. In Figure 6.11(a) we report the evolution of the absorption spectra with the anneal-

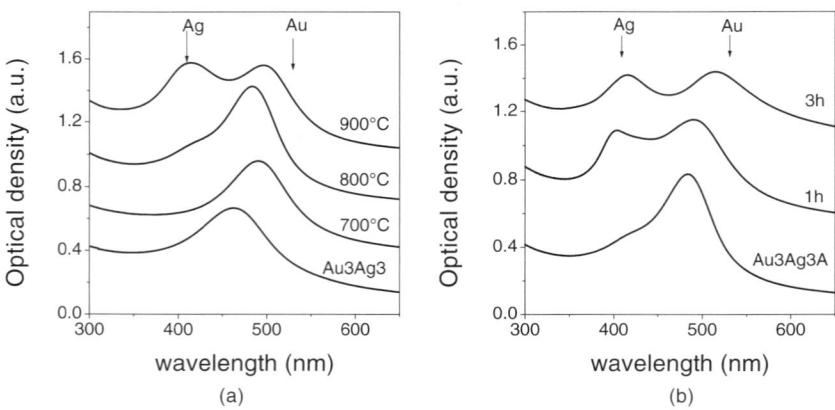

Figure 6.11 (a) Temperature evolution of the optical density in the Au3Ag3A sample annealed for 1 h in air. (b) Time evolution of the optical density in the sample Au3Ag3A (annealed in air at 800°C for 1 h) annealed at 900°C in air. Vertical arrows indicate the SPR position for pure gold or silver nanoclusters in SiO_2. The spectra were vertically shifted to enhance their visibility.

ing temperature recorded for the sample Au3Ag3. A progressive shift (up to 700°C) is evident, with a splitting (above 800°C) of the SPR absorption band that for the as-implanted sample is located at 480 nm. At 900°C the spectrum exhibits two well-resolved bands centered at 400 nm and at 510 nm: the first is consistent with the SPR of spherical Ag clusters in SiO_2, whereas the latter may indicate either Au-rich alloy clusters, i.e., incomplete alloy separation, or the formation of a core-shell structure. A set of isothermal annealings in air of the Au3Ag3A (annealed in air at 800°C for 1h) samples gave rise to the results reported in Figure 6.11(b): the absorption band centered at 490 nm after 3 h annealing shifted to 510 nm, whereas the shoulder near 410 nm becomes a well-defined band. These results are interpreted with a progressive depletion of Ag from the Au-Ag alloy with consequent formation of separated Ag and Au-rich clusters.

Cluster size and composition can be tuned also by laser annealing. The Au+Ag implanted silica sample annealed in reducing atmosphere Au3Ag3H was irradiated with a pulsed excimer-laser, with 20 ns duration and 248 nm wavelength, at different energy densities. Here we report only some representative results. TEM analysis indicates the presence of clusters with a size of 2–3 nm in the near-surface region (depth <40 nm) and about 20 nm at a depth of 40–80 nm. Small clusters are present also down to 400–500 nm from the surface. In Figure 6.12 the absorption spectra after laser irradiation are reported. The SPR value corresponds to the metal alloy formation. The cluster redistribution with respect to the unirradiated samples and the presence of very large clusters (with diameter of 40–70 nm) after laser irradiation indicate a transition to liquid phase. The heat transport equation has been numerically solved, taking into account the Stefan boundary condition related to the solid–liquid phase transition [65]. To attain a liquid phase, an absorption coefficient larger than 5×10^3 cm^{-1} must enter in the heat source term. This absorption is ascribed to the interband absorption of the clusters present before the laser irradiation. The relevant results of the numerical solution of the heat transport equation for this system [65] indicate that the maximum

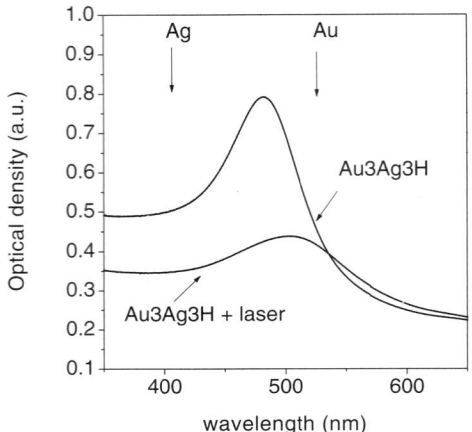

Figure 6.12 Comparison between the OD spectra of the Au3Ag3H sample before and after laser annealing at 1.3 J/cm^2. Vertical arrows indicate the SPR position for pure gold or silver nanoclusters in SiO$_2$. The spectra were vertically shifted to enhance their visibility.

depth involved in the solid–liquid phase transition is about 1 μm, the resolidification velocity is of the order of 2 m/s, and the duration of the molten phase is several hundred ns.

To investigate to what extent the nuclear component of the energy loss can affect this element-selective de-alloying, we irradiated with He and Kr and varied the composition of the irradiated alloy NCs [71], i.e., Au$_\chi$Ag$_{1-\chi}$. The TEM bright-field images of the reference sample Au3Ag3A and of the irradiated one with Kr at 380 keV at a dose of 1.2×10^{16} ions/cm^2 are shown in Figure 6.13. Similarly to the Ne-irradiated Au3Cu3H sample, Kr-irradiation promotes the formation of large (3 to 5 nm in size) satellites around the original clusters (Figure 6.13(b)).

The EDS analysis indicated that in this case there is a preferential extraction of Au atoms from the clusters, the composition of the satellites being characterized by an Au/Ag atomic ratio of 3.7 ± 0.3, to be compared with the corresponding value on the mother cluster of 1.5 ± 0.2. The samples irradiated with He ions (which have a much lower

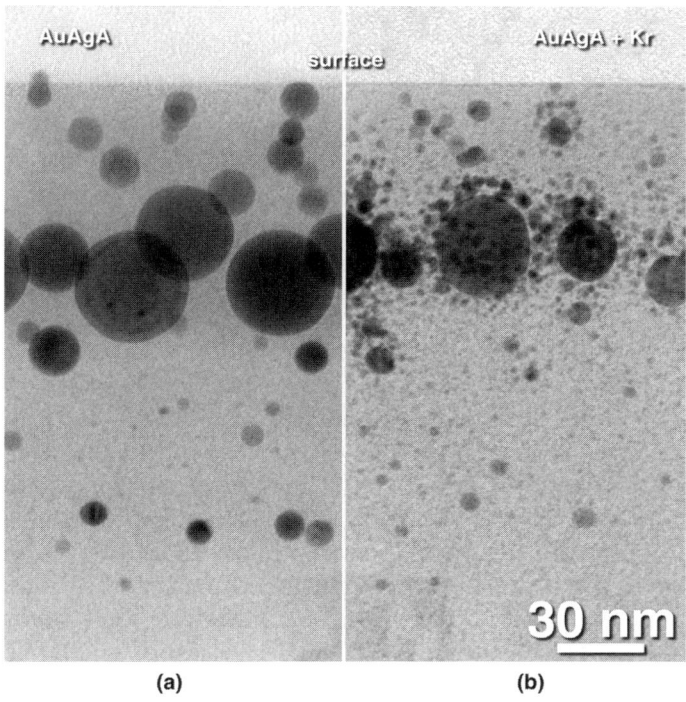

(a)   (b)

Figure 6.13   Bright-field TEM cross-sectional micrographs of the sample Au3Ag3A (annealing in air atmosphere at 800°C for 1h) before (a) and after irradiation at room temperature with 380 keV Kr ions, at a dose of 1.2×10^{16} ions/cm² (b).

nuclear component of the stopping power) exhibit the very early stage of the formation of Au-rich satellites. Due to the size of the satellites, the optical response of the system is even more complicated. Figure 6.14 shows the optical density of the reference Au3Ag3A sample, in comparison with that of the samples irradiated with He (25 keV, 3.2×10^{17} ions/cm²) or Kr. The red-shift already observed in the Au3Cu3H irradiated samples (see Figure 6.10) is also present upon He irradiation and is even more pronounced in the Kr-irradiated one, supporting the hypothesis of a relevant contribution of the nuclear component in the satellite formation. Also in this case, a simple Mie theory is not suitable to account for the band near 540 nm, which can be modeled by varying the local electromagnetic environment around the clusters [12].

6.5 NONLINEAR OPTICAL PROPERTIES OF Au-BASED ALLOYS

Very interesting fast nonlinear optical properties are shown by the Au3Ag6A and Au3Cu6H samples. By means of the Z-scan technique, nonlinear refractive indices n_2 of $(-1.6 \pm 0.3) \times 10^{-10}$ cm²/W and $(6.3 \pm 1.2) \times 10^{-11}$ cm²/W respectively were measured for these samples at 527 nm [72]. The measurements were realized in experimental conditions to avoid cumulative heating effects, i.e., by using a single 6-ps-long laser pulse at 1 Hz repetition rate. The peculiarities of this experimental finding are the very large modulus value of n_2 and the sign change that depends on the cluster composition. Such a large value of the fast n_2 coefficient has never been detected in MNCGs up to now [41]. An explanation of this

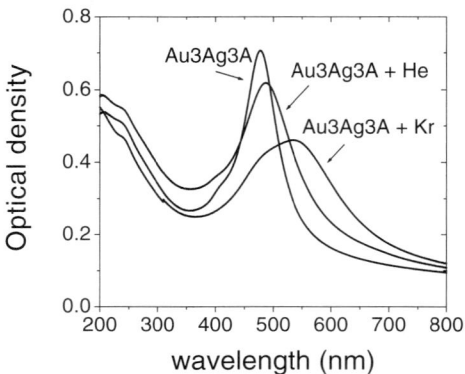

Figure 6.14 Optical density of the sample Au3Ag3A (annealing in air at 900°C for 1h) before, after irradiation at room temperature with 25 keV He ions, at a dose of 3.2×10^{17} ions/cm² and with 380 keV Kr ions, at a dose of 1.2×10^{16} ions/cm².

can be given by considering that for metal volume fractions (i.e., filling factors) p around 0.1 and near the SPR wavelength, the mutual electromagnetic interactions among nanoparticles begin to induce deviations from the low-p approximation of the third-order optical Kerr susceptibility [72]. The effect of this mutual influence has been estimated for a linear chain of gold nanoparticles embedded in silica [73]. The modulus of the local-field enhancement factor results larger than that predicted by the low-concentration approximation, and the difference is much more marked at the SPR wavelength. Mutual interaction among nanoparticles could then explain the large values of n_2 coefficient found in Au3Ag6A and Au3Cu6H MNCG samples. As far as the sign reversal of the nonlinear refractive index as a function of the cluster composition is concerned, this evidence in a MNCG is related to the relative position in wavelength of the nanoparticle surface plasmon band with respect to that of the laser used (527 nm). The local-field enhancement factor for alloy NCs embedded in silica glass has been computed within the framework of Mie theory, starting from experimental bulk dielectric functions of a metallic Au-Ag alloy and correcting them for the finite size of the particles [72]. Indeed, the negative sign of n_2 is well explained by considering the wavelength dependence of the nanoparticles local-field factor.

6.6 MAGNETIC ALLOY NANOCLUSTER FORMATION

The control of NC formation is expected to open new possibilities in the field of magnetic nanocomposite materials, since magnetic properties are strongly dependent on the cluster composition, crystalline structure, size and also on the cluster separation distance. Indeed, cluster correlation distance may play an important role in the magnetic behavior of the material, for example in the shape of its hysteresis loop and especially in determining the coercive field. Besides the cluster size distribution and the cluster distribution in the matrix, another promising parameter for tuning magnetic properties is the cluster composition. In this respect, the possibility of realizing binary alloy nanoclusters with at least one magnetic element is particularly promising. Ferromagnetic Fe-Pt nanoparticles have been produced in SiO_2 and Al_2O_3 single crystals by implantation. Implantation at elevated temperatures (500°C) followed by annealing gives rise to small nanoparticles (up to 25 nm) that exhibit multiple orientations, whereas implantation at 200°C for the same dose gives rise to a buried layer of very large particles oriented with the (111) direction parallel to the Al_2O_3 c-axis [44]. In this section we present the structural results obtained in the synthesis of Co-Ni, Cu-Ni and Cu-Co NCs by sequential ion implantation. For the magnetic properties of these systems the reader is addressed to the cited references.

6.6.1 Co-Ni Nanocluster Formation

To study Co-Ni alloy nanocluster formation, different samples were prepared with different total implanted amount of dopants (from 15 to 40×10^{16} ions/cm^2) [43] and different Co/Ni ratios; the implantation energy was 180 keV for both ion species. Figure 6.15 presents the cross-sectional BFTEM micrographs for the Co8Ni8 sample. It is found that, in general, both the average particle size and particle size distribution of the Co+Ni implanted samples increase with the total implanted dose; for example, the particle sizes for the Co12Ni3, Co24Ni6, Co20Ni20 are 2.6 ± 1.2 nm, 5.2 ± 3.0 nm and 10.0 ± 6.0 nm, respectively [65]; moreover, the particle size increases from the slide surface to the depth. For samples with a total dose of 15×10^{16} ions/cm^2, the particle size and the corresponding size distribution of the pure metal samples are slightly larger than the alloy-based composites observed (the maximum is observed for Ni, 4.6 ± 2.4 nm and the minimum is for Co12Ni3, 2.6 ± 1.2 nm) [42]. The crystalline structure of the NCs formed was investigated by GIXRD [74]. The 2D diffraction patterns (not shown) indicate the presence of randomly oriented crystals. The diffraction intensity profiles for the different samples, as obtained by radial integration, are shown in Figure 6.16. Besides the intense halo due to the silica matrix at $2\theta \sim 7$ deg, the reflections from the nanocrystals are clearly visible. For the

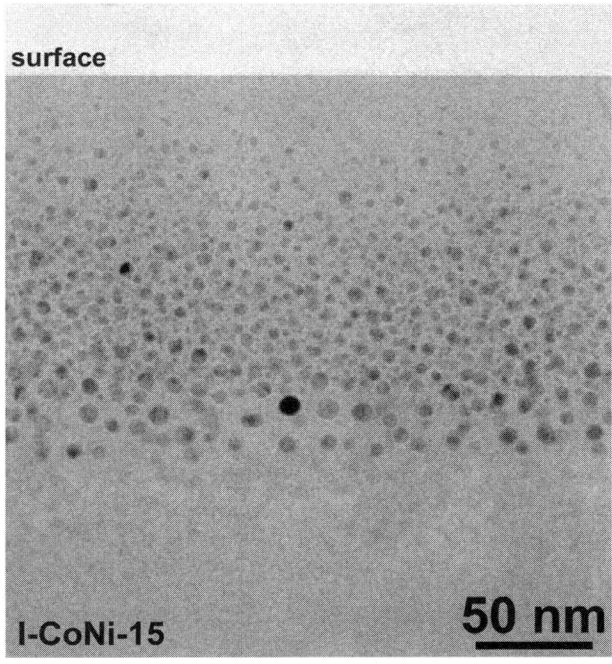

Figure 6.15 Bright-field cross sectional TEM micrograph of the sample Co8Ni8, sequentially implanted with Co+Ni ions at equal dose of 8×10^{16} ions/cm^2.

Co-implanted sample (Co15) the nanocrystal structure is hcp, and this phase is found also in Co-Ni clusters with a Ni concentration less than 30 at %. Upon increasing the Ni content a coexistence of hcp and fcc phases is found. This last phase becomes predominant in the Co8Ni8 sample. SAED analysis found that the lattice parameters are similar to those of the corresponding bulk alloys. The observed variation of the crystal structure is in good agreement with the phase diagram of the bulk alloy.

To properly investigate the alloy formation, EDS measurements were performed with a nanometric electron probe along the diameter of one particle in the Co15Ni15 sample, which is composed of nanoparticles with a larger size distri-

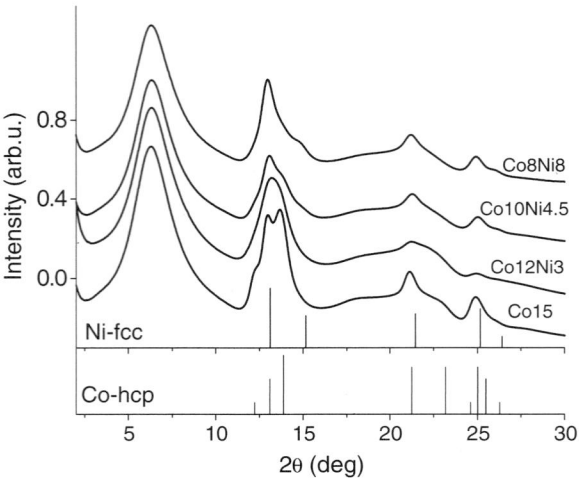

Figure 6.16 Radially integrated diffraction profiles from the Co+Ni implanted samples as a function of the scattering angle (the x-ray beam energy was 27 keV). Besides the matrix contribution (evident in the peak at $2\theta \sim 7$ deg), the reflections from the cluster crystalline structure are clearly visible; vertical bars indicate the reflection positions from an fcc Ni and from an hcp Co bulk crystals.

bution of 10 ± 6 nm and with an fcc structure. The evolution of the intensities of the Co and Ni K_α and K_β absorption peaks was evaluated and considered as representative of the Co and Ni content and is reported in Figure 6.17. This result unambiguously demonstrates the simultaneous presence of Co and Ni in the particles, thus confirming the alloy structure.

Grazing-incidence small-angle x-ray scattering (GISAXS) technique, using a synchrotron x-ray beam, provided information on the cluster size distribution and intercluster correlations in a nondestructive way [74–76]. For these magnetic systems we obtained 2D isotropic scattering patterns characterized by an intercluster correlation ring [74]. With the adopted data processing method, [74, 76] it was found that the intercluster correlation distance is generally less than three times the mean cluster radius. The effect of these cluster–cluster interactions on magnetic properties is

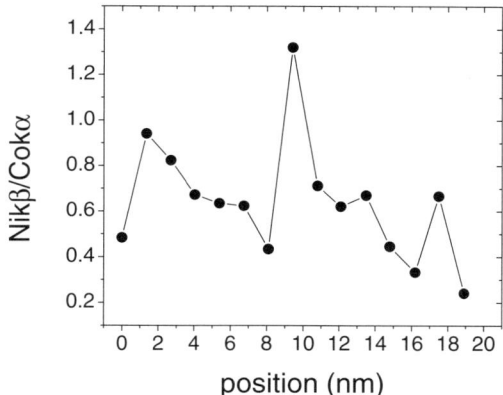

Figure 6.17 Ni/Co concentration profile ratio from the EDS peaks measured by FEG-STEM on a single cluster in the sample Co15Ni15, implanted with Co+Ni ions with equal dose of 15×10^{16} ions/cm^2.

quite complex and could give rise to collective states. In particular, while GISAXS results indicate that the intercluster correlation is isotropic, the analysis of the hysteresis loops shows that the coercive field is weaker if the magnetic field is parallel to the sample surface [74]. This suggests that in the Co+Ni implanted samples the long range interparticle interactions can be sensitive to the macroscopic planar extent of the implanted phase, determining a stronger in-plane demagnetization effect.

6.6.2 Cu-Ni Nanocluster Formation

Cu-Ni alloy nanostructure formation was investigated by implanting Cu and Ni ions into silica slides. The dose was 6 $\times 10^{16}$ ions/cm^2 for each element, while the implantation energy was 90 keV for Cu and 100 keV for Ni.

EXAFS spectroscopy was performed to investigate the local order around the implanted atoms. In Figure 6.18 are represented the k^3-weighted Fourier transform moduli of the EXAFS spectra recorded for the double-implanted sample for both Cu and Ni K-edges; the peaks at about 2.2 Å correspond

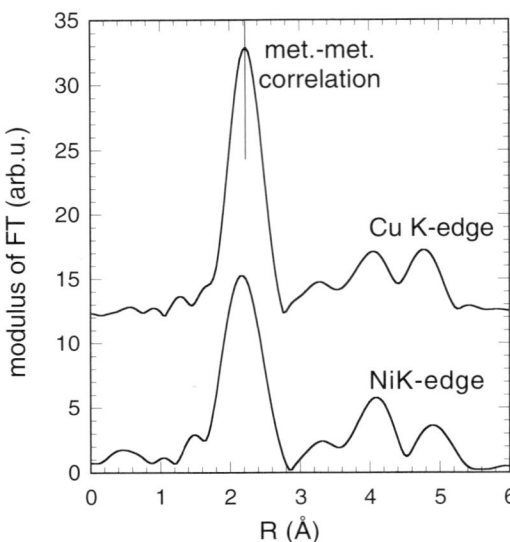

Figure 6.18 Moduli of the Fourier transforms of k^3-weighted spectra of the Cu+Ni implanted sample, measured at Cu and Ni K-edges. The transformation range is $k = 3\text{-}13$ Å^{-1}.

to the metal–metal correlation, while the signals centered at about 1.5 Å are due to the metal–oxygen coordination. This is explained by assuming the coexistence of a metallic cluster phase and an oxidized phase, consisting of copper (nickel) ions at the cluster/glass interface or dispersed in the glass matrix [38]. The calculated metal–metal nearest-neighbor distance was 2.52 Å, for both Cu and Ni K-edge measurements [38, 39]. The distance corresponds in the fcc phase to the lattice parameter of 3.56 Å, between those of the two metal phases, indicating the formation of a CuNi alloy phase (almost 1:1 Cu/Ni ratio). This finding is supported by GIXRD analysis on this sample [38], which indicates the presence of randomly oriented fcc NCs, whose measured lattice parameter is 3.565 ± 0.01 nm, confirming the alloy formation.

6.6.3 Co-Cu Alloy Nanocluster Formation

The sequential ion implantation of Cu+Co ions is particularly interesting for the possibility of obtaining Co-Cu alloy nanoclusters, despite their immiscibility in the bulk phase in the concentration range used in the present work. In fact, the Co-Cu alloy can be produced only by non-equilibrium techniques [77]. For example, Childress and Chien [78] prepared Co-Cu alloys in the whole composition range by dc-sputtering and studied the magnetic diagram phase showing spin-glass, ferromagnetic and paramagnetic phases, while Noetzel et al. [79] have obtained Co/Cu solid solution by implanting up to 25% of Co in Cu targets. Indeed, the Co-Cu bulk alloys can present different magnetic behavior depending on the concentration of the ferromagnetic species. The use of ion implantation in particular gives rise to very thin layers in which different local situations can occur due to the non-equilibrium nature of this process. Co^+ and Cu^+ ions were sequentially implanted at room temperature into fused silica, at energies of either 90 or 180 keV, with projected ranges overlapping for each pair of implants; the total dose was 1.2×10^{17} ions/cm^2 for the doubly implanted samples (Cu+Co and Co+Cu).

An EXAFS analysis was first performed to detect the relative presence of metal-metal and metal-oxygen coordination. In particular, a preliminary investigation by 4-shell fits of EXAFS spectra shows that the percentage of dopant atoms that aggregate in clusters is not lower than 60%. The results from fitting procedures of the first shell indicate that the intermetallic distances are the Co bulk ones for Co-metal correlation [45]. On the other hand, the values obtained for Cu-metal correlation suggest a shortening of the distance, possibly induced by the presence of Cu-rich alloy clusters thus having structural characteristics different from pure metal. In the case of the Cu6Co6 sample, the Co-O coordination distance (1.98 ± 0.02 Å) was shorter than the corresponding one in CoO (2.13 ± 0.01 Å, from EXAFS analysis). This behavior was also observed in similar systems obtained by sequential implantation of Cu and Ni ions [38].

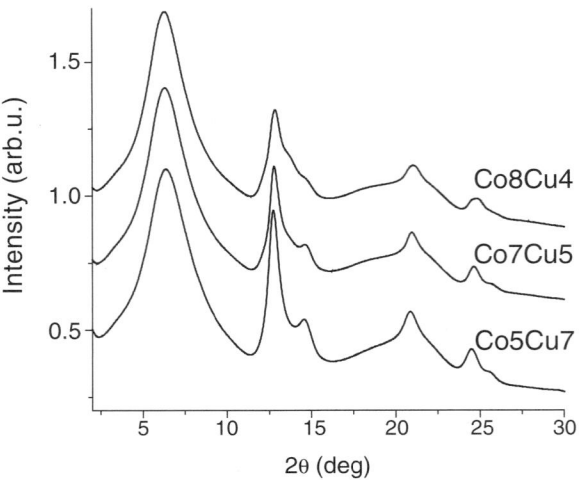

Figure 6.19 Radially integrated diffraction profiles from the Co+Cu implanted samples as a function of the scattering angle (the x-ray beam energy was 27 keV). Besides the matrix contribution (evident in the peak at 2θ ~ 7 deg), the reflections from the cluster crystalline structure are clearly visible.

Figure 6.19 reports the radially integrated profiles from GIXRD scattering patterns in the case of Co+Cu implanted samples. In addition to the first peak, due to the silica substrate [80], a signal from crystalline clusters is clearly visible. Starting from the hcp order of Co nanoclusters (Figure 6.16), increasing the relative amount of copper in the total amount of implanted metal (Co+Cu) leads to the progressively increasing peak heights associated with an fcc crystalline structure. Cobalt aggregates in hcp NCs is introduced in the matrix by ion implantation, while Cu does so in fcc structures [38]. On the other hand, following Childress and Chien [78], a Co–Cu alloy has an fcc structure at up to 80% cobalt. The features observed in Figure 6.19, supported by EXAFS results, suggest that by sequential implantation of Co+Cu two family of NCs are formed: one of Co hcp NCs and the other of Cu-rich crystalline NCs. No oxides are detected.

6.7 CONCLUSIONS

Some case studies have been reported concerning the formation of metal alloy nanoclusters in SiO_2 matrix obtained by sequential ion implantation of Au+Cu, Au+Ag, Co+Ni, Cu+Ni and Co+Cu.

The most relevant technological parameters for ion implantation (dose, energy) and for the subsequent treatments (thermal annealing, ion-irradiation with light ions or laser annealing) were investigated to control cluster composition, size and stability, by exploiting their elemental selective interaction.

Particularly intriguing is the modification induced by ion irradiation on the topology and composition of the Au-Ag and Au-Cu alloy NCs. Around each original cluster a halo of small satellites develops with a preferential outcoming of Au, indicating an elemental selective de-alloying.

As far as the technological properties of the systems investigated is concerned, very interesting results have been obtained about the nonlinear optical properties with the Z-scan technique. High values of the third-order nonlinear refractive index have been measured with an interesting change of its algebraic sign as a function of the NC composition.

As a final remark, we would like to underline the fruitful application of synchrotron-based techniques (GISAXS and EXAFS), which, coupled with more traditional techniques like TEM, allowed a complete structural and compositional characterization of NCs embedded in SiO_2.

REFERENCES

1. M. Respaud, J.M. Broto, H. Rakoto, A.R. Fert, L. Thomas, B. Barbatra, M. Verelst, E. Snoek, P. Lecante, A. Mosset, J. Osuna, T. Ould Ely, C. Amiens, and B. Chaudret, *Phys. Rev.* B 57 (1998) 2925.

2. F. Tihay, J.M. Pourroy, A.C. Roger, and A. Kienneman, *Appl. Catal.* A 206 (2000) 24.

3. M. Faraday, *Phil. Trans. R. Soc.* 147 (1857) 145.

4. C. Flytzanis, F. Hache, M.C. Klein, D. Ricard, and Ph. Roussignol, *Progr. Opt.* 29 (1991) 321.

5. F. Gonella and P. Mazzoldi, Metal Nanocluster Composite Glasses, in: *Handbook of Nanostructured Materials and Nanotechnology,* H.S. Nalwa, Ed., vol. 4, pp. 81–158, Academic Press, San Diego (2000).

6. P. Mazzoldi and G.C. Righini, Glasses for optoelectronic devices, in *Insulating Materials for Optoelectronics* (F. Agulló-López, Ed.), pp. 367–392, World Scientific, Singapore (1995).

7. R.F. Haglund, Jr. Quantum-dot composites for nonlinear optical applications, in *Handbook of Optical Properties II: Optics of Small Particles, Interfaces, and Surfaces* (R.E. Hummel and P. Wissmann, Eds.), vol. 2, p. 191, CRC Press, New York (1997).

8. J.A.A.J. Perenboom and P. Wyder, *Phys. Rep.* 78 (1981) 173.

9. W.P. Halperin, *Rev. Mod. Phys.* 58 (1986) 533.

10. F. Hache, D. Ricard, and C. Girard, *Phys. Rev.* B 38 (1988) 7990.

11. J.W. Haus, N. Kalyaniwalla, R. Inguva, M. Bloemer, and C.M. Bowden, *J. Opt. Soc. Am.* B 6 (198) 797.

12. U. Kreibig and M. Vollmer, *Optical Properties of Metal Clusters, Vol. 25*, Springer-Verlag, Berlin (1995).

13. R.H. Doremus and P. Rao, *J. Mater. Res.* 11 (1996) 2834.

14. L. Yang, D.H. Osborne, Jr., R.F. Haglund, Jr., R.H. Magruder III, C.W. White, R.A. Zuhr, and H. Hosono, *Appl. Phys.* A 62 (1996) 403.

15. C.N. Ironside, *Contemp. Phys.* 34 (1993) 1.

16. E. Cattaruzza, *Nucl. Instr. and Meth.* B 169 (2000) 141.

17. G. Mattei, *Nucl. Instr. and Meth.* B 191 (2002) 323.

18. F. Gonella, *Nucl. Instr. and Meth.* B 166-167 (2000) 831.

19. R.F. Haglund, Jr., L. Yang, R.H. Magruder III, C.W. White, R.A. Zuhr, L. Yang, R. Dorsinville, and R.R. Alfano, *Nucl. Instr. and Meth.* B 91 (1994) 493.

20. N. Skelland and P.D. Townsend, *Nucl. Instr. and Meth.* B 93 (1994) 433.

21. A.L. Stepanov and D.E. Hole, *Recent Res. Devel. Applied Phys.* 5 (2002) 1.

22. A. Meldrum, R.F. Haglund, L.A. Boatner, and C.W. White, *Adv. Mater.* 13 (2001) 1431.

23. F. Gonella, G. Mattei, P. Mazzoldi, C. Sada, G. Battaglin, and E. Cattaruzza, *Appl. Phys. Lett.* 75 (1999) 55.

24. G. Battaglin, E. Cattaruzza, F. Gonella, G. Mattei, P. Mazzoldi, C. Sada, and X. Zhang, *Nucl. Instr. and Meth.* B 166-167 (2000) 857.

25. J.P. Biersack, *Nucl. Instr. and Meth.* 182/183 (1981) 199.

26. E. Valentin, H. Bernas, C. Ricolleau, and F. Creuzet, *Phys. Rev. Lett.* 86 (2001) 99.

27. A. Perez, G. Marest, B.D. Sawicka, J.A. Sawicki, and T. Tyliszczac, *Phys. Rev.* B 28 (1983) 1227.

28. A. Perez, *Nucl. Instr. and Meth.* B 1 (1984) 621.

29. T. Futagami, Y. Aoki, O. Yoda, and S. Nagai, *Nucl. Instr. and Meth.* B 88 (1994) 261.

30. H. Hosono, *Jpn. J. Appl. Phys.* 32 (1993) 3892.

31. H. Hosono and N. Matsunami, *Phys. Rev.* B 48 (1993) 13469.

32. H. Hosono and H. Imagawa, *Nucl. Instr. and Meth.* B 91 (1994) 510.

33. R. Kelly, *Mater. Sci. Eng.* A 115 (1989) 11.

34. R.H. Magruder III, D.H. Osborne, Jr., and R.A. Zuhr, *J. Non-Cryst. Solids* 176 (1994) 299.

35. R.H. Magruder III and R.A. Zuhr, *J. Appl. Phys.* 77 (1995) 3546.

36. R.H. Magruder III, T.S. Anderson, R.A. Zuhr, and D.K. Thomas, *Nucl. Instr. and Meth.* B 108 (1996) 305.

37. T.S. Anderson, R.H. Magruder III, D.L. Kinser, R.A. Zuhr, and D.K. Thomas, *Nucl. Instr. and Meth.* B 124 (1997) 40.

38. F. D'Acapito, S. Mobilio, F. Gonella, C. Maurizio, P. Mazzoldi, G. Battaglin, E. Cattaruzza, and F. Zontone, *Eur. Phys. J.* D 10 (2000) 123.

39. F. Gonella, E. Cattaruzza, G. Battaglin, F. D'Acapito, C. Sada, P. Mazzoldi, C. Maurizio, G. Mattei, A. Martorana, A. Longo, and F. Zontone, *J. Non-Cryst. Solids* 280 (2001) 241.

40. M. Falconieri, G. Salvetti, E. Cattaruzza, F. Gonella, G. Mattei, P. Mazzoldi, M. Piovesan, G. Battaglin, and R. Polloni, *Appl. Phys. Lett.* 73 (1998) 288.

41. E. Cattaruzza, G. Battaglin, F. Gonella, R. Polloni, G. Mattei, C. Maurizio, P. Mazzoldi, C. Sada, C. Tosello, M. Montagna, and M. Ferrari, *Phil. Mag.* B 82 (2002) 735.

42. C. de Julián Fernández, C. Sangregorio, G. Mattei, G. De, A. Saber, S. Lo Russo, G. Battaglin, M. Catalano, E. Cattaruzza, F. Gonella, D. Gatteschi, and P. Mazzoldi, *Mater. Sci. Eng.* C 15 (2001) 59.

43. C. de Julián Fernández, C. Sangregorio, G. Mattei, G. Battaglin, E. Cattaruzza, F. Gonella, S. Lo Russo, F. D'Orazio, F. Lucari, G. De, D. Gatteschi, and P. Mazzoldi, *J. Magn. Magn. Mat.* 226-230 (2001) 1912.

44. C.W. White, S.P. Withrow, J.D. Budai, L.A. Boatner, K.D. Sorge, J.R. Thompson, K.S. Beaty, and A. Meldrum, *Nucl. Instr. and Meth.* B 191 (2002) 437.

45. E. Cattaruzza, F. D'Acapito, C. de Julián Fernandez, A. De Lorenzi, F. Gonella, G. Mattei, C. Maurizio, P. Mazzoldi, S. Padovani, B.F. Scremin, and F. Zontone, *Nucl. Instr. and Meth.* B 191 (2002) 406.

46. G. Battaglin, E. Cattaruzza, G. De Marchi, F. Gonella, G. Mattei, C. Maurizio, P. Mazzoldi, M. Parolin, C. Sada, and I. Calliari, *Nucl. Instr. and Meth.* B 191 (2002) 392.

47. E. Borsella, S. Dal Toè, G. Mattei, C. Maurizio, P. Mazzoldi, A. Saber, G. Battaglin, E. Cattaruzza, F. Gonella, A. Quaranta, and F. D'Acapito, Mater. *Sci. Eng.* B 82 (2001) 148.

48. E. Borsella, E. Cattaruzza, F. D'Acapito, M.A. Garcìa, F. Gonella, G. Mattei, C. Maurizio, P. Mazzoldi, G. Battaglin, and A. Quaranta, *J. Appl. Phys.* 90 (2001) 4467.

49. E. Borsella, C. de Julián Fernández, M.A. Garcìa, G. Mattei, C. Maurizio, P. Mazzoldi, S. Padovani, C. Sada, G. Battaglin, E. Cattaruzza, F. Gonella, A. Quaranta, F. D'Acapito, M.A. Tagliente, and L. Tapfer, *Nucl. Instr. and Meth.* B 191 (2002) 447.

50. F. D'Acapito and F. Zontone, *J. Appl. Cryst.* 32 (1999) 234.

51. G. Mie, *Ann. Phys.* (Leipzig) 25 (1908) 377.

52. H. Yasuda and H. Mori, *Z. Phys.* D 31 (1994) 131.

53. G. Battaglin, M. Catalano, E.Cattaruzza, F. D'Acapito, C. de Julián Fernández, G. De Marchi, F. Gonella, G. Mattei, C. Maurizio, A. Miotello, P. Mazzoldi, and C. Sada, *Nucl. Instr. and Meth.* B 178 (2001) 176.

54. P. Mazzoldi, G. W. Arnold, G. Battaglin, F. Gonella, and R.F. Haglund Jr., *J. Nonlin. Opt. Phys. Mat.* 5 (1996) 285.

55. H. Okamoto, D.J. Chakrabarti, D.E. Laughlin, and T.B. Massalski, *Phase Diagrams of Binary Gold Alloys*, Monograph Series on Alloy Phase Diagrams, ASM International, Materials Park, Ohio (1987).

56. H. Yasuda, K. Mitsuishi, and H. Mori, *Phys. Rev.* B 64 (2001) 094101.

57. Ph. Buffat and J-P. Borel, *Phys. Rev.* A 13 (1976) 2287.

58. G.W. Arnold and J.A. Borders, *J. Appl. Phys.* 48 (1977) 1488.

59. P. Mazzoldi, L. Tramontin, A. Boscolo-Boscoletto, G. Battaglin, and G.W. Arnold, *Nucl. Instr. and Meth.* B 80–81 (1993) 1192.

60. F. Caccavale, G. De Marchi, F. Gonella, P. Mazzoldi, C. Meneghini, A. Quaranta, G. W. Arnold, G. Battaglin, and G. Mattei, *Nucl. Instr. and Meth.* B 96 (1995) 382.

61. C. Maurizio, F. Gonella, E. Cattaruzza, P. Mazzoldi, and F. D'Acapito, *Nucl. Instr. and Meth.* B 200 (2003) 126.

62. C. Maurizio, G. Mattei, P. Mazzoldi, S. Padovani, E. Cattaruzza, F. Gonella, F. D'Acapito, and F. Zontone, *Nucl. Instr. and Meth.* B 200 (2003) 178.

63. B. Massalski, *Binary Alloy Phase Diagrams*, American Society for Metals, Metals Park, Ohio 44073 (1987).

64. G. De, M. Gusso, L. Tapfer, M. Catalano, F. Gonella, G. Mattei, P. Mazzoldi, and G. Battaglin, *J. Appl. Phys.* 80 (1996) 6734.

65. G. Battaglin, E. Cattaruzza, C. de Julián Fernández, G. De Marchi, F. Gonella, G. Mattei, C. Maurizio, P. Mazzoldi, A. Miotello, C. Sada, and F. D'Acapito, *Nucl. Instr. and Meth.* B 175-177 (2001) 410.

66. J.C. Pivin and G. Rizza, *Thin Solid Films* 366 (2000) 284.

67. G. Rizza, M. Strobel, K.H. Heining, and H. Bernas, *Nucl. Instr. and Meth.* B 178 (2001) 78.

68. C.A. Coulter and D.M. Parkin, *J. Nucl. Mater.* 88 (1980) 249.

69. S.M. Foiles, M.I. Baskes, and M.S. Daw, *Phys. Rev.* B 33 (1986) 7983.

70. A. Miotello, G. De Marchi, G. Mattei, P. Mazzoldi, and C. Sada, *Phys. Rev.* B 63 (2001) 075409.

71. G. Mattei, G. De Marchi, C. Maurizio, P. Mazzoldi, C. Sada, and V. Bello, *Phys. Rev. Lett.* 90 (2003) 085502-1.

72. E. Cattaruzza, G. Battaglin, P. Calvelli, F. Gonella, G. Mattei, C. Maurizio, P. Mazzoldi, S. Padovani, R. Polloni, C. Sada, B.F. Scremin, and F. D'Acapito, *Comp. Sci. Technol.* 63 (2003) 1203.

73. N. Pinçon-Roetzinger, D. Prot, B. Palpant, E. Charron, and S. Debrus, *Mat. Sci. Eng.* C 19 (2002) 51.

74. C. Maurizio, A. Longo, A. Martorana, E. Cattaruzza, F. D'Acapito, F. Gonella, C. de Julián Fernández, G. Mattei, P. Mazzoldi, S. Padovani, and P. Boesecke, *J. Appl. Cryst.* 36 (2003) 732.

75. E. Cattaruzza, F. D'Acapito, F. Gonella, A. Longo, A. Martorana, G. Mattei, C. Maurizio, and D. Thiaudière, *J. Appl. Cryst.* 33 (2000) 740.

76. A. Martorana, A. Longo, F. D'Acapito, C. Maurizio, E. Cattaruzza, and F. Gonella, *J. Appl. Cryst.* 34 (2001) 152.

77. T. Nishizawa and K. Ishida, *Bull. Phase Diagrams* 5 (1984) 1.

78. J.R. Childress and C.L. Chien, *Phys. Rev.* B 43 (1991) 8089.

79. J. Noetzel, A. Handstein, A. Mucklich, F. Prockert, H. Reuther, J. Thomas, E. Wieser, and W. Moller, *J. Magn. Magn. Mat.* 205 (1999) 183.

80. F. Zontone, F. D'Acapito, and F. Gonella, *Nucl. Instr. and Meth.* B 147 (1999) 421.

Chapter 7

Intrinsic Residual Stress Evolution in Thin Films During Energetic Particle Bombardment

A. MISRA AND M. NASTASI

CONTENTS

ABSTRACT

Many novel thin film or surface treated materials are synthesized by non-equilibrium processes such as those based on physical vapor deposition and energetic ion bombardment. An inherent component of these processes are the residual stresses that are induced during synthesis. Since these stresses strongly influence the properties, performance and long-term reliability of these materials, understanding the mechanisms of stress generation, relaxation and methods of tailoring residual stresses is of great importance. In this article, we present a review of the current literature on stress generation in ion-beam synthesized thin films. In addition, we discuss a unified model of stress evolution in such materials with the intent of developing ways of tailoring stress to meet the needs of future novel materials.

7.1 INTRODUCTION

A variety of surface modification and surface coating techniques are widely used in industry to modify the near-surface properties of the substrate materials. In the surface modification by ion implantation process, a surface alloy composed of a combination of the substrate elements and the implanted ions is formed. In the ion implantation process the substrate not only provides a backing for the surface alloy but also contributes the material that makes up part of the surface alloy. In this process, there is a slow transition between the surface-modified zone and the substrate. The range of applications for such ion implantation-based surface engineering includes automotive and aerospace components, orthopedic implants, textile-manufacturing components, cutting and machining tools (e.g., punches, tapes, scoring dies, and extrusion dies), etc. [1].

In the surface coatings area we consider the physical vapor deposition (PVD) processes such as sputtering and e-beam evaporation used to deposit thin films in a variety of applications such as microelectronics, microelectromechanical systems (MEMS), nanoelectromechanical systems (NEMS), integrated optoelectronics, etc. The significance of these applications can be appreciated from the simple fact that the PVD equipment market alone is over US$150 billion per year [2]. The primary function of the PVD-synthesized thin films in engineering applications is often non-load bearing, e.g., electrical interconnects in integrated circuits, diffusion barriers, adhesion layers, seed layers to promote texture or epitaxy, coatings in magnetic recording media and heads, optical coatings, wear-resistant coatings, etc. However, the presence of residual stresses can adversely affect the properties, performance and long-term reliability of PVD thin films or ion-synthesized surfaces. Some common deleterious effects of residual stresses are film cracking, delamination from the substrate, stress-induced voiding (e.g., in the electrical interconnect lines in integrated circuits) and undesired modifications of physical properties (e.g., increases in the electrical resistivity). Hence, the need to understand the fundamental origins of residual stresses in these far-from-equilibrium materials and the ability to tailor residual stresses is critical to the integrity and performance of these materials and devices.

One reason for the catastrophic effects of residual stress on film failure is the large magnitudes of stresses typically encountered, as shown in Figure 7.1. It should be pointed out that the residual stresses are elastic and can build up to the level of the film yield strength, provided the yield strength is higher than the film fracture or delamination stress. It is well established that thin films with nanoscale grain sizes have much higher yield strengths as compared to bulk materials [3]. Note from Figure 7.1, taken from old literature [4,5] (much of the earlier work is reviewed by Koch [6]), that maximum residual stresses in metal films can be on the order of 1–2 GPa. Specifically, for the case of Mo films shown in Figure 7.1b, the peak residual stress is $\approx$ 2 GPa. If we assume that this peak residual stress is of the same order as the film yield strength, we infer a yield strength of $\approx$ 2 GPa for nanocrystalline Mo thin films, which is equivalent to Y/150 where Y is the Young's modulus, and hence, only a factor of 5 lower than the theoretical limit of strength of materials of $\approx$ Y/30. Typically, bulk metals have strengths that are lower than the theoretical strength limit by two to three orders of magnitude. Besides the high magnitudes of stresses, Figure 7.1 also shows the strong dependence of stress on the synthesis parameters. For e-beam evaporated W films (Figure 7.1a), bombardment with 400 eV Ar ions resulted in a rapid buildup of tensile stress and transition to compressive stress with increasing ion-to-atom flux ratio. For sputtered Mo films (Figure 7.1b), a similar transition in residual stress from tensile to compressive with reducing Ar pressure during deposition is observed. As discussed later in this chapter, decreasing Ar pressure results in an increase in bombarding particle energy. The effect of an ion beam on film growth and its resultant physical properties will depend on the ion species, energy, and the relative flux ratio of the ions, J_I, and deposited atoms, J_A, customarily defined as R_i. Data showing the effect of ion bombardment on film properties are either expressed simply

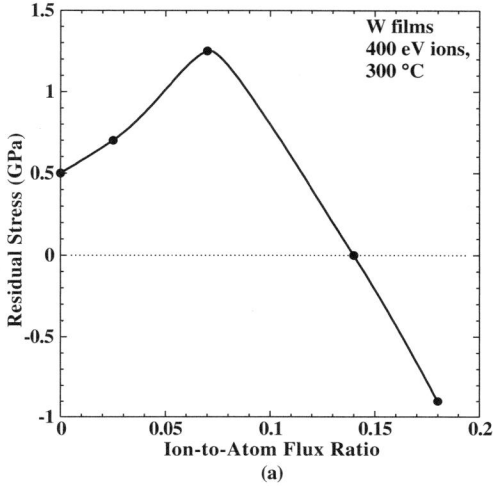

(a)

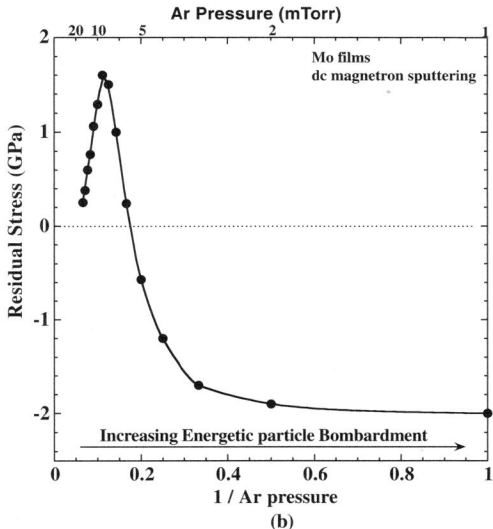

(b)

Figure 7.1 Typical trends in the evolution of thin film residual stresses during energetic particle bombardment; (a) IBAD W films (R.A. Roy, J.J. Cuomo and D.S. Yee, *J. Vac. Sci. Tech.*, A6, 1621 (1988)) and (b) sputtered Mo films (D.W. Hoffman, *Thin Solid Films*, 107, 353 (1983)). Positive values of stress indicate tensile stress and negative values compressive stress.

as ion flux (assuming constant deposited atom flux), as a relative ion/metal atom flux at the substrate (i.e., R_i), or as the average energy deposited per atom, E_{ave}, in eV/atom, which is simply the product of the relative ion/atom flux and the average ion energy, E_{ion}. The mathematical description of these two ion-beam assisted deposition (IBAD) parameters is given by [1]

$$R_i = \frac{\text{ion flux}}{\text{flux of deposited atom}} = \frac{J_I}{J_A} \tag{7.1}$$

and

$$E_{ave} = E_{ion} \cdot \frac{J_I}{J_A} = E_{ion} \cdot R_i \tag{7.2}$$

In this chapter, we present a review of the physical mechanisms of residual stress evolution in thin films subjected to energetic particle bombardment either during or post-deposition. Following Equations (7.1) and (7.2), energetic particle bombardment may be achieved by increasing ion (or atom) energy at a constant R_i, or increasing R_i at a constant E_{ion}. Before describing the models of stress evolution, we present some related general concepts as background.

7.2 MECHANICS OF BIAXIAL STRESS IN THIN FILMS RIGIDLY BONDED TO A SUBSTRATE

Residual stresses exist in thin films because of the constraint from the massive substrate, i.e., the film is rigidly bonded to the substrate. Thus, any change in the in-plane dimensions of the film that is not matched exactly by an equal change in the substrate dimensions will result in a stress in the plane of the film, with no stress in the direction normal to the film plane (i.e., biaxial stress state). It is important to distinguish between *intrinsic* and *extrinsic* residual stresses in thin films. *Intrinsic* stresses (sometimes also referred to as growth stresses) arise due to microstructural relaxation processes that tend to change the stress-free length of the film during deposition. *Extrinsic* stresses occur because of external factors that tend to change the stress-free length of the film subsequent to deposition, e.g., thermal expansion or contraction, or application of an external force to the film. This chapter is focused on the physical mechanisms of intrinsic stress evolution. Needless to say, the basic mechanics and methods of measurement of thin film stresses are the same, irrespective of the origins (intrinsic or extrinsic) of the residual stress.

To express the biaxial stress in thin films [7], consider a 3-D isotropic system in which the stress σ and strain ε are related by the following elasticity equations (x and y are the in-plane directions, and z is the out-of-plane direction):

$$\varepsilon_x = [\sigma_x - \upsilon(\sigma_y + \sigma_z)]/Y$$

$$\varepsilon_y = [\sigma_y - \upsilon(\sigma_x + \sigma_z)]/Y$$

$$\varepsilon_z = [\sigma_z - \upsilon(\sigma_x + \sigma_y)]/Y \qquad (7.3)$$

For thin films in plane stress, there is stress in the plane directions of the film (x and y) but not out of the plane ($\sigma_z = 0$). Therefore,

$$\varepsilon_x = (\sigma_x - \upsilon\sigma_y)/Y$$

$$\varepsilon_y = (\sigma_y - \upsilon\sigma_x)/Y$$

$$\varepsilon_z = (-\upsilon/Y)(\sigma_x + \sigma_y) \qquad (7.4)$$

Rearranging Equation (7.4), we get,

$$\varepsilon_x + \varepsilon_y = (1-\upsilon/Y)(\sigma_x + \sigma_y)$$

and

$$\varepsilon_z = (-\upsilon/1-\upsilon)(\varepsilon_x + \varepsilon_y) \qquad (7.5)$$

For isotropic systems where $\varepsilon_x = \varepsilon_y$,

$$\varepsilon_z = (-2\upsilon/1-\upsilon)(\varepsilon_x) \qquad (7.6)$$

Equation (7.6) gives the simple relation between the in-plane (ε_x) and out-of-plane (ε_z) strains for isotropic case.

The effect of in-plane residual stress on a wafer substrate is schematically illustrated in Figure 7.2 [8]. Consider the case shown in Figure 7.2a where a film if detached from the substrate would have smaller lateral dimensions than the

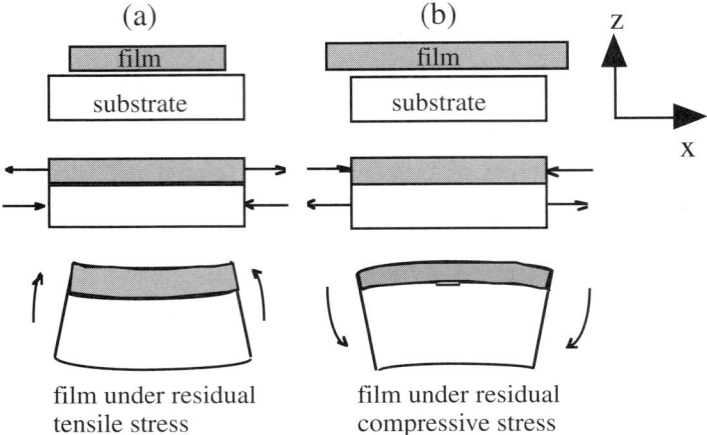

(a)   (b)   z

film under residual
tensile stress

film under residual
compressive stress

Figure 7.2   Schematic illustration of the residual stresses in thin films and the resulting substrate curvatures. A film that tends to shrink develops tensile stress and a film that tends to expand develops compressive stress due to substrate constraint (M. Ohring, *The Materials Science of Thin Films*, Academic Press, New York, 1992).

substrate. To maintain the same in-plane dimensions as the substrate, the film will need to be stretched biaxially, resulting in an induced curvature of the wafer substrate. Similarly, for the case shown in Figure 7.2b, a film with larger in-plane dimensions than the substrate, if unattached to the substrate, would need to be compressed biaxially to remain rigidly bonded to the substrate, resulting in a concave curvature of the substrate.

The bending in the substrate can be used not only to determine the nature of the stress, compressive or tensile (Figure 7.2), it can also be used to quantify the amount of stress. Consider a film of thickness t_f on a substrate of thickness t_s where the in-plane film stress (σ_f) resulted in a curvature of the substrate. A cross-section detail of the film/substrate couple, and the resulting stress distribution for this condition is shown in Figure 7.3. From the theory of elasticity for the bending plates and beams, there is always

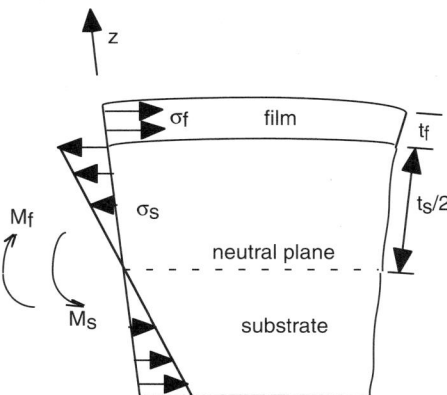

Figure 7.3 Schematic diagram showing the stress distribution in film and substrate and the corresponding bending moments (K.N.Tu, J.W. Mayer and L.C. Feldman, *Electronic Thin Film Science for Electrical Engineers and Materials Scientists*, Macmillan, New York, 1992).

a location in the beam called the neutral plane where the stress is zero. For the condition $t_f \ll t_s$ the neutral plane can be taken as the middle of the substrate. Expressions for the stress in the film and substrate can be derived from the equilibrium requirement that the sum of the moments produced by the stress in the film, M_f, and in the substrate, M_s, be zero. For a stress σ_f, which is uniform across the film thickness, the moment (force times the perpendicular distance) due to the stress in the film with respect to the neutral plane is (Figure 7.3):

$$M_f = \left(\sigma_f W t_f\right)\frac{t_s}{2} \tag{7.7}$$

where W is the width of the film normal to t_f. The moment produced by the stress in the substrate is given by

$$M_s = W \int_{-t_s/2}^{t_s/2} z\sigma_s(z)dz \tag{7.8}$$

Assuming biaxial stress in the substrate

$$\sigma_s(z) = \frac{Y_s}{1 - \upsilon_s} \varepsilon_s(z) \tag{7.9}$$

From Figure 7.3 and geometry it can be shown that

$$\varepsilon_s(z) = z/r \tag{7.10}$$

where r is the radius of curvature of the substrate. From Equations (7.10) and (7.9), we get:

$$\sigma_s(z) = \left(\frac{Y_s}{1 - \upsilon_s} \right) \frac{z}{r} \tag{7.11}$$

The moment from this stress is now written as (substituting Equation (7.11) in (7.8)):

$$M_s = W \frac{Y_s}{r(1 - \upsilon_s)} \int_{-t_s/2}^{t_s/2} z^2 dz = \frac{WY_s t_s^3}{12r(1 - \upsilon_s)} \tag{7.12}$$

Imposing the equilibrium condition, $\Sigma M = 0$, gives the stress in the film

$$\sigma_f = \left(\frac{Y_s}{1 - \upsilon_s} \right) \frac{t_s^2}{6rt_f} \tag{7.13}$$

which is also known as Stoney's equation [9]. This equation can also be derived using an energy minimization approach [10].

7.3 METHODS OF MEASUREMENT OF RESIDUAL STRESSES AND STRAINS IN THIN FILMS

The Stoney equation derived above is the basis for the most commonly used methods for residual stress measurements in thin films on wafer substrates. In the first type of measurement, the wafer is shaped as a cantilever beam with one end clamped (Figure 7.4), and the deflection of the free end (δ in Figure 7.4) is used to infer the film stress. The deflection is

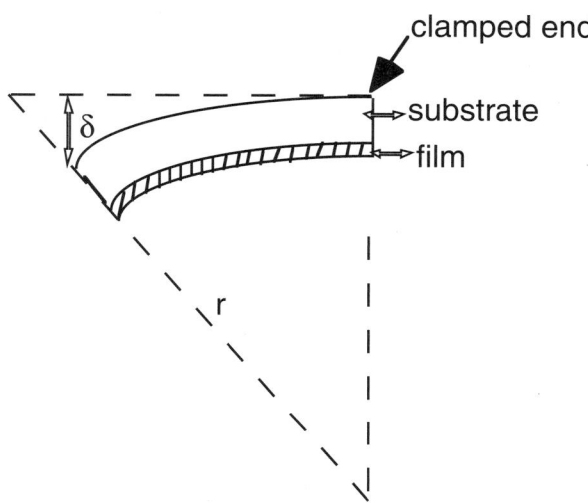

Figure 7.4 Schematic illustration of the principle of film stress measurement from bending of a cantilever beam substrate induced by film residual stress (M. Ohring, *The Materials Science of Thin Films*, Academic Press, New York, 1992).

typically measured using a position-sensitive photo-detector that monitors a laser beam focused on the backside of the free end of the substrate [11,12], although early measurements used a light microscope equipped with an ocular micrometer for measuring δ [13]. Stress in the film is calculated from the following equation, which is the Stoney equation modified for the geometry shown in Figure 7.4:

$$\sigma_f = \left(\frac{Y_s}{1-\upsilon_s}\right)\frac{\delta t_s^2}{3t_f l^2} \tag{7.14}$$

where l is the length of the cantilever beam.

This geometry, with suitable modifications as needed, has been used for *in situ* measurements of film stress during deposition [13] or post-deposition ion irradiation [11,14,15].

The other widely used approach is to measure the radius of curvature (r) using a laser scanning setup and deduce stress from Equation (7.13) [12]. The basic principle of the laser scanning technique to measure wafer curvature is quite simple, and is schematically illustrated in Figure 7.5. A laser beam is reflected from the surface of the wafer, and the displacement of the reflected beam is determined as the wafer is scanned. The change in displacement of the reflected beam is proportional to the change in the angle between the incident laser beam and the wafer surface. In a typical setup, the incident laser beam passes through a galvanometer mirror that is oscillated to have the laser beam scan vertically across a lens. The light that comes through the lens remains parallel to the optic axis as it hits the wafer specimen (Figure 7.5). By symmetry, if the sample is perfectly flat, the laser beam is reflected back to a single spot on the photo-detector throughout the scan. This reflected spot is displaced by a distance L (= 2fθ) on the detector, where f is the focal length of the lens, and θ is the angle by which the flat wafer is tilted (Figure 7.5a). If the sample is curved (Figure 7.5b), the reflected laser spot on the detector (L) moves as the laser is scanned along the sample, s. Mathematically,

$$\frac{\Delta L}{\Delta s} = 2f \frac{\Delta \theta}{\Delta s} = \frac{2f}{r} \tag{7.15}$$

From Equation (7.15), it follows that a plot of the reflected beam displacement (L) as a function of the scan distance (s) will be a straight line with a slope proportional to the inverse of the radius of curvature (r), f being a known constant [10, 12].

Since the commercially wafer substrates typically have a finite curvature in the as-received state, equation (7.13) is modified as follows:

$$\sigma_f = \left(\frac{Y_s}{1 - \upsilon_s} \right) \frac{t_s^2}{6t_f} \left(\frac{1}{r_d} - \frac{1}{r_s} \right) \tag{7.16}$$

where r_s is the curvature of the bare substrate, and r_d is the curvature subsequent to film deposition. For *ex situ* stress measurements, r_s is measured first, then the substrate is loaded in the vacuum chamber for film deposition, and r_d is measured after the deposited substrate is taken out of the

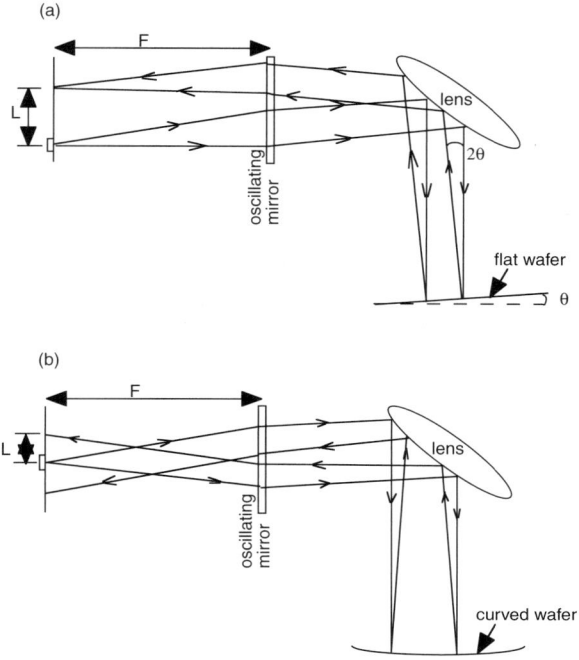

Figure 7.5 Optical ray diagram showing that (a) a parallel incident beam on a flat wafer is reflected back to the same point on the photo detector, and (b) for a curved wafer, the position of the reflected spot varies linearly with the scan position on the sample with a slope that is inversely proportional to the radius of curvature of the sample (P.A. Flinn, D.S. Gardner and W.D. Nix, *IEEE Trans. Electron Devices*, ED-34, 689 (1987); A.I. Van Sambeek and R.S. Averback, *Mater. Res. Soc. Symp. Proc.*, 396, 137 (1996); C.A. Volkert, *J. Appl. Phys.*, 70, 3521 (1991)).

vacuum chamber. Thus, the film is exposed to ambient environment prior to curvature measurement. Several investigators have developed *in situ* measurement capability where the laser scanning setup is coupled to a film deposition chamber [16,17]. Hence, r_s is measured at time zero and r_d is measured in real time, thereby measuring the stress evolution as a function of film thickness, while the film is still in vacuum. Besides the real time measurement capability in vacuum (almost instantaneous measurement of curvature using lasers) that avoids any stress relaxation issues, the other obvious advantage of this technique is that the modulus of the film is not needed to know the stress in the film. However, this technique always gives the total stress in the film. Thus, if the film is multilayered, the respective residual stresses in the constituent layers and the interface are not measured, only the average is measured (for *ex situ* measurements). With *in situ* measurements, stresses in the individual layers can be measured [16].

X-ray diffraction (XRD) methods have also been used widely for measuring the residual strains in the thin films. Elastic stresses change the spacing of crystallographic planes in crystals by amounts easily measured by XRD [18]. Having measured the elastic strains by XRD, the stress can be calculated using Hooke's law provided the elastic constants of the thin film are known. In symmetric reflection geometry (Figure 7.6a) the scattering vector, which is the vector difference between the incident and diffracted beams, is held normal to the film surface while its length is scanned by changing the scattering angle 2θ. Thus, the lattice spacings measured are from planes parallel to the sample surface, i.e., the out-of-plane strain is measured. Assuming biaxial strain state and isotropic material, the in-plane strain may be calculated from Equation (7.6), provided the Poisson's ratio of the film is known and the out-of-plane strain has been measured. To measure in-plane strains, the lattice spacings of planes that are approximately vertical, i.e., normal to the film surface, need to be measured. In grazing incidence x-ray scattering (GIXS) (Figure 7.6b), the incident and diffracted x-ray beams are at a very small angle α to the film surface so that the planes that diffract are nearly perpendicular to the film surface [19]. The XRD techniques have been applied to epitaxial as well as films with a fiber texture (i.e., preferred growth direction but random in-plane orientations). However, due to low diffracted intensities from thin film specimens, strong x-ray beams (rotating anode x-ray source or synchrotron) are often needed. For more details on the XRD technique for residual strain measurement in thin films, the reader is referred to review articles and books on this topic [18–21].

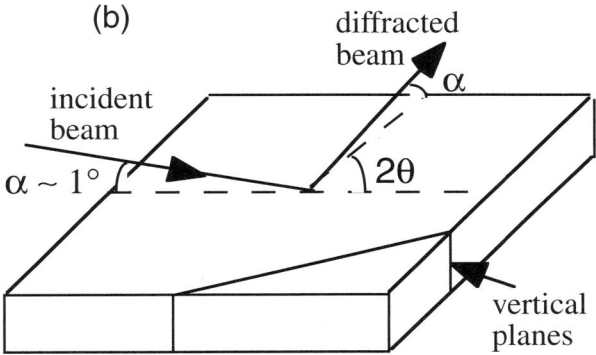

Figure 7.6 Schematic illustration of diffraction geometries used for (a) out-of-plane and (b) in-plane lattice spacing measurements in thin films (M. Clemens and J.A. Bain, *MRS Bull.*, July, 46 (1992); I.C. Noyan and J.B. Cohen, *Residual Stress: Measurement by Diffraction and Interpretation*, Springer-Verlag, New York, 1987).

7.4 SOURCES OF ENERGETIC PARTICLE BOMBARDMENT IN PVD

Before we address the evolution of intrinsic residual stress in thin films during energetic particle bombardment, it is important to briefly describe the sources of hyper-thermal particles that may bombard a growing film during PVD. The two common PVD processes we will consider are e-beam evaporation and magnetron sputtering.

In evaporation, atoms are removed from the source by thermal means, i.e., the source is heated, typically with a focused e-beam, until it melts and starts vaporizing due to the high vacuum. For some materials, semi-melting or sublimation are sufficient for evaporation as opposed to complete melting. In evaporation, the flux of atoms condensing onto the substrate from the vapor phase is typically "thermalized," i.e., atoms have energies on the order of kT where k is the Boltzman constant, and T is the substrate temperature. At low homologous temperatures (T/T_m, where T_m is the melting point), the evaporated films are often under-dense resulting in tensile stress as discussed in the subsequent sections. Energetic ion bombardment of evaporated films during deposition is achieved via an ion source, as schematically illustrated in Figure 7.7, and the process is commonly referred to as ion-beam assisted deposition (IBAD). The ion species is typically produced by a low energy (0.2–2 keV) broad-beam gridded ion source producing beam currents up to 1–2 mA/cm² ($\approx 10^{16}$ ions/s cm²). For more details on the design of IBAD systems, the reader is referred to Refs. [1,22–24] as well as to Chapter 4 of this volume.

In sputtering, the atoms to be deposited are dislodged from a solid target (source) surface through impact of ionized gas. Consider the simplified geometry shown in Figure 7.8a for dc sputtering. The target (cathode) is a plate of the material to be deposited onto the substrate (anode). A dc voltage is applied to the cathode while the anode is at ground potential. After evacuation of the chamber, an inert gas (e.g., Ar) is introduced in the chamber. The electrons emitted from the cathode ionize the inert gas and the Ar^+ ions are accelerated to the negatively charged cathode. The collision of the Ar^+ ions with the cathode knocks-off the target atoms that condense onto the substrate. This process leads to a visible glow discharge between the electrodes. During sputtering, some of the Ar^+ ions also get back reflected from the cathode as neutral atoms and travel toward the substrate. In the simple dc sputtering case shown in Figure 7.8a, a large fraction of the elec-

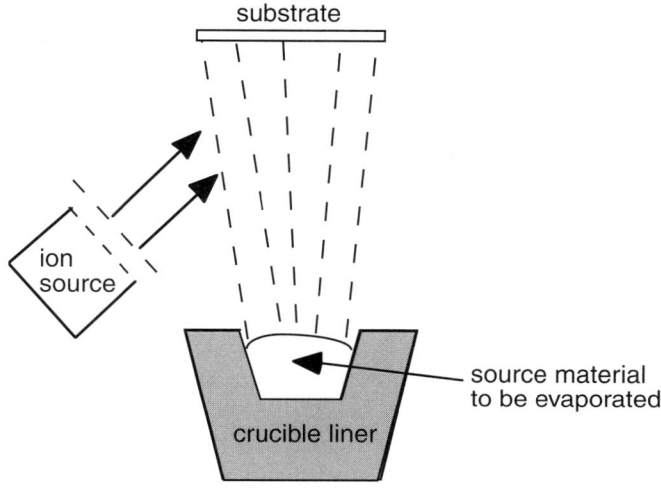

Figure 7.7 Schematic of the geometry employed in energetic ion bombardment of thin films deposited by e-beam evaporation (M. Nastasi, J.W. Mayer and J.K. Hirvonen, in *Ion-Solid Interactions: Fundamentals and Applications*, Cambridge University Press, Cambridge, 1996).

trons emitted from the cathode may escape toward the walls of the chamber; hence, a high base pressure of inert gas (~ 50–100 mtorr) is needed to initiate the glow discharge. The sputtered atoms from the target and the reflected neutral Ar atoms lose most of their energies during transport from the cathode to anode by collisions with the gas particles in the chamber. Higher gas pressure implies higher density of gas particles and therefore, shorter mean free path between collisions. Thus, the flux of atoms condensing on the substrate is essentially "thermalized" (similar to e-beam evaporation). The dc sputtering case is, therefore, a non-energetic deposition. Bombardment of the growing film by energetic ions may be achieved by adding an ion source, similar to the IBAD case shown in Figure 7.7. The development of magnetron sputtering in 1970s allowed the operation of the glow discharge at low Ar base pressures (typically, 1–10 mtorr). This is achieved by adding magnets behind the cathode (Figure 7.8b) such that the magnetic field will keep the emitted electrons confined to the cathode region, thereby making the ionization process more efficient. The low inert gas pressure implies that sputtered atoms from the target and the reflected neutral Ar atoms lose little energy via collisions in the transport from cathode to anode. Thus, the depositing flux is hyper-thermal and lead to energetic particle deposition. Most modern sputtering equipment uses dc-magnetrons for deposition of electrically conducting materials. The energetic particle bombardment during sputter deposition may be further enhanced by applying a negative dc bias voltage to the substrate (Figure 7.8c) that is otherwise kept at ground potential. Typically, dc bias to the substrate is much less than the dc voltage applied to the cathode. The negative dc bias attracts some of the Ar$^+$ ions from the plasma to the anode leading to energetic ion bombardment of the film. In summary, there are three sources of energetic particle bombardment during magnetron sputtering:

1. Sputtered atoms that are ejected from the target with average energies of a few to a few tens of eV with a significant high-energy tail in the energy distribution.
2. Energetic neutral atoms of inert gas that are back-reflected from the target with average energies that scale with the ratio of target atomic mass to incident ion mass and to the energy of the incident ions.
3. Energetic ions bombarding the film due to application of a negative substrate bias. For more quantitative details on the energies of particles bombarding the film during magnetron sputtering, the reader is referred to the literature on measurement of such energy distributions by energy-resolved mass spectrometry [26–28] and theoretical calculations [29].

Finally, thin films on substrates may be subjected to post-deposition ion irradiation in a high-energy ion implanter. Typically, ion energies on the order of hundreds of keV may be needed to penetrate the entire film thickness. While the low-energy bombardment during growth is preferred, there may be situations where subsequent processing after film deposition (e.g., thermal anneal) introduces undesirable film stress

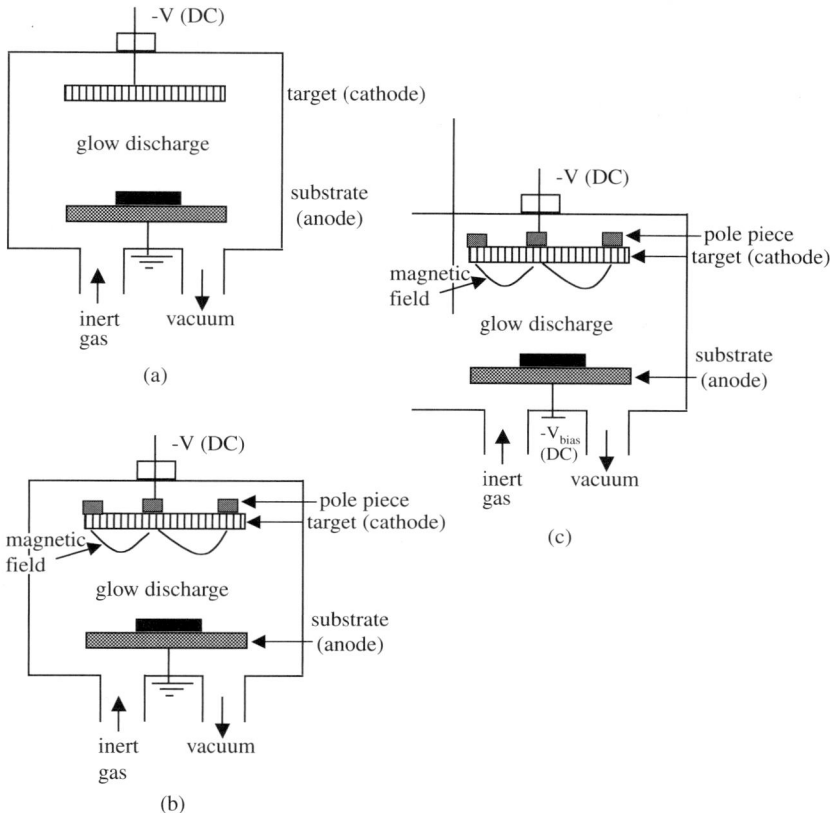

Figure 7.8 Schematic of the geometry used in sputtering: (a) dc sputtering, (b) dc magnetron sputtering and (c) dc magnetron sputtering with negative substrate bias.

that may be modified by a high-energy ion-irradiation of the film.

7.5 REVIEW OF KEY CONCEPTS RELEVANT TO THE ATOMIC ORIGINS OF INTRINSIC RESIDUAL STRESS IN THIN FILMS

In the next section, the physical mechanisms of the intrinsic residual stress evolution in thin films during energetic particle bombardment will be discussed, and models will be developed to interpret these stresses in terms of atomic scale defects. Some basic concepts relevant to the description of atomic origins of stress are briefly reviewed here. First, we review the universal binding energy relation that describes the interatomic potential energy (and hence, the interatomic force) as a function of interatomic distances in solids. Second, we review the ion-solid interactions concepts, particularly the formation of lattice defects during ion irradiation of solids.

7.5.1 Universal Binding Energy Relation

Smith and co-workers [30–32] have developed a universal description of the atomic binding energy as a function of atomic separation in solids. The shape of the curve can be obtained from a simple two-parameter scaling of a universal function $E^*(a^*)$,

$$E(r_{WS}) = \Delta E E^*(a^*) \tag{7.17}$$

where E is the binding energy per atom, r_{WS} is the radius of the Wigner–Seitz sphere containing an average volume per atom, and ΔE is the equilibrium binding energy. The parameter a^* is a scaled length given by:

$$a^* = (r_{WS} - r_{WSE})/\lambda \tag{7.18}$$

The parameter r_{WSE} is the radius of the equilibrium Wigner–Seitz sphere and can be calculated from the expression:

$$r_{WSE} = \left(\frac{3}{4\pi N_o}\right)^{1/3} = \left(\frac{3V_o}{4\pi}\right)^{1/3} \tag{7.19}$$

where N_o and V_o are the equilibrium atomic density and volume, respectively. The length scale λ describes the width of the binding energy curve and is dependent on the isothermal bulk modulus B and the equilibrium binding energy ΔE, and can be expressed as:

$$\lambda = \left(\frac{\Delta E}{12 B \pi r_{WSE}} \right)^{1/2} \tag{7.20}$$

Finally, the form of the universal function $E^*(a^*)$ is given as:

$$E^* = (a^*) = (1 + a^*) \exp(-a^*) \tag{7.21}$$

Substituting Equation (7.21) into (7.17) gives the final form of the universal binding energy relation:

$$E(r_{WS}) = -\Delta E (1 + a^*) \exp(-a^*) \tag{7.22}$$

The values of λ, r_{WSE}, and ΔE are listed in Table 7.1 for several elements. Essentially, the potential can be obtained using the bulk modulus, lattice parameter and cohesive energy of elements. Figure 7.9a shows the interatomic energy-distance curve for Cr, plotted using Equation 7.22 and the characteristic material parameters listed in Table 7.1. While the exact features of this curve are model-dependent, the general shape and trends are universal for all materials. The minimum energy in this curve corresponds to the most stable configuration for Cr atom spacing and represents the maximum energy that must be applied to pull a Cr atom free from the crystal, i.e., the cohesive energy of Cr. The interatomic distance corresponding to the minimum in the interatomic potential energy corresponds to the equilibrium distance between Cr near neighbors. As evident from Figure 7.9a, any departure from the equilibrium interatomic distance results in an increase of the energy of the material (less negative) and makes the material less stable. An alternate way to look at Figure 7.9a is to recall that $-dE/dr = F$, where F is the interatomic force, and r is the interatomic distance. Note that under this convention, a positive force occurs when dr is negative and a negative force results when dr is positive, which is opposite to the stress convention applied in materials science. To maintain the stress convention, i.e., increasing the distance between atoms produces a positive restoring force and compressing the spacing produces a repulsive force we plot $F = dE/dr$ in Figure 7.9b. These data show that there are no forces acting on Cr atoms that have the equilibrium spac-

TABLE 7.1   Values of Parameters λ, r_{WSE}, and ΔE Used in the Universal Binding Energy Relation

Metal	λ (nm)	r_{WSE} (nm)	ΔE (eV)
Li	0.0553	0.1719	1.65
Be	0.0312	0.125	3.33
Na	0.0562	0.2080	1.13
Mg	0.0316	0.177	1.53
Al	0.0336	0.158	3.34
Si	0.0344	0.168	4.64
K	0.0651	0.2573	0.941
Ca	0.0483	0.218	1.825
Ti	0.0340	0.162	4.86
V	0.0310	0.149	5.3
Cr	0.0254	0.142	4.10
Fe	0.0274	0.141	4.29
Co	0.0262	0.139	4.39
Ni	0.027	0.138	4.435
Cu	0.0272	0.141	3.50
Zn	0.0215	0.154	1.35
Ge	0.0348	0.176	3.87
Rb	0.0658	0.2750	0.858
Y	0.047	0.199	4.39
Zr	0.0395	0.177	6.32
Nb	0.0336	0.163	7.47
Mo	0.0265	0.155	6.810
Ru	0.0245	0.148	6.62
Pd	0.0237	0.152	3.936
Ag	0.0269	0.160	2.96
Cd	0.0214	0.173	1.16
In	0.0360	0.184	2.6
Cs	0.0714	0.2977	0.827
Ba	0.0558	0.246	1.86
Ce	0.0648	0.202	4.77
Eu	0.0478	0.227	1.80
Gd	0.0467	0.199	4.14
Dy	0.0404	0.196	3.1
Er	0.0393	0.194	3.3
Yb	0.0506	0.199	1.6
Hf	0.0373	0.174	6.35
Ta	0.0330	0.163	8.089
W	0.0274	0.156	8.66
Re	0.0247	0.152	8.10
Ir	0.0230	0.150	6.93
Pt	0.0237	0.153	5.852
Au	0.0236	0.159	3.78
Tl	0.0331	0.190	1.87
Pb	0.0331	0.193	2.04
Th	0.0483	0.199	5.926

Note: see Equations (7.17)–(7.22) for details

ing of ≈ 0.249 nm. However, if the interatomic distance is greater than equilibrium, a positive or tensile force results. On the contrary, if the interatomic distance is decreased below equilibrium, repulsive or compressive force tries to return the atoms to their equilibrium position.

7.5.2 Fundamentals of Ion Bombardment-Induced Lattice Defects

It has been known for many years that bombardment of a crystal with energetic ions produces regions of lattice disorder. This lattice disorder results from the physical processes responsible for slowing the ion down and allowing it to come to rest in the crystal. There are two energy loss mechanisms that contribute to the slowing down processes: nuclear collisions, in which energy is transmitted as translatory motion to a target atom as a whole, and electronic collisions, in which the moving particle excites or ejects atomic electrons. The total energy-loss rate dE/dx can be expressed as:

$$\frac{dE}{dx} = \frac{dE}{dx}\bigg|_n + \frac{dE}{dx}\bigg|_e$$

where the subscripts n and e denote nuclear and electronic collisions, respectively.

Nuclear collisions can involve large discrete energy losses and significant angular deflection of the trajectory of the ion. It is this process that is responsible for the production of lattice disorder by the displacement of atoms from their positions in the lattice. Electronic collisions involve much smaller energy losses per collision, negligible deflection of the ion trajectory, and negligible lattice disorder. The relative importance of the two energy-loss mechanisms changes rapidly with the energy E and atomic number, Z_1, of the ion: nuclear stopping predominates for low E and high Z_1, whereas electronic stopping takes over for high E and low Z_1.

The nuclear stopping power of an ion of mass M_1 and atomic number Z_1 incident on target atoms of mass M_2 and atomic number Z_2 can also be calculated with a high level of

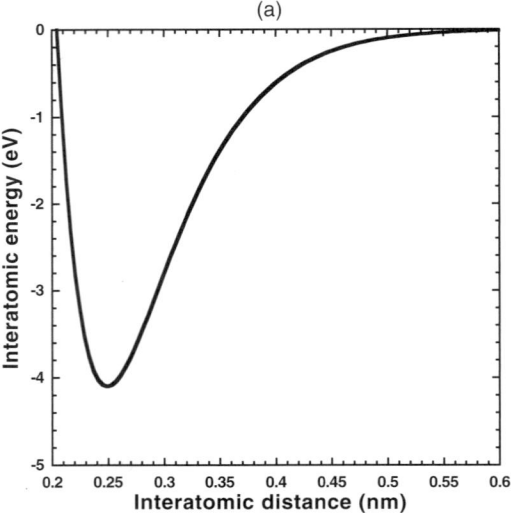

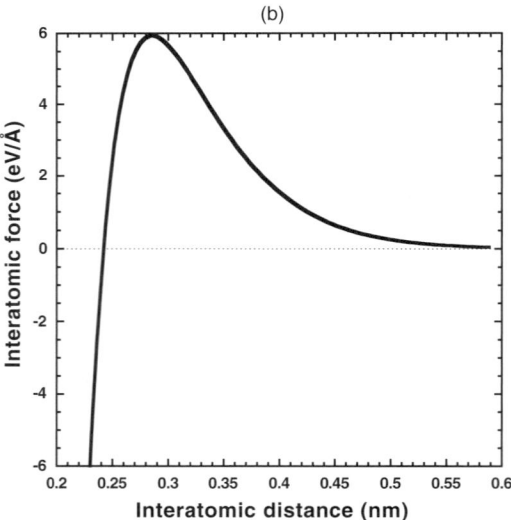

Figure 7.9 Plots of (a) interatomic energy and (b) interatomic force as a function of interatomic distance for Cr using the universal binding energy relation [32].

accuracy using the Ziegler, Biersack and Littmark (ZBL) formula [33]:

$$S_n(E_0) = \frac{8.462 \times 10^{-15} \, Z_1 Z_2 M_1 \, S_n(\varepsilon)}{(M_1 + M_2)(Z_1^{0.23} + Z_2^{0.23})} \quad \frac{\text{eV cm}^2}{\text{atom}} \quad (7.23)$$

where the reduced nuclear stopping cross-section is calculated using

$$S_n(\varepsilon) = \frac{0.5 \ln(1 + 1.1383\varepsilon)}{(\varepsilon + 0.01321 \, \varepsilon^{0.21226} + 0.19593 \, \varepsilon^{0.5})} \text{ for } \varepsilon_{ZBL} \leq 30 \quad (7.24)$$

and

$$S_n(\varepsilon) = \frac{\ln(\varepsilon)}{2\varepsilon} \text{ for } \varepsilon_{ZBL} > 30 \quad (7.25)$$

The term e_{ZBL} is the ZBL reduced energy and is given by

$$\varepsilon_{ZBL} = \frac{32.53 \, M_2 E_0}{Z_1 Z_2 \, (M_1 + M_2)(Z_1^{0.23} + Z_2^{0.23})} \quad (7.26)$$

where E_0 is in keV. The advantage of the reduced energy notation is that a single expression can define nuclear stopping cross-section for all ion/target-atom combinations.

For most ion implantation and ion assisted deposition processes, nuclear collisions dominate the energy losses experienced by the ion. As an ion slows down and comes to rest in a crystal, it makes a number of collisions with the lattice atoms. In these collisions, sufficient energy may be transferred from the ion to displace an atom from its lattice site. Lattice atoms that are displaced by the incident ion are called *primary knock-on atoms* or PKA. If the PKA has been given sufficient energy it can in turn displace other atoms, secondary knock-on atoms, and so on, tertiary knock-ons, etc. — thus creating a cascade of atomic collisions (Figure 7.10). This process leads to a distribution of vacancies, interstitial atoms, and other types of lattice disorder in the region around the ion track. The vacancy-interstitial defect formed by this pro-

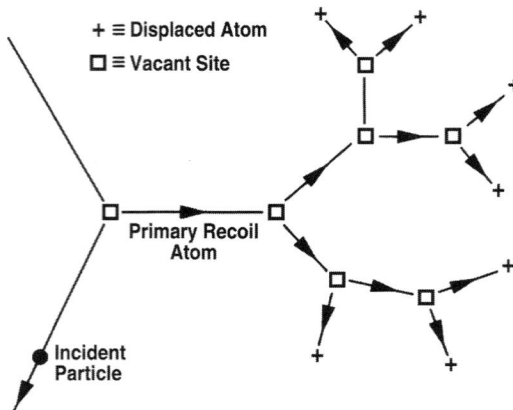

Figure 7.10 Schematic of a collision cascade (M. Nastasi, J.W. Mayer and J.K. Hirvonen, in *Ion-Solid Interactions: Fundamentals and Applications*, Cambridge University Press, Cambridge, 1996).

cess is commonly referred to as a *Frenkel-pair* or a *Frenkel defect*.

In the collision process it is also possible that the PKA has sufficient energy to displace an atom off its lattice site, but after the collision is left with insufficient energy to continue its flight and falls into the vacated site it just created. This process is called a *replacement collision*. While such events have little influence on monatomic materials, replacement collisions can produce considerable disorder in ordered polyatomic materials.

One of the principal effects of introducing point defects into a crystal is the long-range displacement field they produce around themselves. An example of this effect is schematically presented in Figure 7.11 for a vacancy and an interstitial. As can be seen from this example, the defects displacement field results in a change in the crystal's volume, which is commonly referred to as the defects *relaxation volume*, V^{rel}. In the example presented in Figure 7.11 the relaxation volume of the vacancy, V_v^{rel}, is negative and the relation volume of the interstitial, V_i^{rel}, is positive.

Point Defects: Vacancies and Interstitials

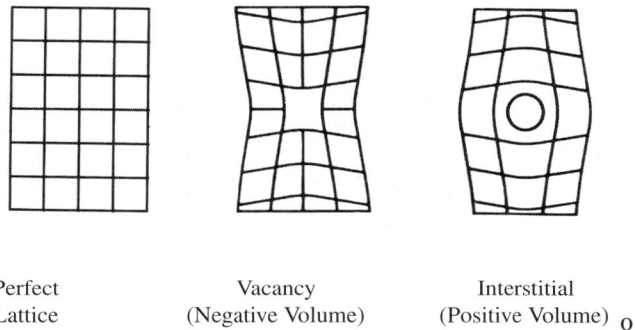

| Perfect | Vacancy | Interstitial |
| Lattice | (Negative Volume) | (Positive Volume) $_0$ |

Figure 7.11 Schematic of lattice strains around point defects.

TABLE 7.2 Relaxation Volumes for Interstitials (V_i^{rel}), Vacancies (V_v^{rel}) and Frenkel Defect (V_{FD}^{rel}), and Young's Modulus (Y) and Poisson's Ratio (ν) for Several Elements

Crystal Structure	Element	V_i^{rel}	V_v^{rel}	V_{FD}^{rel}	Y (GPa)	n
fcc	Cu	1.55	−0.25	1.30	123.6	0.35
	Al	1.9	−0.05	1.85	71.0	0.34
	Ni	1.8	−0.2	1.6	193.2	0.30
	Pt	1.8	−0.2	1.6	170.6	0.38
bcc	Mo	1.1	−0.1	1.0	327.6	0.30
	Fe	1.1	−0.05	1.05	209.9	0.28
					243.21	0.21
hcp	Zn	3.5	−0.6	2.9	397.2	0.28
	Co	1.5	−0.05	1.45	104.9	0.35

The relaxation volume for vacancies and interstitials has been experimentally determined for a number of metals [34]. A summary of these data is presented in Table 7.2. The relaxation volumes presented in Table 7.2 are given in terms of the atomic volume of one atom in the perfect crystal, V_{atom}, and the relaxation volume of a Frenkel defect, V_{FD}^{rel}, is taken as the sum of vacancy and interstitial relaxation volumes. Also presented in Table 7.2 are values of the material's Young's modulus, Y, and Poisson's ratio, v. Examining Table 7.2, we see that for the metals listed, the relaxation volume of the interstitials, V_i^{rel}, is always positive and greater than one atomic volume while the relaxation volume of the vacancies, V_v^{rel}, is always negative and a small fraction of an atomic volume. Therefore, when vacancies and interstitials are created in equal numbers by an irradiation process, positive volume will be added to the crystal causing it to swell.

7.6 STRESS EVOLUTION MODELS

7.6.1 Stress Evolution as a Function of Film Thickness in Non-Energetic Deposition

As discussed in earlier sections, non-energetic deposition primarily applies to the deposition conditions when the depositing flux is essentially thermalized. Since the focus of this article is on stress evolution during energetic particle deposition conditions, only a brief description of stress evolution during non-energetic deposition will be presented here. For more details, the reader is referred to other literature [16,17,35–45]. In the absence of bombardment from hyperthermal particles, the general trend of the evolution of intrinsic stress in island growth thin films as a function of film thickness is as shown in Figure 7.12 [38,39]. For materials with high atomic mobility at room temperature [16,17,38–42] such as Ag, Al, Au, Cu, etc., stress evolution with increasing film thickness has three stages: (i) initial compressive stress, (ii) rapid buildup of tensile stress to a peak, and (iii) relaxation of tensile stress and eventual buildup of compressive stress (Figure 7.12a). Often these measurements are performed *in situ* and some relaxation of the compressive stress is observed when the deposition flux is stopped [44]. For refractory metals with high melting points (e.g., Cr, Mo, W), diffusion at room temperature is limited and the transition from tensile to com-

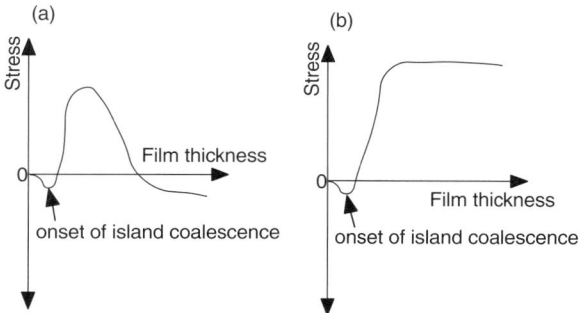

Figure 7.12 Typical trends of growth stress evolution as a function of film thickness for non-energetic deposition: (a) materials with high atomic mobility, and (b) materials with low atomic mobility at the deposition temperature (C.V. Thompson and R. Carel, *J. Mech. Phys. Solids*, 44, 657 (1996); R.C. Cammarata, T.M. Trimble and D.J. Srolovitz, *J. Mater. Res.*, 15, 2468 (2000)).

pressive stress with increasing film thickness is typically not observed during deposition, Figure 7.12b) [38,45].

The initial compressive stress that occurs when the film is discontinuous has been attributed to surface stress [39]. First, we need to define surface stress. Due to the differences in the nature of bonding between surface and bulk atoms in a solid, surface atoms tend to have an equilibrium interatomic spacing different from bulk. However, the surface atoms must remain in atomic registry with the bulk, and this constraint results in a force per unit length acting on a solid surface, referred to as surface stress. For low-index surfaces of metals, the atoms would prefer to adopt a lower equilibrium spacing than bulk to increase the local electron density. In this case the atomic registry between the surface and underlying bulk atoms would result in stretching of the atomic bonds at the surface (i.e., surface stress is tensile) and shrinkage of the bonds in the underlying bulk (i.e., induced compressive stress in the bulk of the island). It should be emphasized that the surface stress represents the work done to elastically *strain* the surface atoms, and hence is different from surface energy that is the work done in *creating* a free surface [47–49].

Similarly, there is interface stress related to the elastic strain-ing of atoms at interfaces between two solids. The compressive stress induced in early stages of island growth is thus attrib-uted to the surface stress. For fcc metals with high surface mobility (Figure 7.12a), island coalescence occurs early on in the deposition process and hence, large compressive stresses typically do not develop. However, in bcc refractory metals (Figure 7.12b) such as Mo, compressive stress on the order of 1 GPa has been observed in early stages of deposition [50].

Next we consider the evolution of tensile stress as the islands coalesce to form a continuous film. A model for this has recently been proposed by Nix and Clemens [36]. Consider that the film has square cross-section islands of height t and width d. The driving force for coalescence is the removal of two free surfaces at the expense of a grain boundary (note that grain boundary energy is typically about one-half to one-third of surface energy in fcc metals) and elastic strain energy that results from the biaxial tensile straining of the film as the islands coalesce. This energy balance can be used to esti-mate the maximum stress that could result from the island coalescence. Let the surface energy per unit area be γ_s, grain boundary energy per unit area be γ_{gb}, and the elastic strain energy per unit volume be $\sigma\varepsilon$ where σ is the stress, and ε is the strain in the film. Note that

$$\varepsilon = \frac{\sigma}{\left(\dfrac{Y}{1-\upsilon}\right)} \tag{7.27}$$

where $Y/(1-\upsilon)$ is the biaxial modulus.

The energy balance gives the following relation:

$$2\gamma_s(td) = \gamma_{gb}(td) + \frac{\sigma^2(d^2t)}{\left(Y/(1-\upsilon)\right)} \tag{7.28}$$

Therefore,

$$\sigma = \sqrt{\frac{\left(2\gamma_s - \gamma_{gb}\right)Y}{d(1-\upsilon)}} \tag{7.29}$$

Substituting typical values for low melting point fcc metals such as Au, Ag, Al, etc., Nix-Clemens showed that Equation (7.29) gives an upper bound stress of 5–6 GPa, which is at least an order of magnitude higher than the peak tensile stress typically observed during island coalescence of these metals.

The reason for overestimation of stress from Equation (7.29) was attributed to the assumption in the model that all crystallites are imagined to coalesce at the same time, with the consequence that no shear stress can be developed on the film/substrate interface. A more realistic picture would be to allow different crystallites to coalesce at different times and some sliding at the interface to occur. Furthermore, any strain relaxation by defect motion within the crystallites was also not considered. Phillips et al. [40] described the gradual buildup of tensile stress during film growth by considering cracks in the film (equivalent to incompletely coalesced islands) that would tend to reduce the curvature of the substrate. As the area fraction of the cracks decreased (i.e., extent of coalescence increased) the tensile stress would increase. This hypothesis was supported by TEM observation on Pt films deposited on SiO_2 where the area coverage of the substrate surface by the Pt islands was observed to increase gradually with film thickness [40]. However, the maximum stress was still significantly lower than what Equation (7.29) would predict.

Several investigators have attempted to extend the Nix–Clemens work to come up with quantitative predictions of tensile stress that compare reasonably with experimentally measured stresses [41–43]. The basic idea is to examine the island coalescence process in more detail, as was first suggested by Nix–Clemens. If the islands are assumed to have an elliptical shape, then the initial contact between growing islands will be at a single point. The contact between islands could then be visualized as an elastic crack edge, and the subsequent coalescence process equivalent to crack closing or "zipping." Following the idea that coalescence follows a crack zipping process, Seel et al. [42] used a finite element method to model it. By minimizing the sum of the positive strain energy and associated reduction in the boundary energy, the equilibrium configuration resulting from island coalescence was determined as a function of island radius. For a given island radius at impingement, this approach yielded average stress values that were an order of magnitude lower than that predicted by Equation (7.29). Freund and Chason [43] modeled the zipping process as a Hertzian contact mechanics problem. Their model is based on the theory of elastic contact of solids with rounded surfaces (Hertz contact theory). This analytical approach also predicted stresses on the order of a couple hundred MPa (as opposed to several GPa from Equation (7.29)) from island coalescence. Floro et al. [41] and Seel et al. [42] also considered the effect of stress relaxation during growth, pointing out that the observed stress is due to a dynamic competition between stress generation due to island coalescence and stress relaxation due to surface diffusion.

Finally, we consider the evolution of compressive stresses during film growth for conditions of non-energetic deposition. Chason et al. [44] have developed a model for compressive stress generation based on the idea that an increase in the surface chemical potential caused by the deposition of atoms from the vapor drives excess atoms into the grain boundaries. Plating of extra atoms at grain boundaries would lead to an in-plane expansion of the film were it not rigidly bonded to the substrate, and the constraint from the substrate leads to compressive stress in the film. Since the compressive stress raises the chemical potential of atoms in the grain boundary, the driving force for additional flow of atoms decreases with increasing stress and eventually a steady state is reached. The model also explains the relaxation of compressive stress, often observed during *in situ* measurements of stress of metals such as Ag with high mobility of surface atoms at room temperature, as the reverse flow of excess atoms from boundaries to surfaces.

The mechanisms of stress evolution and relaxation under conditions of non-energetic deposition described above are schematically shown in Figure 7.13.

7.6.2 Stress Evolution during Energetic Particle Bombardment

Typical trends of residual stress evolution during energetic particle bombardment were shown in Figure 7.1. In this section, we present experimental data of residual stress evolution in thin metal films subjected to energetic particle bombardment in different ways. The corresponding film microstructures are also shown and then models are presented for generation and relaxation of film residual stresses and limits to these stresses.

7.6.2.1 Correlation between Stresses and Microstructures

For simplicity, we will take the example of Cr films deposited on Si substrates, investigated in detail by the authors [14,46,51–56], to show how tailoring of stress in thin films may be achieved by different types of energetic particle bombardment. In these studies, stress evolution is studied as a function of energetic particle bombardment for a constant thickness, as opposed to the non-energetic deposition case discussed in the previous section where stress evolution as a function of film thickness was reported. The thickness of the films in these studies will be on the order of a couple hundred nm or more. Hence, the films will be continuous (i.e., islands have coalesced). Since Cr represents a low atom mobility case (Figure 7.12b), the intrinsic stress is not expected to be compressive in the absence of any energetic deposition. The mechanisms of stress generation developed later in this section will apply to high atom mobility case as well (Figure 7.12a). As discussed in Section 7.4, sputtered films have two sources of hyper-thermal (tens to hundreds of eV) particle bombardment: (i) varying the inert gas pressure (lower pressure results in higher particle bombardment energies) and (ii) negative substrate bias. A third kind of bombardment considered here is post-deposition ion irradiation with high-energy ions (hundreds of keV).

The evolution of tensile stresses in sputtered 150 nm nominal thickness Cr films as a function of Ar gas pressurein

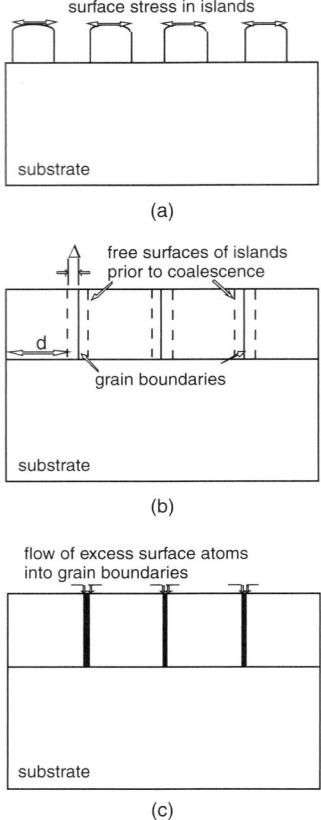

Figure 7.13 Schematic of the mechanisms of stress evolution as a function of film thickness during non-energetic (corresponding to the trend shown in Figure 12a): (a) compressive stress arises in the film due to constraint from the positive surface stresses, in the initial stages of growth when the islands are not coalesced (R.C. Cammarata, T.M. Trimble and D.J. Srolovitz, *J. Mater. Res.*, 15, 2468 (2000)), (b) generation of tensile stress as free surfaces of adjoining islands coalesce to form grain boundaries, tensile strain = D/d (W.D. Nix and B.M. Clemens, *J. Mater. Res.*, 14, 3467 (1999)), (c) relaxation of tensile stress and buildup of compressive stress due to flow of atoms from surface to grain boundaries (a reverse flow of atoms from boundaries to surfaces leads to compressive stress relaxation) (E. Chason, B.W. Sheldon, L.B. Freund, J.A. Floro and S.J. Hearne, *Phys. Rev. Lett.*, 88, 156103, (2002)).

the 1–7.5 mtorr range is shown in Figure 7.14a. With decreasing Ar pressure, the biaxial tensile residual stress increases to a maximum of ~ 1.7 GPa and then rapidly decreases. At the lowest Ar pressure of ~ 1 mtorr used, the stress was still tensile. With increasing substrate bias in the 150 nm thick film (Figure 7.14b), we observe an increase in tensile stress to a maximum of ~ 1.6 GPa, followed by a complete relaxation of tensile stress, and finally the rapid buildup of compressive stress to a saturation value of ~ 2.1 GPa. The stress evolution with substrate bias for 1 μm thick Cr films is shown in Figure 7.14c. The Ar pressure was kept constant at 2.5 mtorr. Note from Figure 7.14a that 2.5 mtorr is to the left of the peak in tensile stress in Figure 7.14a and hence, no initial increase in the tensile stress is observed with increasing bias at 2.5 mtorr deposition (Figure 7.14c). However, with increasing bias at 5 mtorr deposition (which is to the right of the peak in Figure 7.14a, an initial increase in tensile stress to a maximum is observed (Figure 7.14b). The evolution of stress in 150 nm thick Cr films, deposited at 5 mtorr Ar pressure without any substrate bias, with post-deposition ion irradiation is shown in Figure 7.14d. The irradiation was done with 110 keV Ar ions that penetrate about 80 nm of film thickness. With constant ion energy and increasing dose, the stress evolution is similar to that shown in Figure 7.14a and b, i.e., tensile stress initially increases, then decreases to zero, and then compressive stress builds up. No saturation of compressive stress was observed for the dose range considered here. Irradiation with 300 keV Ar ions, where the depth of ion penetration was greater than the film thickness, gave the same results. This indicates that factors such as modification of interface stress and the non-uniform stress due to irradiation of only a fraction of the film thickness have little influence on the observed stress evolution. The evolution of stress was correlated with the microstructures of the films through transmission electron microscopy (TEM) observations. The microstructures were investigated for four processing conditions: (a) films deposited at high Ar pressures without any substrate bias (Figure 7.15a); (b) films deposited under conditions that produced the maximum tensile stress (Figure

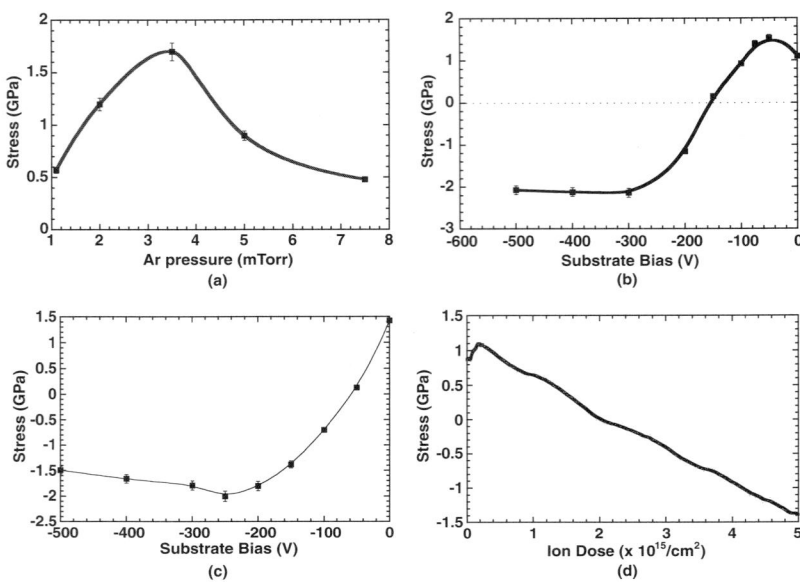

Figure 7.14 Experimental data showing the effects of energetic particle bombardment on the intrinsic residual stress in thin sputtered Cr films: (a) as a function of Ar pressure for 150 nm thick films, (b) as a function of negative substrate bias for 150 nm thick Cr films, (c) as a function of negative substrate bias for 1 mm thick Cr films, and (d) as a function of post-deposition ion-irradiation in 150 nm thick Cr films irradiated with 110 keV Ar ions (deposition was at 5 mtorr Ar pressure with no substrate bias).

7.15b); films deposited under conditions that produced almost no stress (Figure 7.15c); and (d) films deposited under conditions that produced the maximum compressive stress (Figure 7.15d). Typically, nanocrystalline columnar microstructures are observed as shown in the through-focus cross-sectional TEM micrograph in Figure 7.15a from a 150 nm thick Cr film deposited at 7.5 mtorr Ar pressure without substrate bias. The intercolumnar regions exhibited Fresnel fringes in through-focus imaging: bright fringes for under focused images, dark fringes for over focused images and no contrast for exact focus images. While an exact focus image gives the impression that the film is fully dense and all islands have completely coalesced to form grain boundaries, the observation of Fresnel fringes in over- and under-focused images indicate voided intercolumnar regions that may only be a few atomic layers thick. This Fresnel contrast imaging (phase contrast due to differences in atomic numbers) technique has been used to highlight nanoscale voids or films at grain boundaries in bulk materials [57].

As the bombardment energy is increased, the intercolumnar voids tend to shrink and at the tensile stress maximum, no Fresnel fringes were observed (Figure 7.15b). At near-zero stress, clear grains are observed with no evidence of open volume or radiation-induced defects in the film (Figure 7.15c). At compressive stress maximum, clear evidence of radiation-damage-type defect structures is observed (Figure 7.15d), although the details of the radiation-induced point defects cannot be analyzed from these TEM images. However, by comparing Figures 7.15c and d, it is obvious that the major fraction of the grains is covered with a high density of radiation damage in Figure 7.15d that may include vacancies (either isolated or clustered), self-interstitials (either isolated or clustered), Frenkel defects and entrapped Ar. Finally, it should be emphasized that the microstructure evolution with bombardment correlates well with the stress evolution irrespective of the film thickness. For a given stress, the primary difference in the microstructure of the 150 nm and 1 μm thick Cr films was in the grain size: ~ 16 nm in the former and ~ 50 nm in the latter.

In general, the relationship between the film microstructure and the ion beam conditions could be represented schematically as shown in the structure zone diagram (Figure 7.16). The diagram is similar to the evaporation zone diagram in Grovenor et al. [58], but includes bombarding ion energy and the sputtering gas pressure as a processing parameter in addition to the substrate temperature (expressed as a fraction of the melting temperature in degrees Kelvin) [1]. The sputtering gas pressure is included because the energetic particle bombardment of films increases with decreasing gas pressure as discussed in Section 7.4.

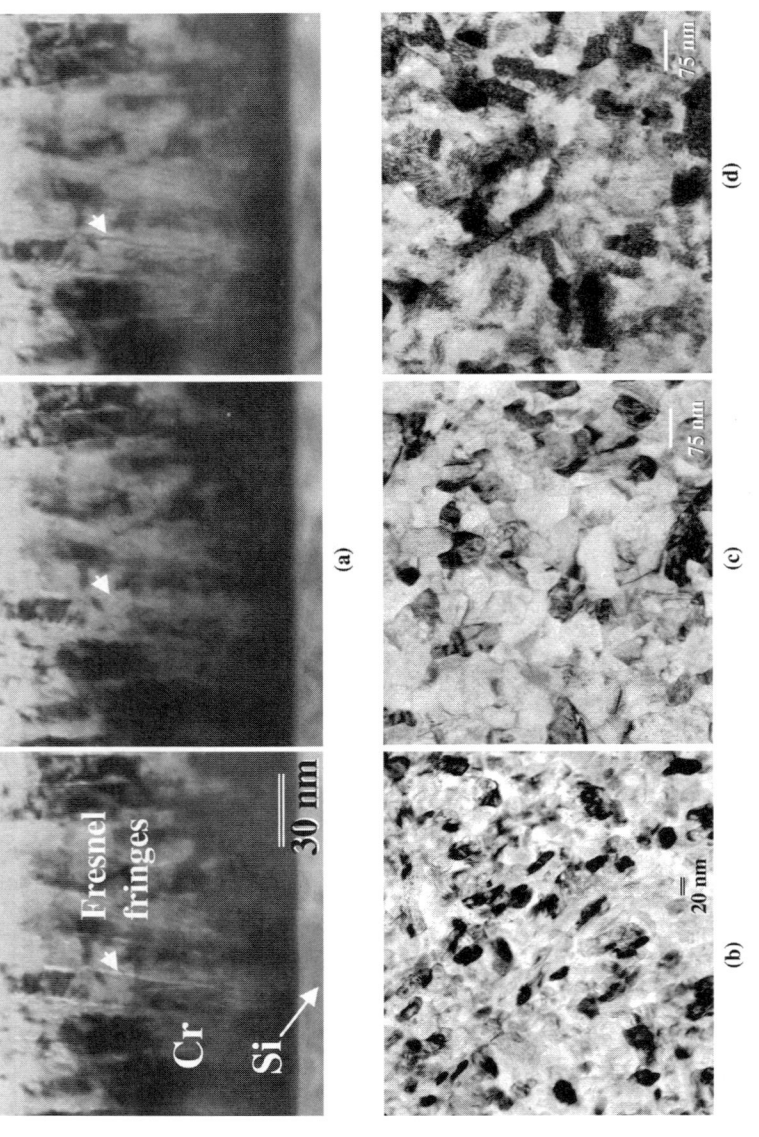

Figure 7.15 TEM micrograph showing (a) cross-sectional view of a 150 nm thick Cr film deposited at 7.5 mtorr Ar pressure without substrate bias (through-focus imaging is used to reveal Fresnel contrast from voided intercolumnar regions), (b) plan view of a 150 nm thick Cr film deposited at 5 mtorr Ar pressure and −50 V bias, (c) plan view of a 1 mm thick Cr film deposited at −50 V bias at 2.5 mTorr Ar pressure, (d) plan view of a 1 mm thick Cr film deposited at −200 V bias at 2.5 mTorr Ar pressure.

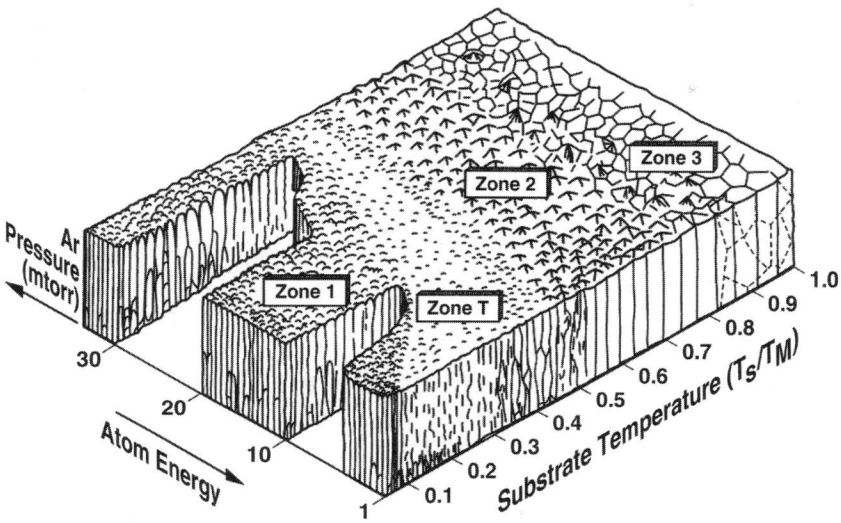

Figure 7.16 Schematic illustration of the relationship between the film microstructure and the energetic particle bombardment conditions (M. Nastasi, J.W. Mayer and J.K. Hirvonen, in *Ion-Solid Interactions: Fundamentals and Applications*, Cambridge University Press, Cambridge, 1996).

Columnar grains with an open structure along the grain boundaries characterize zone 1 in Figure 7.16. The Zone 1 structure results when the deposited atoms diffusion is too limited to overcome the effects of shadowing, and therefore forms at low T_S/T_m, where T_S is the temperature of the substrate and T_m is the melting point of the film material. Shadowing is the process whereby high points on the growing surface receive a higher coating flux than valleys do, and is most prevalent when an oblique deposition flux is present. Shadowing introduces open boundaries that reduce the film density. The addition of bombarding ions during film growth lowers both the temperature window over which Zone 1 structures are stable, and the transition temperature to the formation of Zone T microstructures.

The microstructures in Zone T, or the transition Zone, are also dominated by shadowing effects but have finer structures consisting of a dense array of poorly defined fibrous grains with boundaries that are sufficiently dense to yield respectable mechanical properties. The microstructure in this Zone is attributed to the *onset* of surface diffusion that allows the deposited atoms to migrate on the deposit surface before being covered by the arrival of further material. The boundary of this zone is strongly influenced by substrate temperature and bombarding energy. Intense substrate ion bombardment during deposition can suppress the development of the open Zone 1 structures at low T_s/T_m and ion bombardment yields structures typical of high T_s/T_m depositions.

Zone 2 microstructures nominally form at, T_s/T_m 0·3 – 0.5 where the growth process is dominated by the surface diffusion of the deposited atoms. There is little sensitivity of the microstructure to the ion bombardment energy since temperatures are already high enough to dominate kinetic processes. Columnar grains separated by distinct dense intercrystalline boundaries dominate the structure in this zone.

Zone 3 structures form at $T_s/T_m \geq 0·5$ where bulk diffusion dominates the final structure of the coating. There is little sensitivity of the microstructure to increasing T or to the ion bombardment energy. Recovery and re-crystallization processes typically occur in this temperature regime, driven by the minimization of strain energy and surface energy of the grains.

The effects of energetic particle bombardment on the intrinsic stress evolution and the film microstructures have been described above. In the following sections, we discuss the physical mechanisms that lead to this stress evolution.

7.6.2.2 Tensile Stress

As discussed in Section 7.5, the interatomic forces acting between adjoining islands separated by a small gap (typically, < 1 nm) would lead to an attractive force between islands tending to close the intercolumnar gap. The driving force is the reduction of potential energy of the solid (Figure 7.9) by reducing the interatomic distances in the voided intercolumnar region, at the expense of elastic strain energy from stretching the film in biaxial tension. The elastic strain continues to increase until the islands coalesce completely to form a grain boundary. In other words, significant tensile stress may be generated even before the two free surfaces are replaced by a grain boundary.

This basic concept was first proposed by Hoffman [59] referred to as the grain boundary relaxation model, and discussed in detail recently by Machlin [35] and Nastasi et al. [1]. The idea that energetic particle bombardment leads to the generation of tensile stress via a "densification" process where intercolumnar voids are closed is supported by two kinds of observations: first, TEM observations (Figure 7.15 and Refs. [51,52]) that show closing of gaps on the order of a few atomic layers thick corresponding to an increasing tensile stress; second, molecular dynamics simulation by Müller [60–62] that show the same phenomena at the atomic scale. Figure 7.17 presents the calculated intrinsic stress of a Ni film versus the Ar ion bombarding energy for an ion-to-atom flux ratio of $J_I/J_A = 0.16$. In these simulations it is assumed

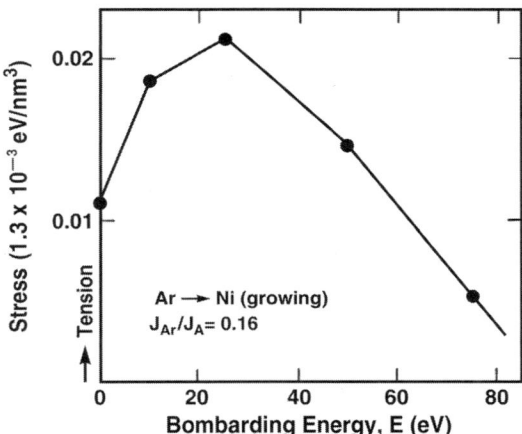

Figure 7.17 Intrinsic tensile stress from MD simulations of an IBAD Ni film as a function of the Ar ion bombarding energy where ion-to-atom flux is 0.16 (R.W. Hoffman, *Thin Solid Films*, 34, 185 (1976)).

that the vapor deposited Ni atoms arrived at the substrate with a kinetic energy of 0.13 eV. The data in Figure 7.17 show that stress state of the Ni film deposited without ion beam assistance is slightly tensile and that the tensile stress initially increases with low energy Ar bombardment, passes over a maximum at an Ar ion energy of about 25 eV and then decreases with increasing bombarding energy. Figure 7.18

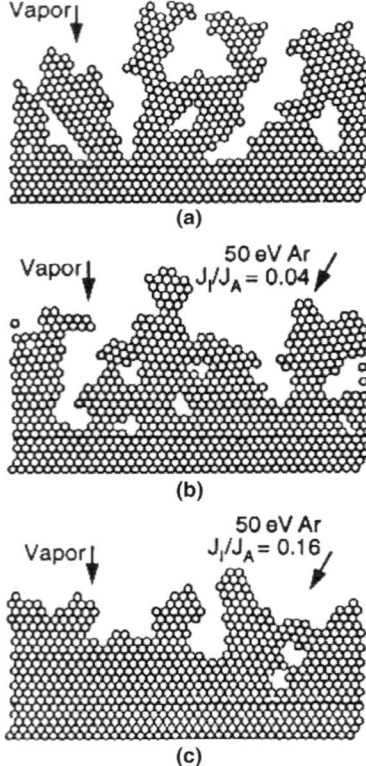

Figure 7.18 Evolution of thin film microstructure in 2-D molecular dynamics simulation for evaporated Ni films: (a) without ion bombardment, (b) with Ar ion bombardment of E = 50 eV, angle of incidence of ions (a) = 30° and ion-to-atom flux of 0.04, and (c) with Ar ion bombardment of E = 50 eV, a = 30° and ion-to-atom flux of 0.16 [61].

contains an MD-generated microstructure for the Ni film grown under the ion bombardment conditions. These microstructures show that the Ni film grown without Ar ion assist contains large micropores and open voids that close (Figure 7.18a–c) with increasing ion bombardment resulting in higher packing density of the film. Since these are very simplistic 2-D simulations, it is really the trends of stress evolution that should be compared with experimental data (e.g., compare Figures 7.17 and 7.14a). The peak value of stress from the MD simulation (Figure 7.17) is about 4 GPa while experiments show about 1–1.5 GPa peak tensile stress in Ni [59, 63].

Using an appropriate interatomic potential, the tensile stress in a thin film can be computed analytically as a function of the gap distance between crystallites as described below. Consider Figure 7.19 that schematically shows interatomic

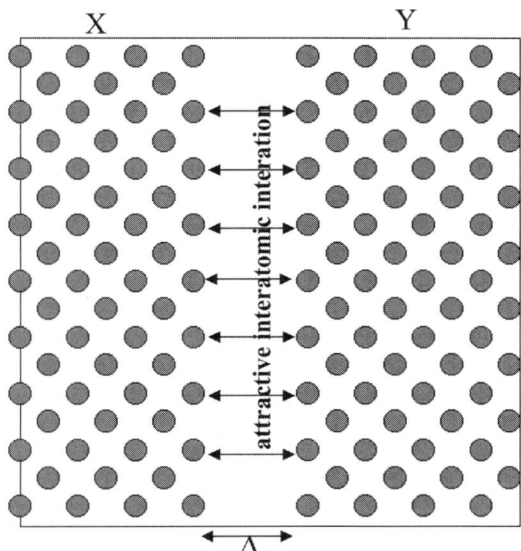

Figure 7.19 Schematic of the attractive interatomic forces between two adjoining grains labeled X and Y separated by a distance Δ.

forces between two crystallites separated by a distance Δ. As a lower bound, consider only the nearest neighbor interatomic forces. In other words, each atom on the surface of crystallite X facing the crystallite Y experiences an attractive force (F) that is readily known from a plot such as Figure 7.9b for a given distance Δ. The tensile stress is then obtained as the net force on each atom times the number of atoms per unit area:

$$\sigma_T = F \ (N_{film})^{2/3} \qquad (7.30)$$

where N_{film} is the atomic density. For bcc refractory metals, the Morse potential has been found to give a reasonable fit to the properties [55,64,65]. Using this potential and $N_{film} = 8.33 \times 10^{22}$ atoms/cm^3 for Cr, σ_T can be plotted as a function of Δ, as shown in Figure 7.20a. The upper bound calculation would include the second and third nearest neighbors interatomic forces as well [55], but is only slightly higher than the lower bound due to rapidly decaying F with Δ relation.

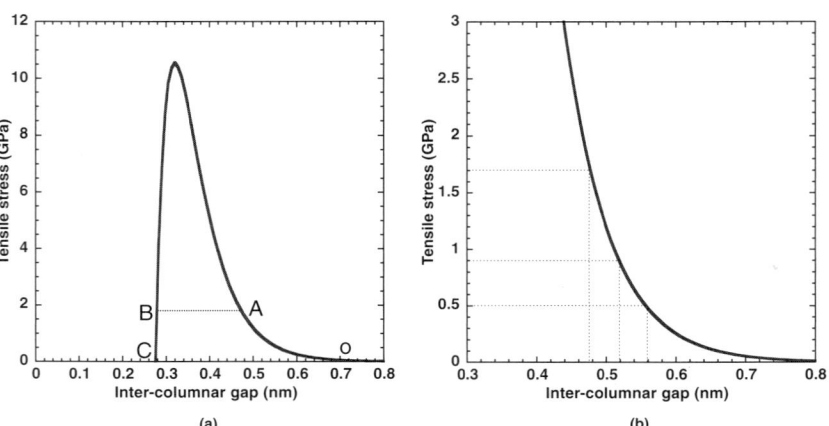

Figure 7.20 Interatomic tensile stress as a function of the inter-columnar gap distance. The line AB in (a) indicates the maximum tensile residual stress limit set by the film yield strength. Further reduction in the gap distance below the point A lead to relaxation of film stress (segment ABC). The segment OA of stress generation matches with the experimental data, shown in detail in (b).

We now compare the model prediction (Figure 7.20a) with the experimental data (Figure 7.14a and b). With increasing bombardment, as the intercolumnar gaps close (TEM data in Figure 7.15), the tensile stress increases to a peak and then rapidly relaxes. This trend is well predicted by the model except that the peak stress is much higher than the experimental observation. We hypothesize that the yield strength for plastic flow sets a limit to the maximum elastic residual stress that can be accumulated. This is discussed further in the next section. The initial rapid buildup of the tensile stress (segment OA in Figure 7.20a) matches with the tensile stress generation in Figure 7.14a and b. The horizontal segment AB in Figure 7.20a shows the film yield strength. Hence, the point labeled A in Figure 7.20a marks the stress and corresponding Δ values at which islands are close enough to spontaneously snap together and in the process relieve some of the stress by dislocation motion (diffusive relaxation may also be possible in high atom mobility materials). This is consistent with recent MD simulation [66]. Reduction in Δ from A to B to C results in relaxation of the stress as Δ approaches the equilibrium value (lowest energy point in the interatomic potential in Figure 7.9a).

To further illustrate the correlation between the model prediction and experimental data, a zoom-in of the OA segment of Figure 7.20a is shown in Figure 7.20b. The experimentally measured stress values (Figure 7.14a) are used to read the corresponding Δ values in Figure 7.20b. For the Cr films deposited with Ar pressures of 7.5, 5 and 3.5 mtorr respectively, we get Δ of ~ 0.56, 0.52 and 0.48 nm, respectively. Two possible approaches for measuring Δ in these films are:

(i) By measuring the film density [55]. Here film density is measured by obtaining the number of atoms per cm^2 by ion-beam analysis methods and the film thickness by profilometry. By knowing the average grain size from TEM and assuming a regular shape (e.g., cylinder or square) of all grains, an average Δ value can be obtained corresponding to the measured film density. This approach [55] yielded good agreement between the Δ values calculated from density measurements and those obtained from Figure 7.20b, considering the crudeness of estimating Δ density measurements and the fact that the interatomic potentials are approximate.

(ii) Δ is measured directly from TEM using the approach described by Page and co-workers [57]. In this approach, Fresnel fringe thickness is measured from the through focus images and plotted as a function of positive (over) and negative (under) defocus. Extrapolating the Fresnel fringe thickness to zero focus gives the gap distance at the boundary. We attempted this approach for the Cr film with 1 GPa tensile stress exhibiting Fresnel fringes at intercolumnar boundaries, and the results are shown in Figure 7.21. This approach shows that Δ is ~ 0.5–0.6 nm for this sample consistent with the model prediction of 0.52 nm from Figure 7.20b.

In summary, starting with non-energetic deposition as the particle bombardment is increased, the generation of tensile stress can be estimated as follows: (i) obtain the interatomic force–distance curve for the material using accurate interatomic potentials (analogous to Figure 7.9b), (ii) multiply the interatomic force with the number of atoms per unit area in the material (Equation (7.30)) to give the stress as a function of interatomic distance (analogous to Figure 7.9b), (iii) obtain the evolution of the intercolumnar gap distance from TEM measurements (Figure 7.21) or from density and grain size measurements — the stress corresponding to a given intercolumnar gap distance can be read from data such as Figure 7.20b to the point of peak in tensile stress. As discussed above, the peak in tensile stress may be set by residual stress increasing to the level of the film yield strength, at which point dislocation motion and spontaneous closure of the intercolumnar gaps relax the tensile stress to zero. More detailed studies using MD simulation to study tensile stress evolution from intercolumnar gap closure have recently been initiated

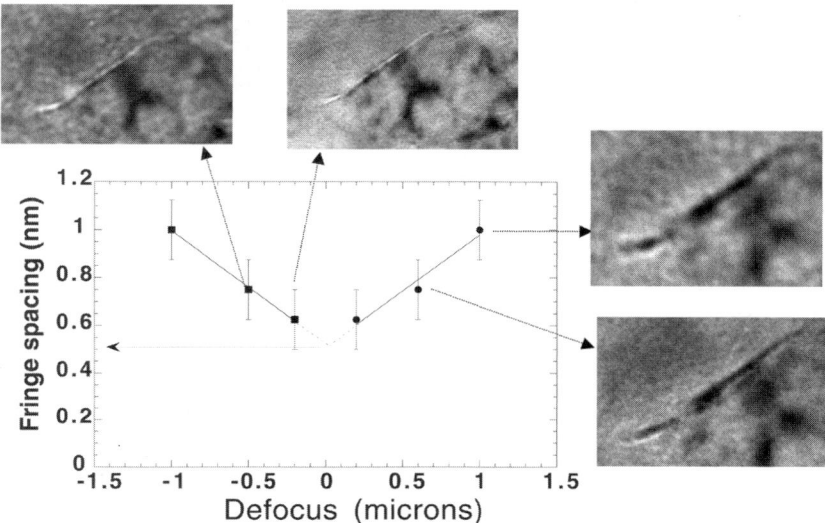

Figure 7.21 Estimation of the intercolumnar gap distance from thickness of the Fresnel fringes in TEM images (plan view imaging from a Cr film).

[67]. The generation of compressive stress at higher bombardment energy/flux is discussed next.

7.6.2.3 Compressive Stress

The hypothesis here is that compressive stress generation during energetic particle deposition is due to the production of irradiation-induced point defects that add positive volume to the film, causing it to swell [53,55,68–70]. However, the constraint that the film must remain rigidly bonded to the substrate prevents any in-plane expansion of the film, leading to compressive stress generation in the plane of the film (Figure 7.2).

Developing a model for compressive stress generation based on the above hypothesis requires knowledge of defect concentrations and the magnitude of local expansion in the lattice introduced by the defects. As described in Section 7.5, interstitials add positive volume to the material and vacancies add a negative volume. However, an interstitial-vacancy pair (Frenkel defect) has a net positive relaxation volume. Thus, compressive stress can be generated if the number fraction of interstitials induced by irradiation either equals or exceeds the number fraction of vacancies generated.

Consider the case of ion-assisted growth of films. The ion bombarding energies used are typically low, on the order of a few hundred eV to a few keV. Under these conditions, interstitials are created by displacing atoms from the surface and driving them into the bulk, leaving vacancies behind at the surface. The maximum amount of energy that can be transferred to a surface atom by the bombarding ion, T_M, occurs for a head-on collision, and can be calculated by [1]:

$$T_M = \frac{4 M_1 M_2}{\left(M_1 + M_2\right)^2} E_0 \qquad (7.31)$$

where M_1 and E_0 are the mass and energy of the ion, respectively, and M_2 is the mass of the target atom.

Equation (7.31) shows that $T_M = E_0$ only for the condition when $M_1 = M_2$ and for all other combinations where $M_1 \neq M_2$, less energy than E_0 will be transferred. The more energy transferred to the surface atom the greater will be its range and the further the interstitial will reside from the surface. Monte Carlo simulations using the TRIM code show that surface atoms that recoiling with energies in the 100 eV to 2 keV range will be implanted to depths between 0.5 and 5 nm (TRIM) [1]. However, the probability of a head-on collision is extremely low with the majority of the collisions occurring at glancing angles where significantly less energy is transferred, which in turn will result in shallower recoil implantation depths. Therefore the interstitials created by this process will reside within a few monolayers of the surface, while the vacant site left at the surface will most probably be filled by a subsequent depositing atom. Therefore it is expected that there will be an imbalance in the number of vacancies and interstitials formed by the bombardment process during ion beam enhanced deposition, with interstitials and interstitial-derived extended defects being dominant. However, it should be noted that the high stress associated with these interstitials, along with their close proximity to the surface, can result in the athermal migration of the interstitial back to the surface, thereby reducing the number of interstitials that are actually retained in the growing film. Therefore, for any given processing condition it will be difficult to know the exact number fractions of interstitials and vacancies retained in the film. We proceed with developing a model with the assumption that all defects induced by irradiation are Frenkel pairs, to obtain a limit of the stress that can be developed by Frenkel pairs.

In general, the biaxial stress equation for the addition of a defect that changes the volume of the crystal by ΔV_{def} is given by [3]:

$$\sigma_{xx}^{defect} = -\frac{Y}{1-\nu} F_{defect} \frac{\Delta V_{defect}}{3} \tag{7.32}$$

where Y is Young's modulus, ν is Poisson's ratio of the material wherein the defect resides, and F_{defect} is the number of defects per unit volume in the material. Generalizing this equation to the Frenkel defect yields

$$\sigma_{xx}^{FD} = -\frac{Y}{1-\nu} F_{FD} \frac{V_{FD}^{rel}}{3} \tag{7.33}$$

where V_{FD}^{rel} is the relaxation volume of the Frenkel defect given in units of atomic volumes, and F_{FD} is the number fraction of Frenkel defects. The data in Table 7.1 show that V_{FD}^{rel} is always greater than zero, which when applied to Equation (7.33) shows that the sign of stress will be negative. Therefore, the formation of Frenkel defects adds compressive stress to the irradiated material.

Given that V_{FD}^{rel} is known or can be estimated, the ultimate magnitude of the stress that results from Frenkel defects will depend on the atomic fraction of Frenkel defects in the material, F_{FD}. The average number of displaced atoms (N_d) produced by an ion of energy E can be estimated using Monte Carlo computer simulations, such as TRIM [33] or can be calculated using the Kinchin–Pease displacement damage function [71–74], which is given by:

$$
\langle N_d(E) \rangle =
\begin{cases}
0 & \text{for } 0 < E < E_d \\[2mm]
1 & \text{for } E_d \leq E < 2E_d / \xi \\[2mm]
\dfrac{\xi v(E)}{2E_d} & \text{for } 2E_d / \xi \leq E < \infty
\end{cases}
\tag{7.34}
$$

where E_d is called the *displacement energy*, and is the energy that a target atom has to receive in order to leave its lattice site and form a stable interstitial, thereby making a Frenkel defect. This energy depends on the direction of the momentum of the target atom. Therefore, a range of displacement energies exists for the creation of a Frenkel pair in a given material. Typical average values of E_d are in the range of 15 to 40 eV [71]. The parameter $n(E)$ in Equation (7.34) is called the *damage energy* and represents the fraction of the ion energy that goes into displacement (nuclear) processes. The damage energy is closely related to the PKA's total nuclear stopping, but is always ($\approx 20 - 30\%$) smaller [1]. Therefore, $n(E)$ can be estimated according to:

$$
v(E) \cong 0.8 \int_0^{E_0} \left. \frac{dE}{dx} \right|_n dx
\tag{7.35}
$$

where $(dE/dx)_n$ is the nuclear energy loss rate, which can calculated from the following relation:

$$
S \equiv \frac{dE / dx}{N}
\tag{7.36}
$$

where N is the atomic density (atoms/cm^3), and S is the nuclear stopping. At very low ion or PKA energies, such as those used in ion assisted deposition, $n(E)$ @ E.

Equation (7.34) describes the case for one ion. For a flux of ions that can penetrate the target to a distance R, the ion range, the atomic fraction of Frenkel defects formed can be estimated from

$$F_{FD} = \frac{\langle N_d \rangle}{N} \cong \frac{0.4 \, \nu(E)}{N \, R E_d} \phi \qquad (7.37)$$

where f is ion dose (ions/cm^2). It is important to note that when the above equations are used to calculate $\langle N_d \rangle$ and F_{FD}, or if these values are determined using Monte Carlo simulation, that the values obtained represent the number or fraction of Frenkel defect generated by an irradiation process as opposed to actual number of defects that ultimately remain in the material. This difference exists because these defect estimates do not take into account defect diffusion, that lead to vacancy-interstitial recombination or the collapse of point defects into other defect structures such as dislocation loops.

The implementation of Equation (7.37) into Equation (7.33) to calculate the stress requires a value for $\nu(E)$, which in turn requires that we can calculate $(dE/dx)_n$ (Equation (7.35)). The nuclear energy loss rate can be calculated using the Ziegler, Biersack and Littmark (ZBL) universal nuclear stopping formula [33], which for an ion with energy E_0 in the laboratory reference frame is given by Equation (7.23).

Equations (7.34)–(7.37) show how F_{FD} can be calculated using simple ion–solid interactions theory. For the data shown in Figure 7.14b and c, where substrate bias was used to produce energetic particle bombardment, some parameters such as ion dose and range, etc., are not known. Hence, we apply Equation (7.33) to the experimental data as described below.

The maximum compressive stress was ≈ 2 GPa in the experimental data shown in Figure 7.14b. Ion beam analysis revealed the presence of entrapped Ar in the compressively stressed Cr film. Hence, we rewrite Equation (7.33) as:

$$\sigma_{comp}^{max} = \left(\frac{1}{3}\right)\left(\frac{Y}{1-\nu}\right)\left(F_{FD} V_{FD}^{rel} + F_{Ar} V_{Ar}^{rel}\right) \qquad (7.38)$$

where F_{Ar} and V_{Ar}^{rel} are the atomic concentrations and relaxation volumes of Ar in Cr, respectively. F_{Ar} was measured by ion-beam analysis to be ~ 0.005 for the Cr films with maximum compressive stress [53], and V_{Ar}^{rel} was estimated to be 1.75 from literature values [34], as described in Ref. [53]. The relaxation volume for Frenkel defects in Cr (V_{FD}^{rel}) is reported to be 1 [34]. Using $(Y/1-\upsilon) \approx 300$ GPa for Cr [53] and σ_{comp}^{max} − 2 GPa as experimentally measured, we calculate F_{FD} from Equation (7.38) as 0.011. Thus, a Frenkel defect concentration of ~ 1% could explain the compressive stress observed in Cr films.

Consider a Frenkel defect concentration of 1%. This would mean that a defect pair must exist every 100 atoms. However, when vacancies and interstitials are in such close proximity to each other athermal recombination is highly probable. Both experimental (at 4.2 K) and computer calculations have found that the spontaneous recombination volume for vacancy-interstitial pairs is between 50 and 100 atomic volumes, and that the Frenkel defect concentration saturates somewhere between 1 and 2% [1]. Since the athermal recombination volume limits F_{FD}, it sets the upper limit to the maximum compressive stress (Equation 7.33) that a material can experience due to the presence of Frenkel defects.

The assumption that all defects produced in thin films during energetic particle deposition are Frenkel defects provides an estimate of the compressive stress generated. For post-deposition ion irradiation at high energies and to high doses, defects may form clusters or even collapse into dislocation loops [74]. These loops preferentially align normal to the tensile stress direction [75] to relax the tensile stress, and continued growth in the size of the loops eventually leads to the development of compressive stress [14,76]. The compressive stress buildup in these high energy ion-irradiated films has been shown to be proportional to $(dpa)^{2/3}$ where dpa is displacements per atom [14,76].

The above description demonstrates that the generation of compressive stress in ion-synthesized films can be interpreted in terms of point defect structures. The correlation of processing parameters with defect structures produced in the film is not described here, and the reader is referred to other texts [1,34].

7.6.2.4 Limits of Residual Stress

The mechanisms of generation of intrinsic elastic residual stresses (tensile or compressive) in thin films are discussed above. Here we show that the maximum values of these elastic residual stresses are set by the plastic yield strength of the film (diffusive relaxation is not considered).

In thin films with the microstructural length scales (in-plane grain size and film thickness) on the order of a few tens of nanometers, continuum-scale dislocation pile-ups cannot be supported and hence, Hall-Petch type models for strengthening due to grain refinement breakdown [77,78]. The plastic flow in these fine-scale structures is accomplished by the motion of single dislocations by bowing between the boundaries (or interfaces) and creating interface dislocations, as shown schematically in Figure 7.22 [79–82]. The yield strength as determined by this Orowan bowing stress is given as, for the case when the in-plane grain size (d) << film thickness (h), [82]:

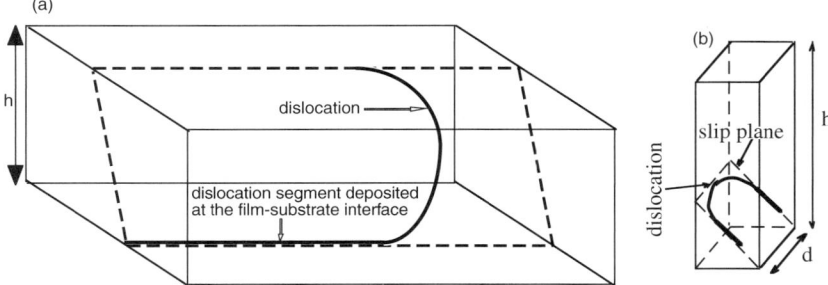

Figure 7.22 Schematic of the Orowan mechanism for dislocation glide in polycrystalline thin films (only one grain is shown for simplicity). The process involves dislocation bowing between the obstacles, film–substrate interface as in (a) or grain boundary as in (b), and creating interface dislocations as the bowed segment advances on the slip plane shown by dotted line. In (a), the film thickness (h) is much smaller than the grain size (d) and hence, strength is primarily determined by h. In (b), d << h and hence, d is used to determine the film yield strength.

$$\sigma_{\text{Orowan}} = M \frac{Gb}{4\pi(1-\upsilon)} d^{-1} \ln\left(\frac{d}{b}\right) \qquad (7.39)$$

where M is the Taylor factor ($\approx$ 2.75 for bcc metals), G is the shear modulus (95 GPa for Cr films), and b is Burger's vector ($\approx$ 2.5 Å for Cr) [53]. In general, grain size and film thickness are additive when applying the Orowan model to estimate the yield strength of thin films [82]. However, if h << d (Figure 7.22a), then h is used in Equation (7.39) instead of d. For the data shown here for columnar-grain Cr films (Figure 7.14), d << h (Figure 7.22b) and hence, Equation (7.39) applies with d instead of h. Substituting these values we obtain σ_{Orowan} = 1.4 ± 0.1 GPa, for d = 21 ± 2 nm, in good agreement with the maximum tensile stress measured (Figure 7.14a and b).

The compressive saturation stress could also be calculated using Equation (7.39), except that an additional term needs to be added to account for hardening due to the radiation-induced point defects (Frenkel pairs or interstitial clusters) in ion-bombarded films having large compressive stress (Figure 7.15d). Hardening due to interstitial solutes is more rapid as compared to substitutional solutes since the strain field around interstitial solutes is usually elliptical while substitutional solutes have spherically symmetric strain field [83]. This rapid hardening ($\Delta\sigma_{\text{defects}}$) due to interstitial solutes is described by the following model [83]:

$$\Delta\sigma_{\text{defects}} = M \frac{G\Delta\varepsilon \sqrt{c}}{\alpha} \qquad (7.40)$$

where α is an empirical parameter, c is the solute concentration, and $\Delta\varepsilon$ is the difference between the longitudinal and transverse strains in the elliptical strain field of interstitials. The $\Delta\sigma \propto \Delta\varepsilon$ dependence implies a diffuse interaction between solutes and dislocations, whereas a localized interaction is more applicable for impenetrable solute clusters leading to a $\Delta\sigma \propto \Delta\varepsilon^{3/2}$ type dependence [83].

We now apply Equation (7.40) to the case of Cr films. From the parameters used in Equation (7.38), we estimate the solute concentration as, c = f_{FD} + f_{Ar} = 0.011 + 0.005 =

0.016. $\Delta\varepsilon = 0.40$ and $\alpha = 20$ for typical interstitial hardening in bcc metals at room temperature [83]. Using these values, we obtain $\Delta\sigma_{\text{defects}}$ from Equation (7.40) as 0.66 GPa. The new yield strength of the film (at high bias deposition), σ_y, is given as:

$$\sigma_y = \sigma_{\text{yield}}^{\text{gb}} + \Delta\sigma_{\text{defects}} \tag{7.41}$$

With $\sigma_{\text{yield}}^{\text{gb}} = 1.56$ GPa (using grain size of ~ 19 nm for films with maximum compressive stress) and $\Delta\sigma_{\text{defects}} = 0.66$ GPa, it follows from Equation (7.41) that $\sigma_y = 2.22$ GPa. This is in good agreement with the experimentally measured saturation in compressive stress, indicating that the biaxial yield strength sets a limit to the maximum elastic intrinsic residual stress in films. The relaxation of compressive stress with increasing bias (Figure 7.14c) has also been observed in another investigation [84]. Since we have shown that the maximum in compressive stress is correlated with point defect buildup and saturation, it is likely that plastic flow in these materials is accomplished, in part, by irradiation enhanced creep [76].

7.7 OTHER PHENOMENA INDUCED BY ENERGETIC PARTICLE BOMBARDMENT THAT MAY INFLUENCE THE INTRINSIC FILM STRESS

7.7.1 Materials Transport and Ion Mixing

It is clear that significant atomic rearrangement occurs in films grown under ion bombardment. In fact it is well known that most materials subjected to ion irradiation will experience significant atomic rearrangement. The most obvious example of this phenomenon is the atomic intermixing and alloying that can occur at the interface separating two different materials during ion irradiation. This process is known as ion beam [85–91] mixing. While many of the examples of ion mixing deal with ion irradiation stimulated interface alloying the fundamental physics that control this phenomena are also responsible for the evolving microstructure, atomic density, and stress that evolve in films grown under ion bombardment.

The interaction of an energetic ion with a solid involves several processes. As an ion penetrates a solid, it slows down by depositing energy to both the atoms and the electrons of the solid. At low ion energies this slowing down process is dominated by the deposition of energy to the target atoms, which occurs by way of nuclear collisions. During the nuclear collision portion of this process, target atoms can be permanently displaced from their lattice sites and relocated several lattice sites away. The displacement mechanism of this atomic rearrangement is the fundamental principle governing ballistic mixing. Ballistic mixing is a purely physical process, dependent only on the kinematics that control nuclear collision events and is independent of thermodynamic driving forces. Two forms of ballistic mixing are recoil mixing and cascade mixing.

7.7.1.1 Recoil Mixing

When an incident ion strikes a target atom some of the incident ion's kinetic energy will be transferred to the target atom. Under such circumstances the struck atom may have sufficient energy to recoil far from its initial location. This process, which results in the transport of atoms through single collision events between the incident ions and target atoms, is the simplest form of ballistic mixing, and is known as recoil implantation, recoil mixing, or sub-implantation. For mixing to be effective by this process, the recoil should travel the maximum range possible; a maximum range will result when the collision between the incident ion and the target atom is head-on ($\theta = 0$). The probability of a head-on collision is very small, with most collisions being *soft* (i.e., $\theta > 0$). The recoils produced in such soft collisions will possess significantly less energy and have reduced ranges relative to head-on collisions, and their trajectory will not be in the forward direction. As a result, the number of target atoms contributing to atomic mixing by the mechanism of recoil implantation will be small.

7.7.1.2 Cascade Mixing

In addition to recoil mixing, other ballistic phenomena are possible during ion bombardment of a solid; enhanced atomic mixing can occur when multiple displacements of target atoms result from a single incident ion. In the multiple displacement process, an initially displaced target atom (primary recoil) continues the knock-on-atom processes, producing secondary recoil atom displacements, which in turn displace additional atoms. The multiple displacement sequence of collision events is commonly referred to as a *collision cascade* (Figure 7.10). Unlike the recoil implantation process, where one atom receives a kinetic energy in a single displacement, atoms in a collision cascade undergo multiple uncorrelated low-energy displacement and relocation events. Atomic mixing resulting from a series of uncorrelated low-energy atomic displacements is referred to as *cascade mixing*.

Calculations of the mean energy of atoms in a cascade show that most recoils are produced near the minimum energy necessary to displace atoms, E_d. Due to the low-energy stochastic nature of these displacement events, the initial momentum of the incident particle is soon lost, and the overall movement of the atoms in a collision cascade becomes isotropic. This isotropic motion gives rise to an atomic redistribution that can be modeled as a random-walk of step size defined by the mean range of an atom with energy near E_d. The effective diffusivity, D_{cas}, for a collision-cascade-induced random-walk process can be expressed as [92]:

$$D_{cas}t = \frac{dpa(x) <r^2>}{6} \qquad (7.42)$$

where $dpa(x)$ is the number of cascade induced displacements-per-atom that occur at a distance x, and $<r^2>$ is the mean squared range of the displaced target atoms. The dpa, resulting from a given dose of ions can be expressed as

$$dpa(x) = \frac{0.4\, F_D(x)\phi}{E_d N} \qquad (7.43)$$

where $F_D(x)$ is the damage energy per unit length at distance x, ϕ is the ion dose, and N is the atomic density. Combining Equations (7.42) and (7.43) gives the effective diffusion coefficient due to ballistic cascade mixing as

$$D_{cas}t = 0.067 \frac{F_D(x)<r^2>}{N\,E_d}\phi \qquad (7.44)$$

A more detailed theoretical formulation of collisional mixing was made by Sigmund and Gras-Marti [93] that accounts for mass difference between the ion, M_1, and the target atom, M_2. However, the primary features the Sigmund and Gras-Marti ballistic cascade mixing equation are the same as Equation (7.44), which indicated that the effective diffusion coefficient under ballistic mixing scales with the dose, f, and the damage energy, F_D. An additional characteristic of the ballistic cascade mixing equations is that they do not contain any temperature-dependent terms. The effective diffusion coefficient described by Equation (7.44) is independent of temperature and can be compared only with experiments in which mixing is also observed to be independent of temperature.

It is easy to envision how these purely physical processes, recoil mixing and cascade mixing, can produce the atomic rearrangements that results in the densification of films grown under ion bombardment. However, a shortcoming of these processes is that they do not take into account chemical gradients, gradients in electric field, pressure, or temperature that will alter the chemical potential of a system and supply a driving force for atomic intermixing.

7.7.1.3 Thermodynamic Effects in Ion Mixing

In certain systems ballistic phenomena, such as cascade mixing, has been found to be inadequate in describing experimental observations. However, these discrepancies could be alleviated if the ballistic description of atomic mixing was modified to account for thermodynamic driving forces. The first work to quantitatively describe the role of thermodynamics during ion mixing was carried out by Johnson et al. [94], who derived a phenomenological expression for the ion-mixing rate that included a Darken term to account for the chemical potential that exists at the interface between different elements. In this way the equation for the effective diffusion coefficient was transformed from one that described a random walk process to one that described a biased random walk.

An extensive set of experiments has shown that ion mixing can be correlated with thermodynamic driving forces [95–99]. The majority of this work has focused on the driving force presented by the gradient in chemical differences (activity coefficient or heat of mixing) that exist between the intermixing species. It is well known in atomic diffusion that in addition to chemical gradients, gradients in electric field, pressure, or temperature alter the chemical potential of a system and can supply a driving force for interdiffusion [100]. Since it has been established that analogs exist between interdiffusion and ion-stimulated ion mixing, it is reasonable to assume that gradients in pressure or stress, both of which can influence the chemical potential, can also influence the ion mixing process.

Nastasi and co-workers have examined the effect of stress in ion mixing [101]. In their analysis the change in chemical potential that an atom would experience upon a change in biaxial stress of $\Delta\sigma$ is given by

$$\Delta\mu_\sigma = -\frac{2}{3}\Omega\Delta\sigma \qquad (7.45)$$

where Ω is the molar volume. This equation indicates that there is a driving force for an atom to move toward regions of more positive stress (e.g., from regions of compressive stress to region of tensile stress). Given that interstitial atoms in most metals are in a state of compressive stress and that voids, pores and grain boundaries are in a state of tensile stress, cascades that form at the boundaries of open volumes will result in atomic transport into these volumes.

The influence of cascades on reducing tensile stress has been examined and modeled by Brighton and Hubler [102], who postulated that stress relief occurs if each atom in the growing film is involved with at least one cascade. If the average volume affected by a cascade is V_{cas}, and the average atomic density in the film is N, then the average number of atoms affected per cascade is NV_{cas}. Therefore, if one ignores cascade overlap, a lower limit for the critical ion-to-atom flux ratio, $\left(J_I / J_A \right)_c$, for stress relief can be expressed as:

$$\left(J_I / J_A \right)_c = \left(N V_{cas} \right)^{-1} \qquad (7.46)$$

With cascade volumes derived from Monte Carlo simulations, Brighton and Hubler found excellent agreement with the experimental data of Hirsch and Varga [103]. Experiments have also shown that cascades and radiation damage can also lead to a reduction in compressive stress. The relief of compressive stress in ion-bombarded amorphous materials has been extensively studied, with the bulk of the work focused on high-energy ion bombardment [12,104–106]. This work shows that under these conditions stress relief is accomplished by radiation-induced plastic/viscous flow. Similarly, plastic flow is also expected [107] and observed [14,51,108] in irradiated metal films. In this case the plastic flow occurs by point defect rearrangement into dislocation loops that plate out in sympathy with existing stress. The effects of ion irradiation induced interfacial mixing on the intrinsic stress have been observed in metallic multilayers such as Ag–Fe [109] and Cu–W [110,111]. Ion beam-induced intermixing at the interfaces, in the otherwise immiscible CuW system, was inferred from XRD measurements [110]. In the case of Ag–Fe [109], ion beam mixing occurred to the extent that the crystal structure of the Fe layer changed from bcc to fcc (Fe–Ag alloy).

7.7.2 Bombardment-Induced Phase Transformation

Another effect of energetic particle bombardment that has a consequence on intrinsic film stress is phase transformation in the film induced by energetic particle bombardment. The effect of phase transformation on stress originates from the volume difference in the product versus the parent phase. If the phase transformation tends to produce in-plane expansion, then the constraint from the substrate would lead to compressive stress. However, if the phase transformation would lead to an in-plane shrinkage of the film were it not attached to the substrate, then the constraint from the substrate would produce tensile stress. A well-known example is amorphization (e.g., in Si [12]) at high doses of ion irradiation. The amorphization and the related radiation-enhanced plastic flow may lead to relaxation of compressive stress [12,112,113]. Similar effects have been observed in ion-irradiated metal silicide films on Si substrates [114,115]. Other examples are discussed in Ref. [35].

7.8 SUMMARY

The intrinsic stress evolution during energetic particle deposition of thin films is summarized in Figure 7.23. Films deposited at low homologous temperatures without energetic ion bombardment, exhibit a low-density porous structure and will in general have a density that is lower than the nominal density and be in a state of tensile stress. As the energetic ion bombardment is increased, the densification of the film starts, leading to tensile stress generation due to attractive interatomic forces across the sub-nanoscale intercolumnar gaps in the film. The film is nearly stress free when the intercolumnar gap distance approaches the equilibrium interatomic distance for minimum potential energy. Under conditions where diffusion is limited and ion bombardment during growth is intensified, the implantation of the assisting ions and/or the formation of subsurface interstitial atoms that result from surface collision process that push deposited atoms into the bulk, can create an excess density and a state of compressive stress that scales with the concentration and relaxation volumes of these point defects. At high bombardment energies and/or dose, compressive stress may relax. While this work focused on correlating stresses to atomic-scale defects in thin films, more work is needed in correlating the processing conditions to the defect structure of ion-bombarded films or surfaces. This is a rich area for further research particular simulations (e.g., molecular dynamics,

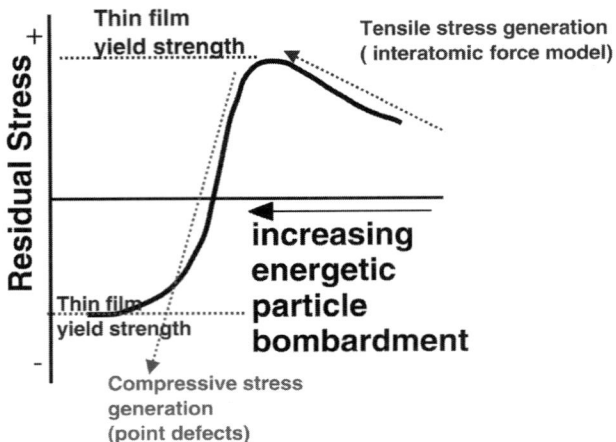

Figure 7.23 Schematic illustration of the stress evolution trend with energetic particle bombardment and the related physical mechanisms of residual stress generation and stress limits.

Monte Carlo, etc.) that can accurately account for the time and length scales of these phenomena.

ACKNOWLEDGMENTS

This research is funded by the U.S. Department of Energy, Office of Basic Energy Sciences. The authors acknowledge discussions with J.P. Hirth, R.G. Hoagland, J. Sprague and H. Kung.

REFERENCES

1. M. Nastasi, J.W. Mayer and J.K. Hirvonen, in *Ion-Solid Interactions: Fundamentals and Applications*, Cambridge University Press, Cambridge, 1996.

2. R.A. Powell and S. Rossnagel, *PVD for Microelectronics*, 26, Academic Press, 1999.

3. M.F. Doerner and W.D. Nix, in *Stress and Deformation Processes in Thin Films on Substrates,* CRC Critical Reviews in Solid State Materials Sciences, 14, 225 (1988).

4. R.A. Roy, J.J. Cuomo and D.S. Yee, *J. Vac. Sci. Tech.*, A6, 1621 (1988).

5. D.W. Hoffman, *Thin Solid Films*, 107, 353 (1983).

6. R. Koch, *J. Phys.: Condens. Matter*, 6, 9519 (1994).

7. K.N. Tu, J.W. Mayer and L.C. Feldman, *Electronic Thin Film Science for Electrical Engineers and Materials Scientists*, Macmillan, New York, 1992.

8. M. Ohring, *The Materials Science of Thin Films*, Academic Press, New York, 1992.

9. G.G. Stoney, *Proc. R. Soc. Lond.* A, 82, 172 (1909).

10. P.A. Flinn, D.S. Gardner and W.D. Nix, *IEEE Trans. Electron Devices*, ED-34, 689 (1987).

11. A.I. Van Sambeek and R.S. Averback, *Mater. Res. Soc. Symp. Proc.*, 396, 137 (1996).

12. C.A. Volkert, *J. Appl. Phys.,* 70, 3521 (1991).

13. K. Kinosita, K. Maki, K. Nakamizo and K.Takeuchi, *Jpn. J. Appl. Phys.,* 6, 42 (1967).

14. A. Misra, S. Fayeulle, H. Kung, T.E. Mitchell and M. Nastasi, *Nucl. Instrum. Methods* B, 148, 211 (1999).

15. Y.S. Kim and S.C. Shin, *Mater. Res. Soc. Symp. Proc.*, 382, 285 (1995).

16. A.L. Shull and F. Spaepen, *J. Appl. Phys.,* 80, 6243 (1996).

17. V. Ramaswamy, B.M. Clemens and W.D. Nix, *Mater. Res. Soc. Symp. Proc.*, 528, 161 (1998).

18. B.M. Clemens and J.A. Bain, *MRS Bull.*, July, 46 (1992).

19. H. Dosch, *Phys. Rev.* B, 35, 2137 (1987).

20. I.C. Noyan and J.B. Cohen, *Residual Stress: Measurement by Diffraction and Interpretation*, Springer-Verlag, New York, 1987.

21. A. Segmuller and M. Murakami, in H. Herbert (Ed.), *Treatise on Materials Science and Technology*, 27, 143, Academic Press, New York, 1988.

22. J.J. Cuomo and S.M. Rossnagel, *Nucl. Instrum. Methods*, B, 19/20, 963 (1987).

23. J.K. Hirvonen, *Mater. Sci. Rep.*, 6, 215 (1991).

24. F.A. Smidt, *Int. Mater. Rev.*, 35, 61 (1990).

25. H. Windishmann, *Crit. Rev. Solid State Mater. Sci.*, 19, 547 (1992).

26. P.Y. Jouan and G. Lemperiere, *Vacuum*, 42, 927 (1991).

27. S. Kadlec, C. Quaeyhaegens, G. Knuyt and L.M. *Stals, Surf. Coat. Technol.*, 89, 177 (1997).

28. I. Petrov, A. Myers, J.E. Greene and J.R. Abelson, *J. Vac. Sci. Technol.* A, 12, 2846 (1994).

29. R.E. Somekh, *J. Vac. Sci. Technol.* A, 2, 1285 (1984).

30. J.H. Rose, J.R. Smith and J. Ferrante, *Phys. Rev.* B, 28, 1835, (1983); *Phys. Rev. Lett.*, 47, 675 (1981).

31. J.H. Rose, J.R. Smith, F. Guines and J. Ferrante, *Phys. Rev. B*, 29, 2963 (1984).

32. A. Banerjea and J.R. Smith, *Phys. Rev. B*, 37, 6632 (1988).

33. J.F. Ziegler, J.P. Biersack and U. Littmark, *The Stopping and Range of Ions in Solids*, Pergamon Press, New York, 1985.

34. P. Ehrhart, P. Jung, H. Schultz and H. Ullmaier, in *Atomic Defects in Metals*, Landolt-Bornstein, Group III, 25, Springer-Verlag, Berlin, 1992, Chapter 2.

35. E.S. Machlin, *Materials Science in Microelectronics — The Relationships between Thin Film Processing and Structure*, vol. 1, GIRO press, New York, 1995, pp 157–184.

36. W.D. Nix and B.M. Clemens, *J. Mater. Res.*, 14, 3467 (1999).

37. C.V. Thompson, *J. Mater. Res.*, 14, 3164 (1999).

38. C.V. Thompson and R. Carel, *J. Mech. Phys. Solids*, 44, 657 (1996).

39. R.C. Cammarata, T.M. Trimble and D.J. Srolovitz, *J. Mater. Res.*, 15, 2468 (2000).

40. M.A. Phillips, V. Ramaswamy, B.M. Clemens and W.D. Nix, *J. Mater. Res.*, 15, 2540 (2000).

41. J.A. Floro, S.J. Hearne, J.A. Hunter, P. Kotula, E. Chason, S.C. Seal and C.V. Thompson, *J. Appl. Phys.*, 89, 4886 (2001).

42. S.C. Seel, C.V. Thompson, S.J. Hearne and J.A. Floro, *J. Appl. Phys.*, 88, 7079 (2000).

43. L.B. Freund and E. Chason, *J. Appl. Phys.*, 89, 4866 (2001).

44. E. Chason, B.W. Sheldon, L.B. Freund, J.A. Floro and S.J. Hearne, *Phys. Rev. Lett.*, 88, 156103, (2002).

45. B.W. Sheldon, K.H.A. Lau, and A. Rajamani, *J. Appl. Phys.*, 90, 5097 (2001).

46. A. Misra, H. Kung, T.E. Mitchell and M. Nastasi, *J. Mater. Res.*, 15, 756 (2000).

47. R.C. Cammarata and K. Sieradzki, *Annu. Rev. Mxater. Sci.*, 24, 215 (1994).

48. F. Spaepen, *Acta Mater.*, 48, 31 (2000).

49. R.C. Cammarata, K. Sieradzki and F. Spaepen, *J. Appl. Phys.*, 87, 1227 (2000).

50. D.P. Adams, L.J. Parfitt, J.C. Billelo, S.M. Yalisove and Z.U. Rek, *Thin Solid Films*, 266, 52 (1995).

51. A. Misra, S. Fayeulle, H. Kung, T.E. Mitchell and M. Nastasi, *Appl. Phys. Lett.* 73, 891 (1998).

52. A. Misra and M. Nastasi, *J. Mater. Res.*, 14, 4466 (1999).

53. A. Misra and M. Nastasi, *Appl. Phys. Lett.*, 75, 3123 (1999).

54. A. Misra and M. Nastasi, *J. Vac. Sci. Technol.* A, 18, 2517 (2000).

55. A. Misra and M. Nastasi, *Nucl. Inst. Methods* B, 175/177, 688 (2001).

56. A. Misra and M. Nastasi, in K.L. Mittal (Ed.), *Adhesion Aspects of Thin Films*, vol. 1, 2001, pp. 17–29.

57. J.N. Ness, W.M. Stobbs and T.F. Page, *Phil. Mag.* A, 54, 679 (1986).

58. C.R.M. Grovenor, H.T.G. Hentzell and D.A. Smith, *Acta Metall.*, 32, 773 (1984).

59. R.W. Hoffman, *Thin Solid Films*, 34, 185 (1976).

60. K.H. Müller, *J. Appl. Phys.*, 62, 1796 (1987).

61. K.H. Müller, *J. Appl. Phys.*, 59, 2803 (1986).

62. K.H. Müller, *Phys. Rev.* B, 35, 7906 (1987).

63. R. Mitra, R.A. Hoffman, A. Madan and J.R. Weertman, *J. Mater. Res.*, 16, 1010 (2001).

64. L.A. Girifalco and V.G. Weizer, *Phys. Rev.*, 114, 687 (1959).

65. M. Itoh, M. Hori and S. Nadahara, *J. Vac. Sci. Technol.* B, 9, 149 (1991).

66. R.G. Hoagland, LANL, unpublished work.

67. J.A. Sprague, NRL, unpublished work.

68. H. Windischmann, *J. Appl. Phys.*, 62, 1800 (1987).

69. C.A. Davis, *Thin Solid Films*, 226, 30 (1993).

70. G. Knuyt, W. Lauwerens, L.M. Stals, *Thin Solid Films*, 370, 232 (2000).

71. M.T. Robinson, *Phil. Mag.* 12, (1965) 741.

72. M.T. Robinson and O.S. Oen, *J. Nucl. Mater.,* 110, 147 (1982).

73. P. Sigmund, *Radiat. Eff.*, 1, 15 (1969).

74. D.J. Bacon, A.F. Calder and F. Cao, *Radiat. Eff. Def. Solids*, 141, 283 (1997).

75. A.B. Lidiard and R. Perrin, *Phil. Mag.*, 14, 49 (1973).

76. A. Jain, S. Loganathan and U. Jain, *Nucl. Inst. Methods* B, 127/128, 43 (1997).

77. B.M. Clemens, H. Kung and S.A. Barnett, *MRS Bull.*, 24, 20 (1999).

78. A. Misra and H. Kung, *Adv. Eng. Mater.*, 3, 217 (2001).

79. J.D. Embury and J.P. Hirth, *Acta Metall.*, 42, 2051 (1994).

80. W.D. Nix, *Scripta Mater.*, 39, 545 (1998).

81. A. Misra, J.P. Hirth and H. Kung, *Phil. Mag.* A, 82, 2935 (2002).

82. C.V. Thompson, *J. Mater. Res.*, 8, 237 (1993).

83. R.L. Fleischer, in D. Peckner (Ed.), *The Strengthening of Metals*, Reinhold Press, New York, 1964, pp. 93–162.

84. B. Window, F. Sharples and N. Savvides, *J. Vac. Sci. Technol.* A, 6, 2333 (1988).

85. J.W. Mayer and S.S. Lau, in J.M. Poate, G. Foti and D.C. Jacobson (Eds.), *Surface Modification and Alloying by Laser, Ion, and Electron Beams*, Plenum Press, New York, 1983, p. 241.

86. B.M. Paine and R.S. Averback, *Nucl. Instrum. Methods* B, 7/8 (1985) 666.

87. F.W. Saris, in C.J. McHargue, R. Kossowsky and W.O. Hofer (Eds.), *Structure–Property Relationships in Surface-Modified Ceramics*, Kluwer Academic Publishers, Dordrecht, 1989, p. 103.

88. Y.-T. Cheng, *Mater. Sci. Rep.*, 5 (1990) 45.

89. A. Gras-Marti and U. Littmark, in R. Kelly and M.F. da Silva (Eds.), *Materials Modification by High-Fluence Ion Beams*, Kluwer Academic, Dordrecht, 1987, p. 257.

90. M. Nastasi and J.W. Mayer, *Mater. Sci. Eng.*, R12, 1 (1994).

91. W. Bolse, *Mater. Sci. Eng.*, R12, 53 (1994).

92. H.H. Andersen, *Appl. Phys.*, 18, 131 (1979).

93. P. Sigmund and A. Gras-Marti, *Nucl. Instrum. Methods*, 182/183, 25 (1981).

94. W.L. Johnson, Y.-T. Cheng, M. Van Rossum and M.-A. Nicolet, *Nucl. Instrum. Methods* B, 7/8, 657 (1985).

95. R. Kelly and A. Miotello, *Surf. Coat. Technol.*, 51, 343 (1992).

96. A. Miotello and R. Kelly, *Thin Solid Films*, 241, 192 (1994).

97. A. Miotello and R. Kelly, *Surf. Sci.*, 314, 275 (1994).

98. P. Sigmund and N.Q. Lam, *Mater. Fys. Medd. K. Dan. Vidensk. Selsk.*, 43, 255, (1993).

99. L.C. Wei and R.S. Averback, *J. Appl. Phys.*,81, 613 (1997).

100. J.L. Bocquet, G. Brebec and Y. Limoge, in R.W. Cahn and P. Haasen (Eds.), *Physical Metallurgy*, North-Holland, Amsterdam, 1983, p. 385.

101. M. Nastasi, S. Fayeulle, Y.-C.Lu and H. Kung, *Mater. Sci. Eng.* A, 253, 202 (1998).

102. D.R. Brighton and G.K. Hubler, *Nucl. Instrum. Methods* B, 28, 527 (1987).

103. E.H. Hirsch and I.K. Varga, *Thin Solid Films*, 52, 445 (1978).

104. H. Trinkaus, *J. Nucl. Mater.*, 223, 196 (1995).

105. M.L. Brongersma, E. Snoeks, T. van Dillen and A. Polman, *J. Appl. Phys.*, 88, 59 (2000).

106. S.G. Mayr and R.S. Averback, *Phys. Rev. Lett.*, 87, 6106 (2001).

107. A. Jain and U. Jain, *Thin Solid Films*, 256, 116 (1995).

108. A. Jain, *Surf. Coat. Technol.*, 104, 20 (1998).

109. H.U. Krebs, Y. Luo, M. Stormer, A. Crespo, P. Schaaf and W. Bolse, *Appl. Phys.* A, 61, 591 (1995).

110. G. Gladyszewski, Ph. Goudeau, A. Naudon, C. Jaouen and J. Pacaud, *Appl. Surf. Sci.*, 65/66, 28 (1993).

111. L. Pranevicius, K.F. Badawi, N. Durand, J. Delafond and Ph. Goudeau, *Surf. Coat. Technol.*, 71, 254, (1995).

112. S. Tamulevicius, I. Pozela and J. Jankauskas, *J. Phys. D: Appl. Phys.*, 31, 2991 (1998).

113. E. Snoeks, K.S. Boutros and J. Barone, *J. Appl. Phys. Lett.*, 71, 267 (1997).

114. C. Hardtke, W. Schilling and H. Ullmaier, *Nucl. Instrum. Methods* B, 59/60, 377 (1991).

115. C. Hardtke, H. Ullmaier, W. Schilling and M. Gebauer, *Thin Solid Films*, 175, 61 (1989).

Chapter 8

Industrial Aspects of Ion-Implantation Equipment and Ion Beam Generation

KOJI MATSUDA AND MASAYASU TANJYO

CONTENTS

ABSTRACT

Ion-implantation equipment is an indispensable tool in the manufacture of semiconductor devices. It also serves in the research and development of surface modification techniques for materials. In this chapter, (1) the desired components, performance and specifications of typical production-line ion-implantation equipment are described and (2) ion beam generation methods for elements ranging from hydrogen to heavy-mass elements are presented. Since these elements are found in both gaseous and solid phases, various methods must be used for producing beams of their respective ions.

8.1 INTRODUCTION

The Ion Engineering Center Corporation (IECC) was established in 1988 in Japan as a joint project of the Japanese government and a number of private companies with the purpose of supporting new technology development and creating new business opportunities in the field of ion engineering. IECC has supplied ion beam services on more than 7400 occasions while operating five ion-implantation systems.[1] The number of ion species available in the service is now up to 68, covering all the elements in the periodic table except for rare and harmful ones. The application fields of the ion beam service include semiconductor device manufacture, metal surface modification, chemistry and microanalysis, among others. Of these applications, semiconductor research and development (R&D) is the most frequent, representing more than 30% of all operations.

Most sales of ion beam technology to industrial clients are for semiconductor device fabrication. The ion-implantation equipment used for such applications is mainly for the fabrication of source, drain and gate electrodes in MOS-FET devices, a key step in their manufacture. In 2000, sales of ion-implantation equipment were in excess of US$1.5 billion worldwide. In this chapter, we will describe the current state of industrial ion-implantation equipment, which impacts strongly on the area of semiconductor device fabrication, and ion beam generation, which constitutes part of the basic technology upon which this equipment is based.

Of all the ion-implantation equipment in production-line use for microelectronic devices, it is medium-current equipment that has the most demanding requirements for control systems and components. This equipment is presented in the first section, followed by a discussion of several ion beam generation techniques in the second section.

8.2 ION-IMPLANTATION EQUIPMENT

8.2.1 Features and Composition

The features and composition of ion-implantation equipment depend on application requirements and trends in technological innovation. Impurity doping of silicon wafers is the most frequent application of this technology. Therefore, this section will focus on production-line ion-implantation systems integrating advanced technology. Ion implanters used for research in areas such as material surface modification and microanalysis consist of almost the same components as production-line equipment, but their features are adapted specifically to those research applications.

The main features of ion-implantation equipment used on the semiconductor production line are the following:

1. Ion beams are chemically and isotopically pure, and spatially separated from those of other elements.
2. Precise measurement and control of the implanted ion dose.
3. Precise measurement and control of implanted ion energy.
4. Uniform implantation flux over the entire surface.
5. Uniform ion beam angle of incidence over the entire surface.

6. Adapted for mass production.
7. Protection of the substrate from contamination by other elements.
8. Made of durable components seldom requiring replacement.
9. Operation of the system can be automated and controlled remotely.
10. Changes in operating conditions, such as ion species, energy or implantation dose, can be made quickly.
11. All operating conditions can be logged and made available for readout at any time.
12. Equipment is compact and easy to maintain.

The main components of an ion-implantation system are:

1. An ion beam generation section for source material ionization, beam extraction from the ionized plasma and control of beam intensity.
2. A purification stage where the desired ion species is selected according to ion mass using a source analyzer magnet and mass analyzer slit.
3. An acceleration section consisting of an acceleration column and ion-energy controls.
4. A beam-line section composed of a final-energy magnet, a beam-scanning magnet, a collimator magnet and Faraday cups for controlling the uniformity of ion implantation into the substrate.
5. An implantation chamber section consisting of the target chamber, transfer robot and wafer cassette, which ensure that implantation into the substrate is accurate and that the substrates are transported smoothly.
6. A control section with cabinets housing automatic and manual controls for the other components.

The layout of a typical ion-implantation system is depicted in Figure 8.1.

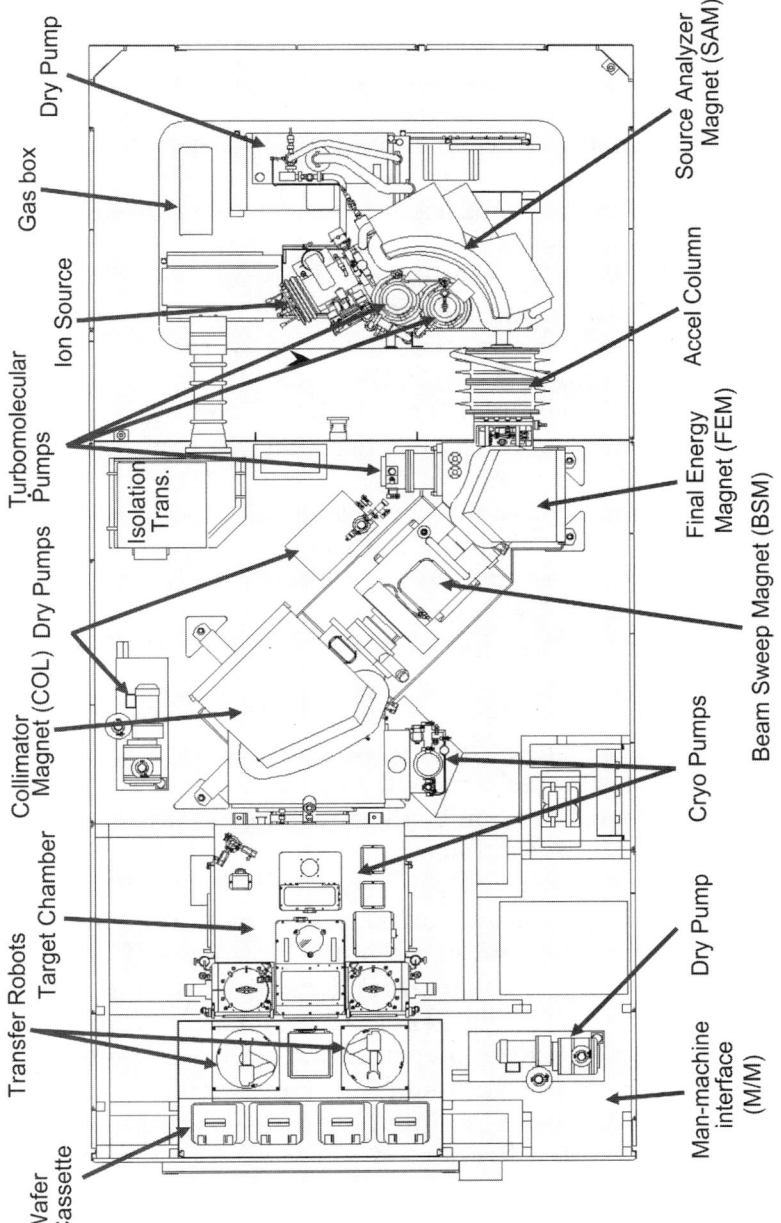

Figure 8.1 Layout of a typical (medium current) ion implantation system.

8.2.2 The Evolution of Ion-Implantation Equipment

The first ion-implantation apparatus was initially conceived and developed at an ion accelerator for studying nuclear reactions in experimental nuclear physics and for isotope separation for the purpose of isotope production.

The use of ion implantation in the manufacture of semiconductor devices was well established by the latter half of the 1960s and companies producing equipment for this application were flourishing. Among notable ion-implantation equipment manufacturers of that period were Ion Physics, High Voltage Engineering and Danfysik. It was recognized that the use of ion implantation allowed for more precise control of threshold voltages in micro-devices than did thermal diffusion. This stimulated further technological development of ion implanters with a focus on production equipment.

Improvements to ion-implantation technology continue to this day, keeping in step with the demands of the microelectronics industry for ever smaller and faster devices and process innovation in that industry. Examples of recent advances are higher intensity ion beam transport at lower energies, decreased ion beam energy, reduced metal contamination, more uniform ion implantation, more accurate beam-scanning parallelism, decreased downtime to set up new implantation conditions, ion beam generation of different species, handling of larger diameter substrates and cleaner environments.

In the ion-implantation market, equipment falls into one of three categories depending on ion beam intensity and ion energy: medium energy (up to around 250 keV) with ion doses up to around 10^{15} cm^{-2}, high energy (ion energies ranging to more than 1 MeV), and high-current ion-implantation equipment with ion doses in excess of 10^{15} cm^{-2} with ion energies up to around 200 keV. The application fields of these three kinds of equipment differ significantly with regard to ion implantation dose and implantation depth.

Further technological developments and changes in micro-device processing may lead to changes in the relative importance of these three kinds of equipment. Medium-current ion-implantation equipment may eventually encompass both high ion-dose processes by developing higher ion beam currents, and high ion-energy processes by developing multi-charged ion beams. For that matter, high-energy ion-implantation equipment may absorb medium-current ion-implantation processes by developing lower ion beam energies.

8.2.3 Medium-Current Ion-Implantation Equipment

8.2.3.1 Outline of the Equipment

Medium-current ion-implantation equipment represents more than half of the ion-implantation processes involved in LSI (large-scale integrated circuit) fabrication. This is because the precision and accuracy requirements of components in these systems are greater than are found in other systems, such as high-current and high-energy equipment. Therefore, the following description of medium-current ion-implantation equipment can suitably represent implantation equipment in general. Typical specifications for such equipment are presented in Table 8.1.

With regard to the main specifications, the ion beam energy can be changed from around 200 to 5 keV, the ion beam current from around the nA up to the mA range and a through-put of up to 200 wafers per hour is possible. The mass-resolving power ($M/\Delta M$, where M is the ion mass number) is greater than 80. The ion-implantation uniformity and ion-implantation reproducibility are both within 0.5%.

This uniformity is obtained by mechanical up-and-down scanning along one substrate axis and electromagnetic right-to-left scanning along the other. The angle of the substrate at the time of ion implantation can be adjusted to the required orientation-flat angle, tilt angle and stepping rotation for quad-mode implantation. (Note: There are two kinds of orientations of the beam-incidence angle to the wafer. One is the orientation-flat angle, meaning the angle to the [1 1 0] plane of the Si-crystal that is indicated as a notch mark or an orientation-flat mark at the side of a wafer. The other is the

TABLE 8.1   Specifications of Typical Medium-Current Ion-Implantation Equipment

Item	EXCEED 2000A/2000AH	EXCEED 2300/2300H
Ion energy	5~200 keV (2000AH, 2300H: 250 keV)	
Beam current	Max. beam 11B+: 2 mA 31P+: 4.5 mA 75As: 3 mA Min. beam 1 μA	Max. beam 11B+: 2 mA 31P+: 4.5 mA 75As: 3 mA Min. beam 1 μA
Wafer size (mm)	200 mm	300 mm
Ion mass analyzing power	M/ΔM ≥ 80	M/ΔM ≥ 80
Implantation uniformity	$\sigma/\times \leq 0.5\%$	$\sigma/\times \leq 0.5\%$
Beam parallelism	≤ ± 0.5°	≤ ± 0.5°
Method of implantation	Horizontal: magnetic parallel scan Vertical: mechanical scan	Horizontal: magnetic parallel scan Vertical: mechanical scan
Implantation chamber	Four cassettes Implantation angle: 0~60° OF angle: 0~359° Step rotation: 2~36 steps	Four cassettes Implantation angle: 0~60° OF angle: 0~359° Step rotation: 2~36 steps
Through-put	200 wafers/h	200 wafers/h
Reliability	Wafer breakage less than 1/50,000	Wafer breakage less than 1/50,000
Wafer cooling power	0.6 W/cm^2 at 100°C	0.6 W/cm^2 at 100°C
Particles	0.05 pcs/cm^2 at ≥ 0.18 μm	0.05 pcs/cm^2 at 0.18 μm
Dimensions (mm)	3000 (W) × 6335 (D) × 2500 (H) mm	3000 (W) × 6500 (D) × 2500 (H) mm

tilt angle, which means the angle to the [1 0 0] plane, the wafer-surface plane. The stepping rotation for quad-mode implantation means that the implantation is done with four orientation-flat angles with rotation of the wafer for different mechanical scanning.)

Substrate purity is compromised by particle adhesion and metal contamination. The number and size of adhering particles are subject to specifications that become more and more stringent as devices become smaller.

A vacuum of less than 10^{-4} Pa is maintained in the ion-implantation equipment. This is to prevent discharging between electrodes operated at high voltage and decreases in ion beam intensity caused by collisions of the beam with residual gas molecules in the vacuum. To prevent contamination such as oil vapors from the vacuum pumps, cryo-pumps and other similar types of pump are used.

If the ion beam collides with residual gas molecules in the vacuum, the beam current will decrease due to decomposition of molecular ions with a concomitant reduction in ion beam energy. Implantation with the ion-energy distribution thus spread out would impact negatively on micro-device fabrication. To prevent this from occurring, an ion-energy analyzing system (usually magnetic) is placed after the accelerating column.

Every step involved in operating the equipment is fully automated, from setup, through all operations, to shutdown. The system can be operated remotely from the production facility's central control station. System operating conditions can be continuously recorded; both information on the current conditions and information that has already been logged are accessible at all times. All data can be password protected.

8.2.3.2 Ion Source Modules

The purpose of the ion-source module is the generation of a stable ion beam. The module consists of an ion source, ion extraction electrode, the ion-source housing and the controller.

The ion source must be able to produce stable plasma with wide ranges of densities, which are equivalent to extracted ion beam intensities from nA to mA ranges. Furthermore, it must maintain the stability of this plasma over a long period and adapt quickly to changes in operation conditions such as a change in ion species. The ion source is designed to be compact and easily handled; since some parts of the source are consumable and one must be able to change them quickly.

Many types of ion sources are used, some examples of which are the radio frequency (RF), electron cyclotron resonance (ECR), and Freeman and Bernas types.[2] The Bernas type of ion source is currently used most often because of its superior controllability, stability and long life in ion-implantation equipment. The typical Bernas ion source is shown in Figure 8.2.

The plasma in the arc chamber is produced by first generating thermionic electrons from a hot filament, then accelerating these to several tens of volts. These will then ionize the source material that has been vaporized by heating in a crucible. For more efficient ionization, an electron repeller is placed facing the hot filament, this induces the electrons to oscillate between the hot filament and the repeller, thereby increasing the number of ionizing collisions. In addition, a magnetic field is applied to the electron flux to create a spiral

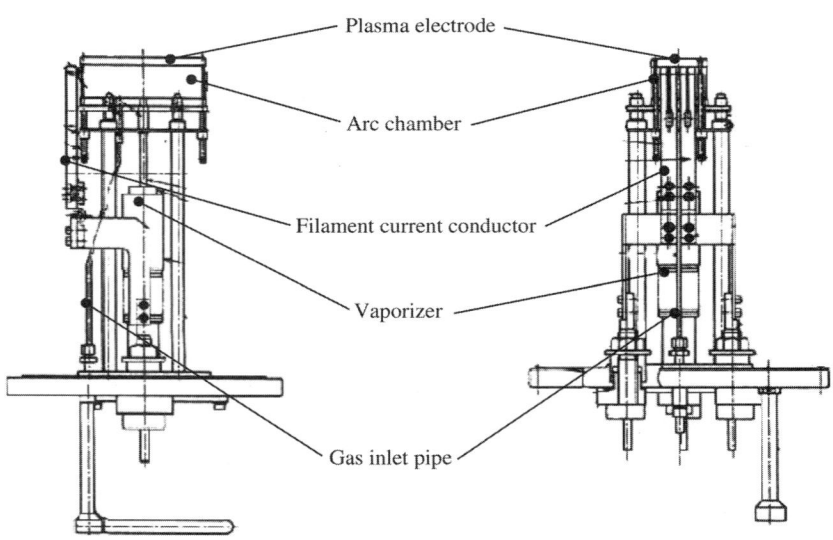

Figure 8.2 Bernas type ion source currently used in ion implantation equipment.

motion of the electrons, resulting in even more electron passes through the vapor. Positioning the filament in the arc chamber so that it is not exposed to the ion-extraction aperture extends filament life, since it is mainly the ions produced at the extraction electrode returning back through this aperture that damage the filament.

The ion-source control is composed of arc and filament power supplies that are necessary for ionization, an electromagnet power supply and input/output of the control console signal.

Gaseous or solid materials can be used as ion sources. For boron ion beam generation, BF_3 gas is normally used. For producing phosphorus ions, PH_3 gas or solid phosphorus is used and for arsenic, AsH_3 gas or solid arsenic. The ion-source material flow in the arc chamber is controlled by the mass flow meter in the case of gaseous materials and by the crucible temperature in the case of solid materials.

The function of the ion beam extraction assembly is to extract the required amount of beam from the source. Applying an electric field between the extraction electrode and the ion-source arc chamber results in extraction from the ion source. A simulation of the extracted ion beam trajectory in the extraction area is shown in Figure 8.3.

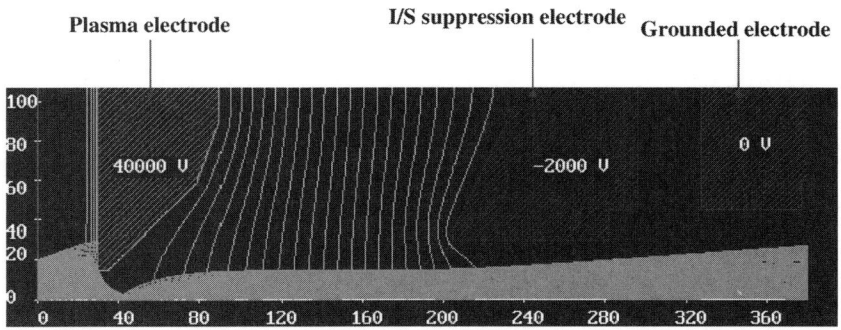

Plasma electrode   I/S suppression electrode   Grounded electrode

Cross Section of Extraction Electrode System

Figure 8.3 Simulated ion beam trajectories in the extraction area.

The ion-source extraction assembly is composed of the ion-extraction electrode and its position control. The ion-extraction electrode can be moved perpendicularly to the ion beam flight direction, along that direction and inclined, so that the amount of ion beam and the extraction angle can be adjusted. Since the ion-extraction voltage applied can range from 5 to 40 kV depending on the desired beam intensity, the insulation on the electrode must be shielded from metal contamination to avoid shorting out the high voltage. Many techniques, such as maze structures, are used to protect the insulation from vaporized and sputtered metals.

8.2.3.3 Ion Mass Selection

The ion species required for implantation should be in the pure atomic state. Ion mass selection is achieved by an electromagnet placed after the ion extraction module, since some atoms other than those required will inevitably be generated in the ion extraction module.

The source analyzer magnet (SAM) will typically have a magnetic rigidity (ion mass number in atomic mass units times ion energy) of at least 3 amu-MeV, since the heaviest ions used for semiconductor applications of arsenic ion (75 amu) are extracted with a 40 kV ion extraction voltage.

The slit aperture is adjustable, since it determines not only the ion mass resolving power but also the beam intensity passing through the slit. Undesirable ion species will collide with the inside wall of the vacuum chamber in the SAM, since the trajectories of these ions in the SAM are different from those of the selected ions and do not pass through the slit aperture. In order that the beam passing through the slit aperture be transferred with a minimum loss of intensity into the acceleration column and the beam line, the SAM serves to adjust the ion beam profile and its focusing angle. Varying the pole-edge angles in the entrance and the exit of the SAM as well as adapting the shape of the magnet pole pieces provides additional beam focusing.

8.2.3.4 Ion Beam Acceleration

The function of the acceleration column is to increase the kinetic energy of the ions passing through the column to the desired level, since that of the ions entering the column is usually insufficient for implantation. The main components of the ion beam acceleration stage are the acceleration column and the high-voltage power supply. The supply voltage on the column is typically around 200 kV. The column has a sandwich structure of electrodes and insulator plates, which helps to prevent high-voltage breakdown. Good focusing of the beam in the acceleration column results from the column having a high voltage gradient.

If any part of the accelerated ion beam hits the material surrounding the exit canal of the acceleration tube, secondary electrons may be generated. These will be accelerated by the electric field back through the acceleration column and hit the material surrounding the column inlet, which could result in the emission of x-rays in lethal doses. To prevent such x-ray emission, using appropriate materials will minimize the generation and therefore backward flux of secondary electrons, using materials with low atomic number where possible will reduce x-ray generation, as will putting a negative electron suppressor at the column exit. X-ray leakage is kept at safe levels by shielding the equipment cabinet with lead to absorb any residual x-rays.

The acceleration power supply must assure the stability and adjustability of the output voltage, provide countermeasures against discharges, be compact and easy to maintain. The power supply is usually a Cockloft–Walton type with possible modifications. Once a high voltage is applied to the acceleration column, the ion-source and mass-selection areas float at high voltage. Electric power is supplied to the latter using an isolation transformer or a motor generator with an insulated shaft. When the ion beam energy used is below the ion-extraction voltage, a negative voltage is applied to the accelerating column and the ion beam decelerated in the column, since there is a minimum extraction energy that is required to extract a significant amount of beam current from the source.

8.2.3.5 Ion Beam Line

The diameter of the ion beam entering the ion beam line is around 30 mm at most. The diameter of the substrate, on the other hand, is 300 mm or so, hence scanning of the ion beam is required to obtain a uniform implantation dose and a uniform implantation angle over the entire substrate. In addition to ion beam scanning, other approaches such as enlarging the ion beam by defocusing or by mechanical scanning of the substrate may be used although good implantation uniformity is difficult to achieve with these methods.

Despite the fact that the pressure in the ion beam line is kept below 10^{-4} Pa, the ion beam may still collide with residual gas molecules in the vacuum chamber, causing breakup into molecules of smaller mass or charge-exchange into neutral molecules, resulting in lower kinetic energy. Furthermore, when the beam intensity is increased, electrostatic repulsive forces within the beam increase and beam divergence can occur. The purpose of the various beam-line components is to resolve these problems caused within the vacuum chamber.

The ion beam increases in energy as it passes through the acceleration column. When the beam collides with the residual molecules in the column, some part of the beam will lose energy. These lower energy ion beams are referred to as being "energy-contaminated."

Use of the energy-separation magnet as shown in Figure 8.1, which is described as the final energy magnet (FEM), allows removal of the energy-contaminated beam. The separation of the neutral beam from the charged ion beam is done by allowing the neutral beam to pass straight through while deviating the charged beam with an electromagnetic field.

Such energy contamination in the case of BF_3 ion-source material is shown in Figure 8.4. Here, with an ion extraction energy of 40 keV and acceleration energy of 60 keV, BF^+ ions can dissociate from BF_2^+ ions before entering the acceleration column with an energy of 85 keV. If B^+ ions dissociate from BF_2^+ ions after exiting the acceleration column, the B^+ ion energy is 22 keV. The removal of the energy-contaminated ions by the final energy magnet (FEM) located after the column is shown in Figure 8.5.

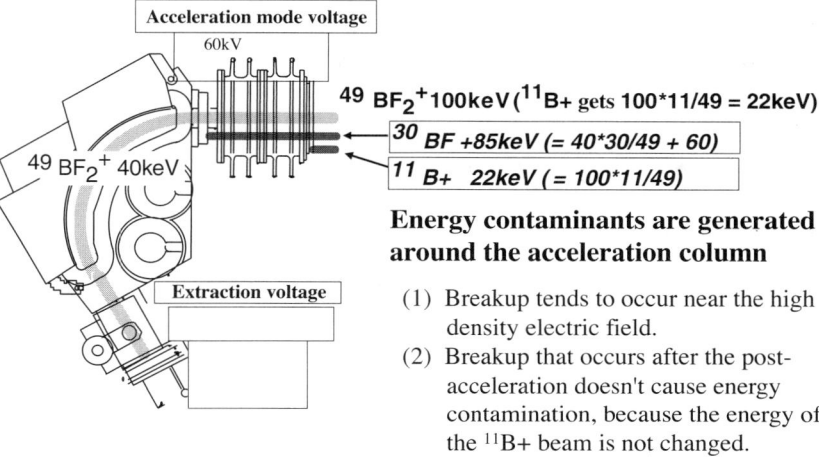

Figure 8.4 Beam energy contamination with BF_3 source material.

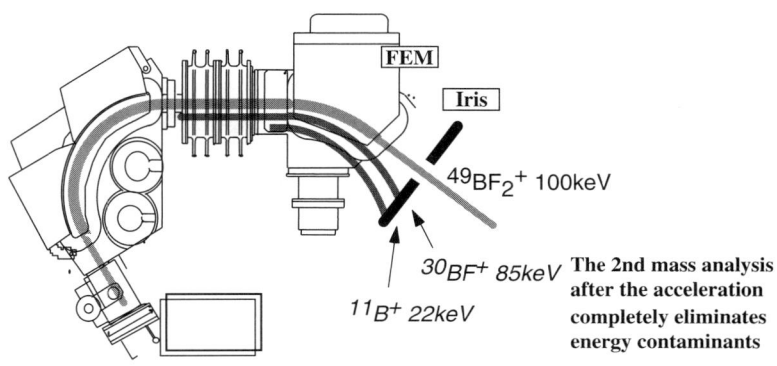

Figure 8.5 Removal of energy contamination by means of the final energy magnet (FEM).

Some data on the secondary-ion mass spectrum (SIMS) are shown in Figure 8.6.

Figure 8.6a–d shows that by converting the doubly charged SIMS profile into the singly charged one for B, P, As and BF_2 ions, both profiles display the same shape, with no

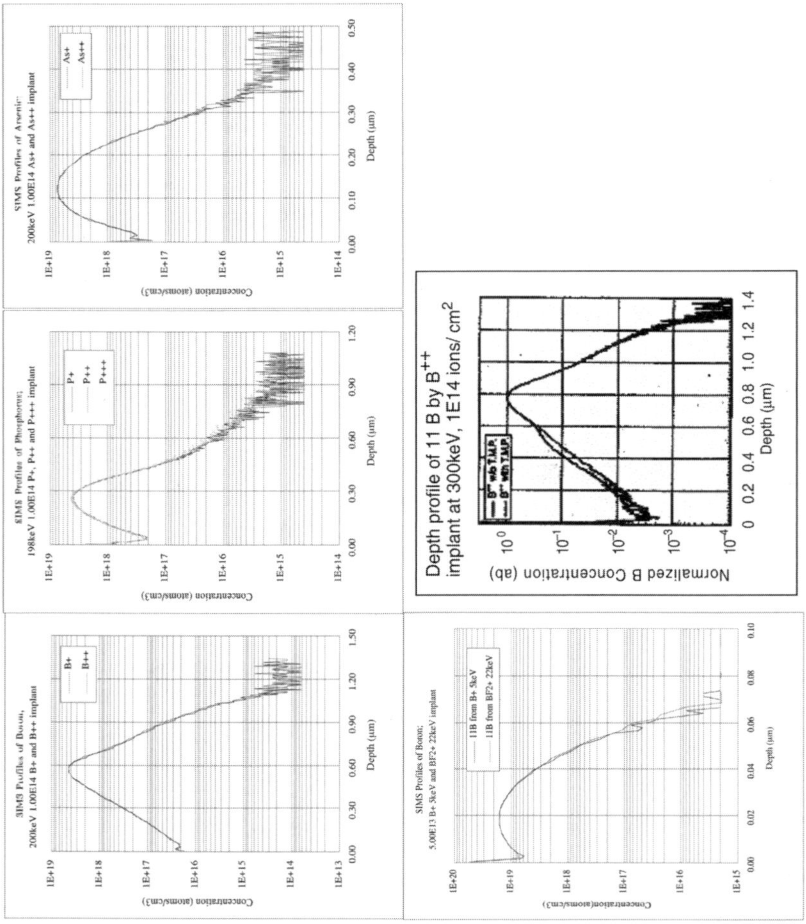

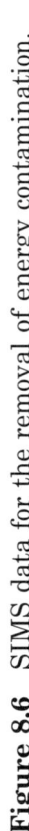

Figure 8.6 SIMS data for the removal of energy contamination.

S, the strength of the space charge effect,
is given by the following equation:

$$S = Al(1-p)\sqrt{(m/qV^3)}$$

1: beam current
m: mass of the ion
q: charge number of the ion
V: acceleration potential
p: electron ratio (amount of electron charge/amount of ion charge)
A: constant

Figure 8.7 Simulation formula of the space-charge effect.

energy contamination. Figure 8.6e shows the SIMS profile for boron taken by the old system without a FEM, and with and without the pumping with a turbo-molecular pump (TMP) in the beam acceleration area. It clearly shows the energy contamination fragments without TMP pumping, which simulates the vacuum level degradation by the resist out-gassing. So, while using a bare wafer, there is energy contamination present, but using the resist-patterned wafer without FEM leads to the appearance of energy contamination. FEM has thus been shown to be an effective way to remove the energy contamination in a semiconductor production line.

As the ion beam propagates through the beam line, it ionizes the residual gas and creates electrons and negative ions. As a result of the negative charge thus created, the repulsive Coulomb force is diminished. This is called "charge neutralization."

Divergence of the ion beam caused by the repulsive force without charge neutralization can be estimated by simulation. The formula is shown in Figure 8.7[3] and an example of the simulation is shown in Figure 8.8.

In Figure 8.8, the repulsive Coulomb force[3] caused by the space-charge effect depends on the beam current density, the ion energy, the ion mass and the residual gas pressure. The existence of electromagnetic fields will also influence the space-charge effect. *Beam current density*: the smaller the

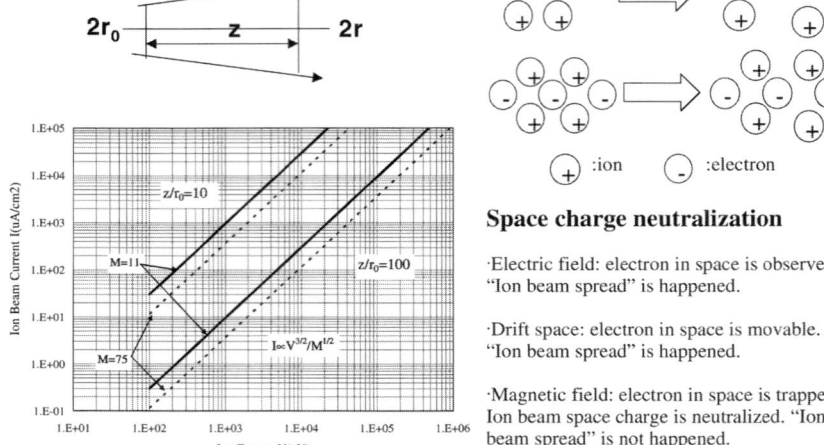

Figure 8.8   Ion beam divergence caused by the space-charge effect.

beam current density, the greater the distance between the ions and the weaker the repulsive force. *Ion energy*: the higher the ion energy, the less time the ions will be in proximity to each other and consequently, the shorter will be the duration of the Coulomb repulsive force. *Ion mass*: the smaller the ion mass given the same ion energy, the higher the speed of the ions and consequently, the less time they spend adjacent to each other, which weakens the effect of the repulsive force. *Vacuum*: the higher the concentration of residual gas, the greater the possibility of collisions between the ions and residual gas molecules, which increases the likelihood of the resulting negatively charged particles causing space-charge neutralization. *Electromagnetic fields*: while the electric field absorbs electrons in space into the electrodes and thus strengthens the space-charge effect of the ion beam, the magnetic field restrains the electrons in space due to Larmor gyration and effectively weakens the effect.

A graph of the beam current density versus the beam energy for the same beam divergence is shown in Figure 8.8.

Enlargement of the ion beam size of up to three times with flight lengths of 10 and 100 times the ion beam radius is shown by using the formula in the same figure in the cases of ion mass numbers 11 and 75. It can be seen that the ion beam is more likely to diverge as the beam density increases.

The ion-implantation equipment shown in Figure 8.1 is equipped with magnets such as the FEM, the beam-sweep magnet (BSM) and the collimator magnet (COL). This allows even higher ion beam currents with lower beam energies to propagate with fewer losses in the beam line.

Since the substrate size is generally bigger than the beam-spot size, the ion beam must be scanned to obtain uniform ion implantation into the substrate. Ion beam scanning in two dimensions may be accomplished by supplying triangular wave voltage signals to two pairs of parallel plates. However, this way of scanning produces non-uniform angles of ion incidence between the center and edges of the substrate and changes in the shadows of micro-patterns if micro-devices are being implanted.

To obtain uniform shadowing, parallel ion beam scanning is needed. This type of scanning may require two pairs of scan units and two pairs of anti-scan units if it is to be accomplished by beam scanning alone, but that would require considerable space in the beam line. To avoid this, compact parallel scanning is used in the horizontal direction, whereas the vertical scanning is achieved by mechanical translation of the substrate.

An electric or magnetic field can be used to control the horizontal scanning of the ion beam. Electric scanning is structurally simpler, but may cause problems with beam-charge neutralization. Magnetic scanning is somewhat bulkier, but does not have the problem of charge neutralization. For parallel ion beam scanning, one scan magnet and one static magnet are used. This approach requires more work in designing the magnet but uses a simpler power supply. The parallel ion beam scan system using a scan magnet and a static magnet is shown in Figure 8.9.

An accurate parallel ion beam scan system must include a means of measuring and controlling the quality of the beam parallelism. One way of measuring beam position is with Faraday cups. Faraday cups monitoring the beam position are set in front of as well as behind the substrate. Each Faraday cup is composed of many small Faraday cups positioned parallel to the beam-scan axis. The way of monitoring the parallelism is to compare the number (A) of small Faraday cups in front of the substrate that catch the scanned beam of a specific phase with the number (B) of small Faraday cups behind the substrate that catch beam of the same phase. The parallelism is measured by calculating the difference between A and B, and the distance between the two Faraday cup positions. The setup for the front and back Faraday cups is shown in Figure 8.10.

An example of the ion beam parallelism attained by the system described above is shown in Figure 8.11. Parallelism of less than 0.2° is shown in the 200 mm diameter substrate. As for the scanning frequencies of the ion beam, those in excess of 100 Hz are normally used for a 10-sec ion-implantation time to obtain an implantation uniformity to within 0.5%, since a higher frequency implies more frequent scanning on a wafer and hence a better uniformity.

8.2.3.6 Ion-Implantation Chamber

The ion-implantation chamber, commonly called the end-station, is where the substrate is implanted by the ion beam and transferred by the transfer robot. The rate at which substrates can be treated, i.e., the through-put, and the implantation characteristics such as implantation angle and step rotation, are related to the design of the ion-implantation chamber.

Substrates are loaded in a cassette that can hold up to 25 at a time. A transfer robot moves the substrate from the cassette to the wafer aligner where the orientation flat of the substrate is aligned, then the substrate is transferred to the preliminary evacuation chamber.

In the preliminary evacuation chamber the pressure is pumped down to 10^{-1} Pa in a few seconds. Evacuation should be done quickly but gently to avoid creating dust in the chamber, which would require additional venting and limit substrate through-put. Once evacuated sufficiently, the sub-

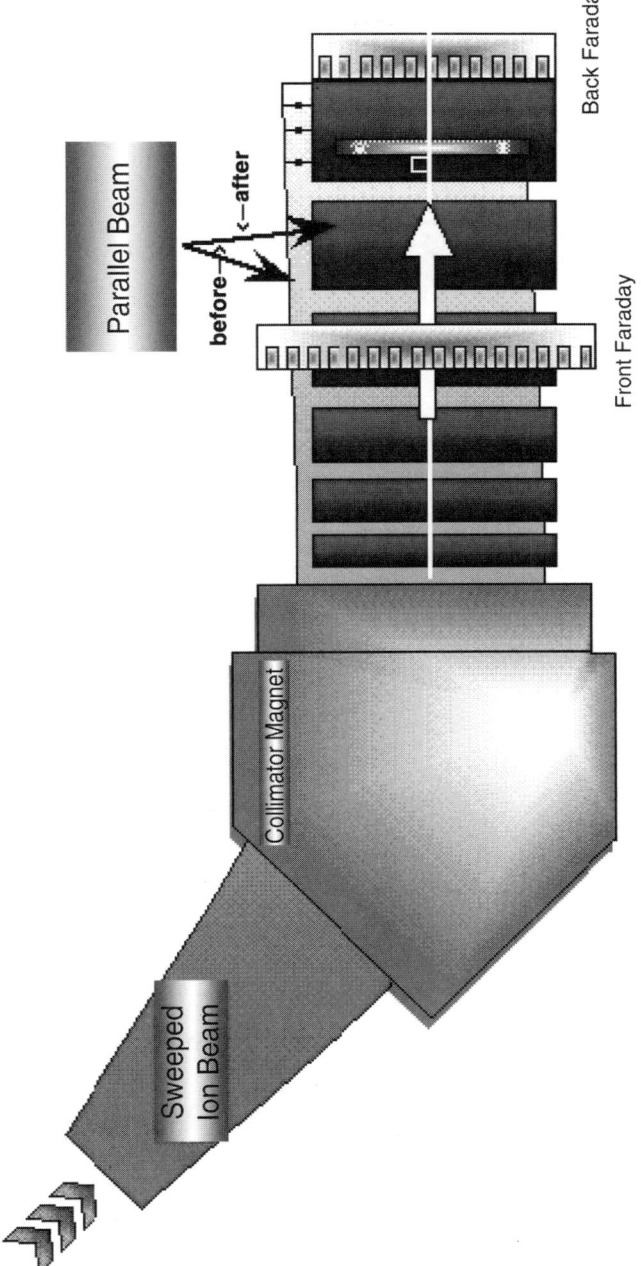

Figure 8.9   Parallel ion beam scan system using a scan magnet and a static magnet.

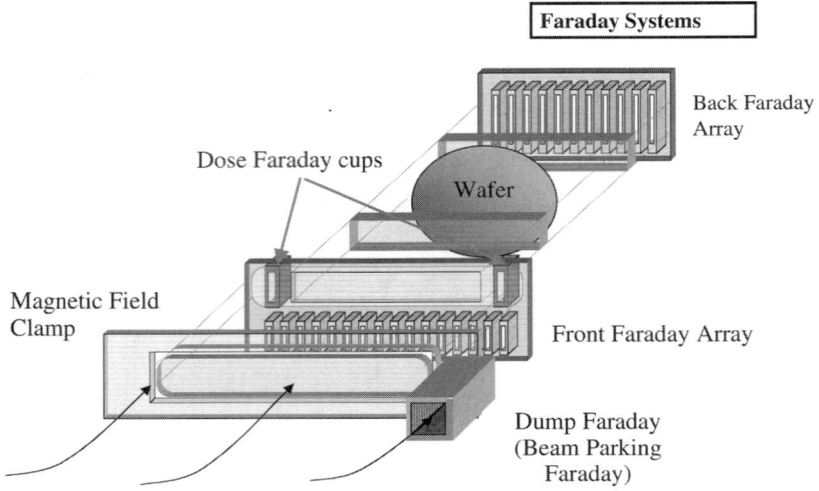

Figure 8.10 The set-up of the "front" and "back" Faraday cups for measuring parallelism.

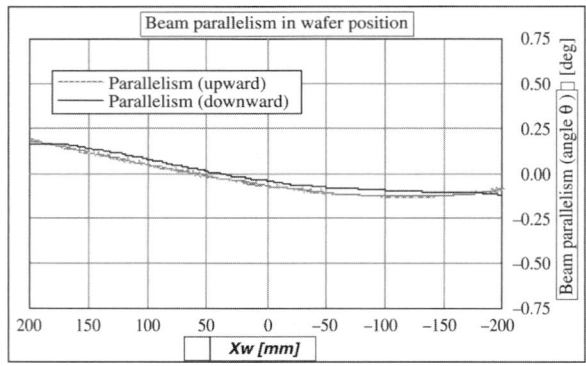

Figure 8.11 An example of the scanned ion beam parallelism.

strate is transferred to the main chamber. Once ion implantation is complete, the treated substrate is then returned through the preliminary evacuation chamber to the cassette in open air. The substrate is then either sent back to its position in the original cassette or to a new cassette. In the former case, a fully automated procedure requires that the details of the ion-implantation treatment applied to the substrates in any given cassette be somehow stored in or associated with the cassette itself, and that the factory process control center issue implantation directions matching those corresponding to each cassette.

The ion-implantation chamber consists of the substrate holder, the platen controlling the implantation angle, the mechanical scan system driving the platen up and down, the beam current monitor, the ion beam charge suppression on the substrate, substrate transfer robot and vacuum pumps. The substrate holder both holds the substrate in place during implantation and prevents excessive heating of the substrate as a result of the procedure. There are two ways of keeping the substrate on the holder: (a) clamp the edge of the substrate mechanically or (b) use an electrostatic chuck.

Mechanical clamping has both advantages and disadvantages. Some of the advantages are that:

1. It can be adapted to any substrate material and shape since the material and shape of the clamp itself are changeable.
2. The effectiveness of methods for suppressing ion beam charge accumulation on the substrate can be gauged using the secondary electrons generated from the clamp material.

On the other hand, disadvantages would be that:

1. Clamped edges of the substrate cannot be implanted.
2. The substrate surface is hard to hold down uniformly and without bending.
3. Contaminating particles may be produced by rubbing between the substrate and the clamp.
4. Metal contamination of the substrate results from sputtering of the clamp materials by ion impingement.

The electrostatic chuck is made of electrostatic plates covered by an insulating film. Its function is to hold on to the substrate using the electrostatic force induced by a potential difference applied to the plates.

Advantages of this method are that:

1. The substrate surface can be held down uniformly and remain flat.
2. There is no physical contact with the substrate surface, so the risk of contamination is reduced.
3. Since there are no materials in front of the substrate surface, the risk of contamination due to sputtering is also reduced.

Disadvantages would be that:

1. Some additional limitations are imposed on the material of the substrate, such as its being electrically conductive.
2. Provision must be made for quick release of the substrate from the holder, since otherwise the induced static charge can leak into the substrate and the induced force can persist for several seconds.

The electrostatic chuck is superior for use in ion-implantation equipment, since damage to the substrate from contamination and sputtering would appear to be reduced. An example of the electrostatic chuck is shown in Figure 8.12.

The temperature rise of the substrate during ion implantation is kept below approximately 100°C so that the characteristics of the photo-resist covering the substrate are not altered by thermal stresses. Heat transfer from the substrate to the holder takes place by thermal conduction, since the substrate is in a vacuum and the temperature increase is less than around 100°C. Better thermal conduction can be achieved, increasing contact between the substrate and the holder, by:

1. Laying a sheet of soft material such as silicon rubber on the holder
2. Polishing the substrate surface to a mirror finish

Gas cooling, in which helium gas is made to flow through the gap between the substrate and the holder, can be used to decrease the temperature. However, gas cooling needs a pressure of approximately 1 Pa in a high vacuum chamber that

$$F=S/2 \cdot \varepsilon_\gamma \cdot (V/d)^2$$

F: Absorption Power (g/cm²)

S: Area of electrode (cm²)

ε_γ : Specific Dielectric Constant

V: Supply Voltage (V)

d: Thickness of Insulator (cm)

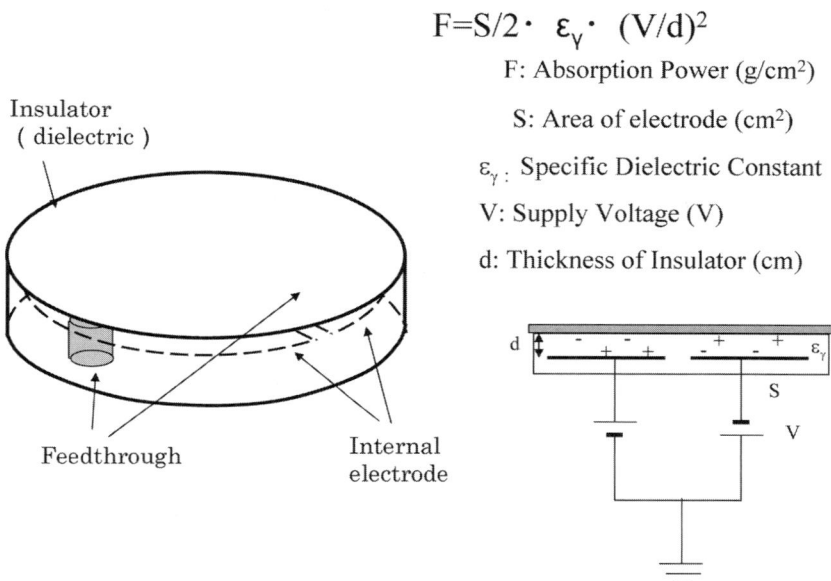

Figure 8.12 Layout of the electrostatic chuck.

is at about 10^{-4} Pa. Therefore, some way of dealing with gas leakage must be provided.

The main function of the platen is to control the angle of incidence of the ion beam and the step angle for ion implantation. The control of the angle of incidence serves to protect the uniform projected range from non-uniformity caused by crystal channeling; the substrate is inclined to an off-channel state. Control of the step angle is required to implant the ion beam to the three-dimensionally structured micro-devices preventing the shadow effects.

The ion beam entering the main chamber is usually scanned in the horizontal direction by using electrical or magnetic fields. Mechanical movement of the substrate takes care of scanning in the vertical direction. Such mixed electromagnetic and mechanical scanning is called hybrid scanning.

The mechanical scanning system, consisting of the substrate holder and the platen, is heavy, weighing over 10 kg, so the maximum driving speed is around 30 cm/sec and the number of repetitions is approximately 2–10 per implantation. Control of the repetition speed is very sensitive to fluctuations in the ion beam intensity, so that the end of the implantation procedure will coincide with the end of the scanning cycle and deviations in uniformity are maintained to within 0.5%.

There are two ways of measuring the beam current entering the substrate, one direct and the other indirect. The direct way is to measure the ion beam current entering the substrate using a current integrator. This has the advantages that:

1. No correction of the measured data is required.
2. No special precaution is required since the beam-current measurement area is the same as the substrate wafer size, which is large compared with the beam size.

On the other hand, with this approach,

1. The substrate is one of the Faraday cage components and the substrate holder needs additional parts to complete the Faraday cage.
2. The substrate needs to be fixed in position.

The indirect approach consists of measuring the beam current entering the implantation chamber using a Faraday cage set near the substrate and applying appropriate corrections. This method has several advantages:

1. Any substrate material can be used since the substrate is grounded and plays no part in the current measurement.
2. No limitation is imposed on movement of the substrate.
3. Ion beam charge accumulation on the substrate can be suppressed by spraying with negative charges such as electrons, since the required components can be placed in front of the substrate without disturbing the ion beam measurement.

Disadvantages of the indirect method are that:

1. Beam-intensity correction on the substrate is necessary since the substrate is grounded.
2. The ion beam collection area of the Faraday cage is considerably smaller than the one on the substrate and a more accurate current integrator is required. The accuracy of the current integrator must be better than 0.5% over a range from 10 nA to 10 mA. Changes in range can be made automatically.

Beam-charge accumulation on the substrate is kept down to protect micro-device from breakdown due to the micro-discharges on the substrate that would otherwise result. Spraying with electrons and/or negatively charged ions neutralizes the charge on the substrate. There are various ways of producing the negatively charged particles:

1. Generation of secondary electrons by the ion beam hitting the edges of the substrate holder
2. Extraction of thermal electrons from a hot filament placed in front of the substrate
3. Scattering of thermal electrons by a metal target to decrease their energy before entering the substrate
4. Extraction of electrons and negatively charged ions from a plasma source placed in front of the substrate

In the micro-device fabrication process, as device density increases and circuits are further miniaturized, the insulating layer of the circuit elements becomes thinner and the voltage tolerance of that layer lower. Therefore, not only ion-charge accumulation but also negative-charge accumulation must be prevented. When using negatively charged particles to suppress positive charging on the substrate, the final energy of the negative charges should be lower than the tolerance voltage of the layer. The fourth method listed above is the most commonly used in current production equipment. An example of a plasma source, called the plasma flood gun, is shown in Figure 8.13.

The time saved in transferring the substrate is important in order to maintain substrate through-put. The speed control of all moving components and the actions of these components are equally important.

Evacuation of any particles released by degassing of the implanted substrate and any gaseous chemical materials emanating from the ion source is important in order to keep the vacuum clean. A cryo- or similar type of pump is usually used for pumping out these by-products. The cold trap of the cryo-pump is automatically and regularly maintained by heating.

A preliminary evacuation is done to save evacuation time in the main chamber. The substrates are evacuated one by one for only a few seconds each. Rapid venting may cause the release of particles by adiabatic expansion and strong currents of gas. To overcome these problems, the preliminary evacuation requires a dry gas such as dry nitrogen, gas filters and multiple exhaust ports.

8.2.3.7 Control Section

The control section runs all of the components in the ion-implantation equipment. Operating commands for the section are communicated from the outside, generally from the central control station. Commands are issued via a menu driven display on a screen using a mouse, a touch-sensitive screen or a light pen. All of the operating data for the ion implantation procedure are automatically stored with password protection, so that access is restricted to authorized individuals.

8.3 TOPICS OF CONCERN IN ION IMPLANTATION

The topics discussed in this section relate to the requirements of ion-implantation technologies for manufacturing the micro-devices.

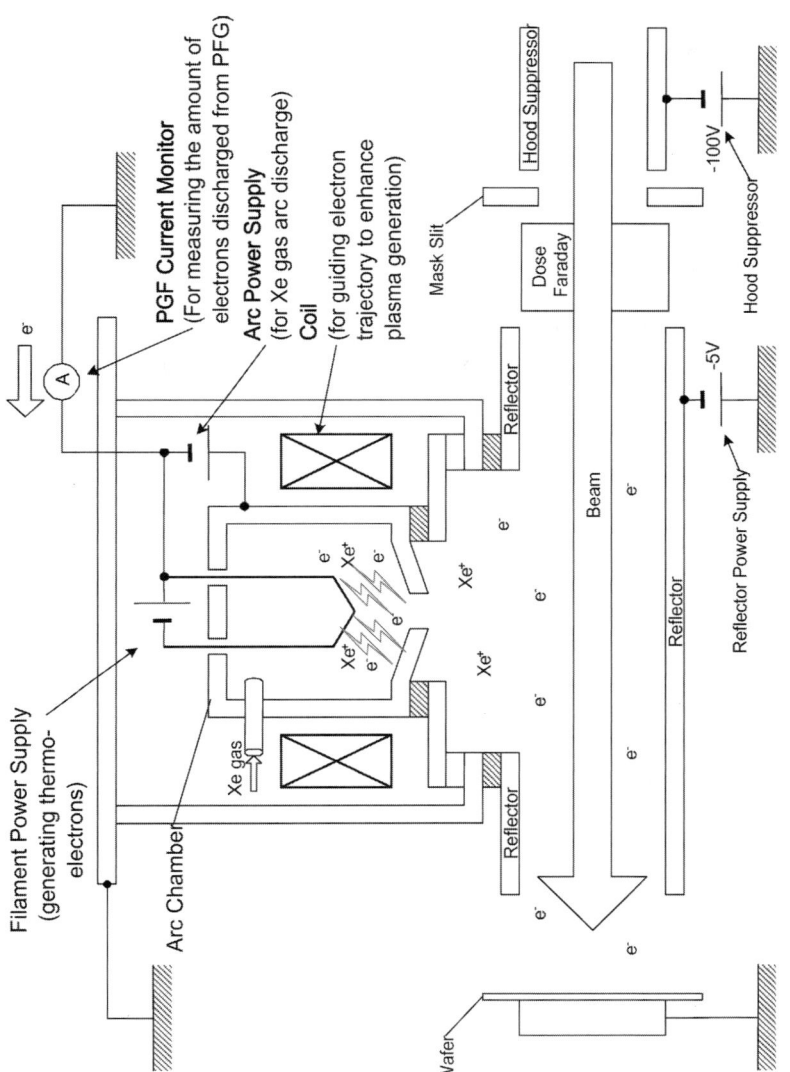

Figure 8.13 The plasma flood gun used to suppress ion beam charge accumulation.

8.3.1   Transient-Enhanced Diffusion (TED)

As micro-device miniaturization continues and designs begin to integrate structures measuring less than 0.1 μm, demand for lower-energy ion implantation increases. The projected range of the implanted ions is specified in the device design, but the high-temperature annealing required to activate the implanted ions causes thermal diffusion of the ions and spreads them outside their specified range. This phenomenon is called transient-enhanced diffusion. It is believed to occur when implanted ions such as boron collide with silicon in the silicon crystal, producing interstitial silicon. Then the boron and the interstitial silicon bond in such a way that thermal diffusion of the interstitial silicon results in diffusion of the boron as well.

Some methods for preventing TED currently under development or in actual use are:

1. Preventing bonding between the boron and interstitial silicon atoms. Nitrogen atoms bond with interstitial silicon readily so that by implanting nitrogen ions, interstitial silicon will bond with these before it can do so with boron.[4]

2. Decreasing the diffusion velocity of boron. The diffusion velocity of a molecule composed of boron and hydrogen (B–H) is lower than that of free boron in crystalline silicon. When boron ions are implanted, if hydrogen is also implanted into the same area, the implanted hydrogen will bond with the boron ions and thus reduce the diffusion velocity of the latter.[5]

3. Quenching interstitial silicon atoms in the surface layer of the wafer. For the low-energy implantation, such as with $B_{10}H_{14}$ ions at 3 keV, an amorphous layer is readily produced on the surface of the wafer, which suppresses defect formation and acts as a sink for interstitial silicon atoms. It is well known that low-energy implantation with $B_{10}H_{14}$ ions does not exhibit TED.[6]

4. Enclosing boron within dense amorphous areas. Boron tends to get trapped in amorphous areas.

Dense amorphous areas are formed by germanium ion implantation, so boron is more likely to be trapped under those conditions than to bond with interstitial silicon.[7]

5. Decreasing diffusion by rapid annealing. To take advantage of the different rates for thermal activation and diffusion, the boron-implanted substrate can be activated at a higher temperature for a shorter time. In this way, sufficient activation results will result with minimal diffusion. A pulsed heating tool called a spike annealer is used, with ramp-up and ramp-down rates of 200°C/sec and 50°C/sec, respectively.

In practice the last of these, spike-annealing technology, is used. Solutions involving nitrogen or germanium ions would require another ion implantation step, in addition to that for boron. In order to avoid such a two-step implantation, implantation with B–Ge molecules may be considered. Ion implantation and annealing have traditionally been treated as separate processes in micro-device fabrication, but the TED effects discussed above are related to both. Therefore, a solution to these effects must also involve both processes.

8.3.2   Metal Contamination

When the energized ion beam enters the vacuum chamber, some of the ions may hit and sputter the materials in the chamber, depositing metal particles onto the substrate. Metal contamination tolerance levels for ion implantation in micro-device fabrication are becoming lower with increasing device miniaturization. In the *International Road Map for Semiconductors*,[7] the maximum acceptable metal particle contamination level was given as $5 \times 10^9/cm^2$ in 2001 but has since been revised downward, making such requirements even more stringent.

Technological advances used in the prevention of metal contamination fall into two categories: analysis technology and protection technology.

1. Analysis technology. At the $5 \times 10^9/cm^2$ tolerance level some solutions do exist, such as secondary ion mass

spectrometry (SIMS). For greater sensitivity, on the order of $10^8/cm^2$, analysis technology is still in the research stage. Other important concerns for these technologies are where they are to be used and whether *in situ* monitoring is possible. The latter is an issue because the substrate to be monitored is not always stationary.

2. Protection technology. Some methods of protecting against metal contamination are:

   a. The use of an electrostatic chuck rather than a clamp to hold the substrate avoids metal contamination resulting from rubbing between the clamp and the substrate.

   b. The use of mechanical shielding or coating with harmless materials such as carbon and silicon diminishes ion sputtering of contaminating materials that would otherwise result from ion bombardment.

   c. A clean vacuum will minimize ion beam scattering by gas molecules in the chamber.

8.3.3   Charging on the Substrate

When implanting in materials that are good insulators, ion beam charge can accumulate on the insulating layer of the substrate. In micro-device fabrication, the insulating layer in question may be a photo-resist coating on the device or a micro-transistor gate oxide thin film. When beam charge accumulation on the insulator occurs and the charging potential exceeds the breakdown threshold of the material, the insulating layer will breakdown and an electrostatic discharge between the charge-accumulated layer and the micro-device layer will effectively destroy the micro-transistors. Several methods exist for suppressing charge accumulation on the substrate, such as:

1. Spraying with electrons from a hot filament, which has been used because it is easy to implement, although the energies involved are rather high, at several tens of electronvolts.

2. Spraying with electrons scattered by a metal target, which does lower their energy but not below the tolerance level of less than 10 eV.

Spraying with electrons and negatively charged particles from a plasma of relatively heavy atoms such as xenon. These negatively charged particles have lower energies than do those of the above methods. This is the approach used primarily in current ion-implantation equipment.

8.4 ION BEAM GENERATION

Over the past ten years of activity in ion beam engineering, the IECC has used ion beams of over 60 different species from the periodic table. These are listed in Figure 8.14.

The initial state of ion-source material used for ion beam generation may be solid, liquid or gaseous. In the ion-source chamber, source materials should be in a gaseous state since their ionization is usually produced by gaseous discharge. Solid ion-source materials are first vaporized by heating in a crucible before entering the ionization chamber, as shown in Figure 8.15. While it is true that there are ion sources such as the sputter source, which can use solid materials directly, these kinds of sources are mainly restricted to specific research applications.

Source materials that are gaseous at room temperature are stored in cylinders. A control valve leading to the ionization chamber controls their flow. When the source material is a liquid at room temperature, the liquid is stored in a cylindrical container, which can be heated and then converted into the gaseous state by evacuation with a vacuum pump. The gas pipe and control valve leading from the cylinder bottle to the ionization chamber are heated to prevent condensation of the source material.

The temperature in the discharge chamber is normally around 400°C and the vapor pressure in the chamber is maintained between 0.1 and 1 Pa to keep the discharge stable. Therefore, if the vapor pressure of the material is high, the crucible should be kept away from the discharge chamber or

Ion Implanter (I/I) :

8MeV I/I

400keV I/I

200keV I/I-4set

1 H ○																		2 He ○
3 Li □	4 Be											5 B ○	6 C ○	7 N ○	8 O ○	9 F ○	10 Ne ○
11 Na □	12 Ma ○											13 Al ○	14 Si ○	15 P ○	16 S ○	17 Cl ○	18 Ar ○
19 K ○	20 Ca ○	21 Sc ○	22 Ti ○	23 V ○	24 Cr ○	25 Mn ○	26 Fe ○	27 Co ○	28 Ni ○	29 Cu ○	30 Zn ○	31 Ga ○	32 Ge ○	33 As ○	34 Se ○	35 Br	36 Kr ○
37 Rb ○	38 Sr ○	39 Y ○	40 Zr ○	41 Nb ○	42 Mo ○	43 Tc	44 Ru ○	45 Rh ○	46 Pd ○	47 Ag ○	48 Cd ○	49 In ○	50 Sn ○	51 Sb ○	52 Te	53 I ○	54 Xe ○
55 Cs ○	56 Ba ○	57-71	72 Hf	73 Ta ○	74 W ○	75 Re	76 Os ○	77 Ir ○	78 Pt ○	79 Au ○	80 Hg	81 Tl	82 Pb ○	83 Bi ○	84 Po	85 At	86 Rn
87 Fr	88 Ra ○	89 Ac	90 Th ○	91 Pa	92 U	93 Np	94 Pu ○	95 Am	96 Cm	97 Bk	98 Cf ○	99 Es	100 Fm	101 Md	102 No	103 Lr	

57-71 Lanthanoide (Rear earth)	57 La ○	58 Ce ○	59 Pr ○	60 Nd ○	61 Pm	62 Sm ○	63 Eu ○	64 Gd	65 Tb ○	66 Dy ○	67 Ho	68 Er	69 Tm ○	70 Yb ○	71 Lu ○

Figure 8.14   The ion species being used in IECC ion beam facilities.

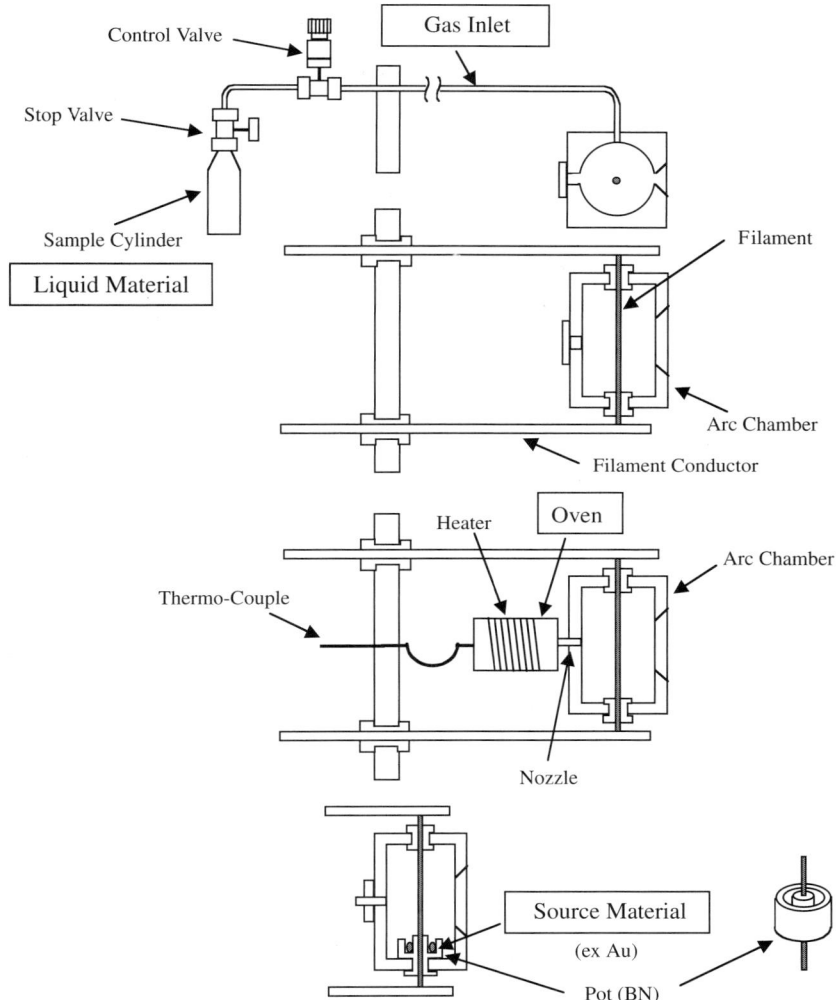

Figure 8.15 Vaporization of solid material by heating in a crucible prior to ionization.

should be set behind a plate to block radiant heat coming from the discharge chamber. If the vapor pressure of the material is low, the temperature of the crucible should be high. However, maximum temperature of the crucible is usually limited to about 1000°C, since above this value there is excessive radiant heat. If the vapor pressure for the material is too low, a crucible cannot be used. Another approach is that shown in Figure 8.15, wherein material placed near the filament in the discharge chamber is vaporized by heat radiated from the hot filament and/or sputtered with ions in the discharge chamber.

Recently, the safety delivery system (SDS), in which the source material is adsorbed on the surface of zeolite or porous carbon, has been developed. It is used with metal-organic chemicals, for which the vapor pressure is usually high and hence the required temperatures not as high. The pressure inside the storage bottle is less than atmospheric pressure, so that toxic materials can be handled rather safely.

8.5 CONCLUSION

Ion-implantation equipment used in a factory environment for the fabrication of semiconductor micro-devices has been described. There are special requirements if ion implantation is to be used in an industrial setting, such as the generation of suitable ion species, the prevention of contamination and the elimination of charging problems. The most important issue is the reliability and stability of the process within an acceptable cost. Ion implantation is already an indispensable tool in the manufacture of semiconductor devices, but the next generation of micro-devices will have much more stringent requirements for which solutions must be developed. In addition to its use in such industrial applications, ion implantation will remain a convenient tool for surface modification of materials in the field of research and development.

REFERENCES

1. IECC Report No. 47, October 2001.

2. See, for example, B. Wolf (Ed.), *Handbook of Ion Sources*, CRC Press, Boca Raton, FL, 1995.

3. T.S. Green, *IEEE Trans. Nucl. Sci.* NS-23, 2, 918, 1976.

4. S. Nakashima, M. Takahashi, S. Nakayama and T. Ohno, in *Proceedings 1998 International Conference on Ion Implantation Technology*, 1999, pp.122–125.

5. K. Yokota, H. Nishimura, K. Terada, K. Nakamura, S. Sakai, M. Tanjyo, H. Takano and M. Kumagai, *Proceedings 1998 International Conference on Ion Implantation Technology*, 1999, pp. 91–94.

6. T. Kusaba, N. Shimada, T. Aoki, J. Matsuo, I. Yamada, K. Goto, and T. Sugii, *Proceedings 1998 International Conference on Ion Implantation Technology*, pp. 1258–1261.

7. International SEMATEC, SEA (Eds.) *International Technology Roadmap for Semiconductors, 2002 Edition*, SEMATEC, 2002.

Chapter 9

Nanostructured Transition-Metal Nitride Layers

DANIEL GALL

CONTENTS

ABSTRACT

This chapter describes the current atomic-level understanding of the growth of transition-metal nitride layers by physical vapor deposition, presents a summary of the growth of 1-nm-wide nanopipes, and describes the effect of atomic shadowing on surface morphological evolution and nanostructure formation. Transition-metal nitrides exhibit a highly anisotropic surface diffusion, which is several orders of magnitude higher on (001) versus (111) surfaces and favors ⟨111⟩-oriented grains during the kinetically limited competitive growth of polycrystalline layers. The texture evolution toward ⟨111⟩ is reversed under the influence of high-flux low-energy N_2^+-ion irradiation that results in a steady-state atomic N coverage on (001) surfaces, increasing the effective adatom binding energy on 002-grains, and leading to a 002-oriented texture. Epitaxial single-crystal layers exhibit self-organized arrays of 1-nm-wide nanopipes that extend through the layers and form due to a combination of anisotropic adatom mobilities and atomic shadowing from periodic surface mound structures. The nanopipe density is controlled by the ion-irradiation flux during deposition. Ion irradiation increases the surface diffusion by a simple momentum transfer process, resulting in smoother surfaces, less atomic shadowing, and the suppression of nanopipe formation. Atomic shadowing causes strong surface roughening and leads to separated columns. The degree of shadowing can be increased by deposition from oblique angles. This is illustrated by the glancing-angle deposition (GLAD) technique that leads to well-separated nanopillars, which themselves exhibit complex engineered shapes, obtained by controlling both azimuthal and polar angles of the deposition flux. Such angular control will in the future allow the creation of complex interconnected nanochannel arrays in epitaxial transition-metal nitrides.

9.1 INTRODUCTION

Transition-metal (TM) nitrides are well known for their remarkable physical properties including high hardness and

mechanical strength, chemical inertness, and electrical resistivities that vary from metallic to semiconducting. As a result,
they are widely studied and have become technologically
important for applications such as hard wear-resistant coatings on cutting tools, as diffusion barriers in microelectronic
devices, and as corrosion and abrasion-resistant layers on
optical components. Transition metals from the left side of
the periodic table, including Sc, Ti, V, Cr, Y, Zr, Nb, Hf, and
Ta, form nitrides with a B1-NaCl structure[1,2] as illustrated
in Figure 9.1a. The excellent mechanical properties of these
materials are due to strong covalent-ionic bonds between the
TM and N ions resulting from the fully occupied N 2p bands.[2,3]
In contrast, optical and electronic transport properties are
primarily determined by the occupation of the metal d-bands,
thus vary dramatically as a function of the atomic number of
the metal atom. Figure 9.1b illustrates the chemical bonding
in these cubic TM nitrides. The N 2p orbitals, which contain
a component of p-d hybridization, are fully occupied by 3 N
valence electrons and 3 Me d-electrons per formula unit,
resulting in strong interatomic bonds. The remaining d-elec-

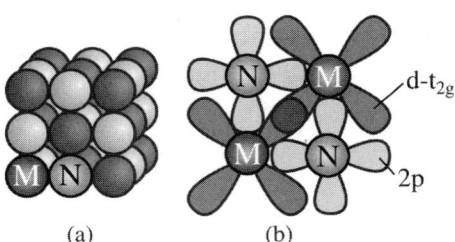

(a) (b)

Figure 9.1 (a) Schematic of the B1-NaCl (rock-salt) crystal structure. The transition-metal atoms (M) form a close packed fcc lattice,
while the N atoms fill in the empty spaces, resulting in a very
compact material. (b) Schematic of the chemical bonding in TM-
nitrides. Bonding is primarily due to the N 2p orbitals and occurs
along ⟨100⟩ crystal directions. The d-orbitals from the metal atoms
have primarily t_{2g} symmetry and determine electronic transport and
optical properties, without strongly affecting the interatomic bonding.

trons from the metal ions occupy d-orbitals with primarily t_{2g} symmetry. The filling of the d-bands, in turn, determines the density of states at the Fermi level and consequently electronic transport and optical properties.

ScN has zero electrons in the 3d conduction bands and has recently been shown to be a semiconductor with a 1.3 eV indirect gap.[4] In contrast, TiN has one more electron per unit cell resulting in metallic conductivity with a room temperature resistivity of only 13 $\mu\Omega$-cm.[5,6] The three electrons per unit cell in the d-band of CrN form local magnetic moments, resulting in a Mott–Hubbard type gap and weakly insulating behavior of stoichiometric CrN.[7,8] The NaCl-structure TM-nitrides represent an ideal materials system to build nanostructures. Their strong interatomic bonding, responsible for the excellent mechanical and temperature stability, is due to the N 2p orbitals and therefore nearly independent of the metal used. However, their optical and electronic transport properties, determined by the metal 3d-bands, vary drastically. They all have comparable lattice constants and can all be grown epitaxial on MgO(001), offering potential for usage of epitaxial superlattices, multilayers, and alloys as high-temperature extreme-environment devices inside furnaces, chemical reactors, or jet engines.

In addition to the great potential for vertical structures on the nanometer scale, these materials also offer considerable promise for laterally confined self-assembled nanostructures due to their highly anisotropic adatom mobilities, adatom binding, and surface free energies. Epitaxial transition-metal nitride layers exhibit, due to atomic shadowing from a kinetically roughened periodic surface mound structure, arrays of self-organized rectangular 1-nm-wide nanopipes. The formation of such nanopipes is described in detail in this chapter. The first section summarizes the large number of recent developments in the understanding of the growth of transition metal nitrides on an atomistic level, including the effects of anisotropic surface diffusion, atomic N-flux, adatom potential energies, and ion irradiation. All of these effects are essential in order to understand and control the formation of nanopipes, as shown in the second section, which contains a summary of experimental results on nan-opipe growth, followed by a detailed discussion on surface morphological evolution and its effect on nanopipe formation. This chapter also contains a review on atomic shadowing, its effect on surface roughening and its potential for building nanostructures, with the example being sculptured thin films obtained by glancing angle deposition. Finally, a discussion on potential methods to control nanopipe shape, size, and arrangement as well as potential applications for such structures is presented.

9.2 ATOMISTIC PROCESSES IN THE GROWTH OF TRANSITION-METAL NITRIDES

The microstructure of transition-metal nitride thin films, including layer density, texture, strain, surface morphology, and ultimately designed nanostructures, determines performance, usefulness, and lifetime of the coatings for given applications. It is therefore of crucial importance to develop a detailed understanding of the atomistic processes involved in the layer growth in order to adjust deposition conditions accordingly to obtain the desired microstructure. An example for the microstructure effect on coating performance is the growth of TiN diffusion barriers for microelectronic device applications, as studied in detail by Chun et al.[9–12] Figure 9.2 shows typical plan-view and cross-sectional transmission electron micrographs (TEM and XTEM) of a TiN layer grown on SiO_2 by ultrahigh vacuum reactive magnetron sputter deposition in pure N_2. The growth temperature $T_s = 450°C$, limited by the diffusion-barrier application, corresponds to only ~ 0.2× (in K) of the melting point $T_m = 2949°C$ for TiN.[1] The layer is underdense due to limited adatom mobilities leading to atomic shadowing which, in turn, results in a columnar microstructure with both inter- and intracolumnar voids as seen in both plan-view and cross-sectional micrographs. The breakdown kinetics of such TiN diffusion barriers for contact with Al interconnects was investigated by *in situ* synchrotron x-ray diffraction annealing experiments.[11] Al was found to rapidly diffuse into the voids of the TiN layer, resulting in a large

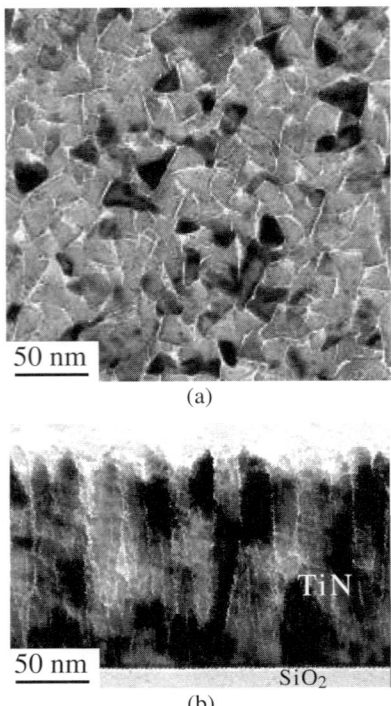

(a)

(b)

Figure 9.2   (a) Plan-view and (b) cross-sectional transmission electron micrographs from a polycrystalline 160-nm-thick TiN layer deposited on SiO_2 by reactive magnetron sputtering at 450°C, showing an underdense porous microstructure. (J.-S. Chun, P. Desjardins, C. Lavoie, I. Petrov, C. Cabral, Jr., and J.E. Greene, *J. Vac. Sci. Technol.* A 19, 2207, 2001. With permission.)

Al/TiN interface area, the formation of tetragonal Al_3Ti, and an overall break-down temperature of 450°C.

The microstructure shown in Figure 9.2 is typical for NaCl-structure TM-nitride layers grown by physical vapor deposition techniques. They exhibit strong 111-texture and are underdense. However, 002-oriented layers can be obtained by careful choice of deposition parameters including high-flux low-energy ion irradiation, as discussed in detail in the following section. 002-textured TiN layers are, in contrast to

111, fully dense and lead to enhanced diffusion barrier properties as recently shown by Chun et al.[9] who reported a breakdown temperature of 560°C, more than 100°C above the 111-value.

9.2.1 Texture Evolution

During the initial stages of growth, primarily 002- and 111-oriented grains nucleate with approximately equal number densities.[13,14] However, grain growth rates are a function of crystalline orientation. This leads, with increasing film thickness, to the development of strong preferred layer orientation that is either 002 or 111, depending upon film deposition conditions.[9,13–16] The resulting film texture has, as described above, a determining effect on the overall coating performance due to, in addition to layer density, highly anisotropic elastic properties and differences in grain sizes. It is therefore of interest to understand the underlying physical processes that favor the growth of 002- and 111-oriented grains.

A model system to study the competition between 111- and 002-oriented grains is the growth of ScN on single crystal MgO(001) substrates. While both layer and substrate crystallize with a cubic structure, there is a 7% lattice mismatch between MgO (a_o = 4.213 Å) and ScN (a_o = 4.501 Å), leading to two possible orientations for ScN nuclei as shown in the schematic in Figure 9.3. ScN grains can grow with 002-orientation and a cube-on-cube epitaxial relationship with $(001)_{ScN} | | (001)_{MgO}$ and $[100]_{ScN} | | [100]_{MgO}$. Alternatively, 111-oriented grains also nucleate with local epitaxy. In that case, strained triangular unit cells of the ScN 111-planes mesh with the squares of the MgO(001) surface, resulting in an epitaxial relationship with $(111)_{ScN} | | (001)_{MgO}$ and $[11\bar{2}]_{ScN} | | [110]_{MgO}$. ScN layers grown on MgO(001) by reactive magnetron sputter deposition in pure N_2 discharges at 750°C exhibit exactly these two grain orientations, as shown by a combination of x-ray diffraction θ–2θ scans and pole figures, as well as electron diffraction patterns.[17]

Figure 9.4 shows typical bright- and dark-field XTEM micrographs from a 180-nm-thick ScN/MgO(001) layer,

ScN(002)/MgO(001)   ScN(111)/MgO(001)

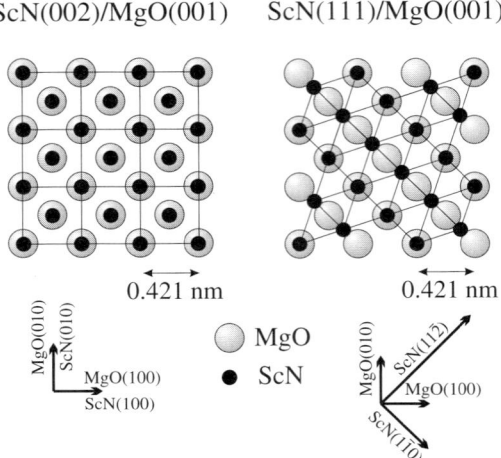

Figure 9.3 Schematic of ScN grain nucleation on MgO(001). 002 oriented grains have a cube-on-cube epitaxial relationship with the substrate (left). Strained triangular unit-cells of the ScN(111) planes mesh with the MgO squares (right).

showing a columnar microstructure with a column width of $\cong 30$ nm. The bright-field XTEM image in Figure 9.4a is obtained with the electron beam oriented along the [100] zone axis of the MgO substrate, resulting in the MgO appearing dark. Approximately half of the ScN columns adjacent to the film/substrate interface are also of uniformly dark contrast and, during tilting experiments, have the same contrast as the substrate. These grains appear bright in the dark-field image in Figure 9.4d, obtained with the 020 diffraction spots from Figure 9.4b labeled in Figure 9.4c, showing that they exhibit $[001]_{ScN} || [001]_{MgO}$ and therefore are the 002-oriented grains. The micrographs reveal that the 002 columns, which have broad bases and peaked tops, are overtaken by columns with 111 orientation. Thus, during nucleation and the early stages of film growth, the layers consist of approximately equal volume fractions of 002- and

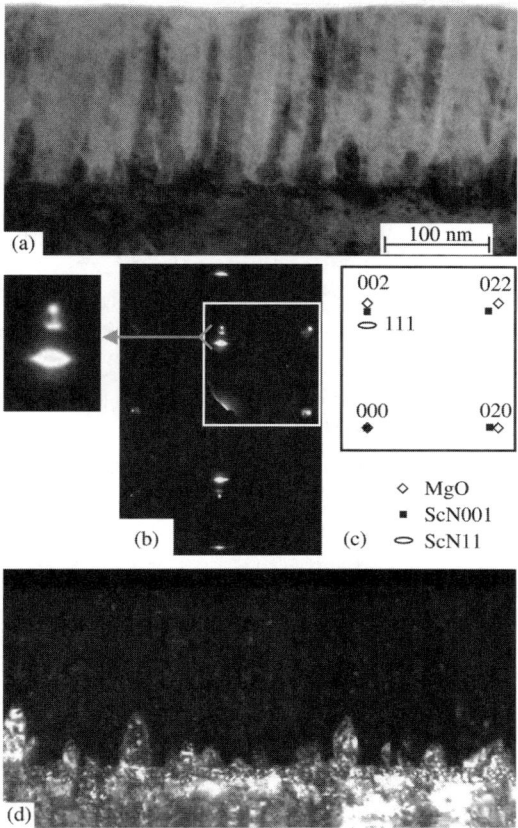

Figure 9.4 (a) Bright-field XTEM micrograph, (b) a corresponding SAED pattern, (c) an indexed SAED pattern, and (d) a dark-field XTEM obtained using the 020 ScN and MgO reflections in (b), from a polycrystalline 180-nm-thick ScN film grown on MgO(001). (D. Gall, I. Petrov, L.D. Madsen, J.-E. Sundgren, and J.E. Greene, *J. Vac. Sci. Technol.* A 16, 2411, 1998. With permission.)

111-oriented grains. However, preferred orientation evolves toward a purely 111 texture within ≈ 40 nm as the 002 grains grow out of existence.

The lowest surface-energy face for cubic TM nitrides is the (001), since metal–nitrogen bonds are along perpendicular $\langle 100 \rangle$ directions (see Figure 9.1). Hence, (001) is expected to be the preferred orientation under growth conditions where adatom mobilities are sufficiently high to favor crystallites bound by low-energy planes. However, the result from Figure 9.4, together with reports on TiN,[13] Ti$_{0.5}$Al$_{0.5}$N,[18] and TaN,[19] all indicate the opposite trend. Thus, the eventual dominance of the 111 versus 002 texture during competitive columnar growth must be due to kinetic limitations during film growth rather than thermodynamic driving forces. Similar results in which the higher-energy surface emerges as the preferred orientation are also reported from studies using atomistic Monte Carlo simulations for fcc-metal film growth.[20]

A simplified schematic of the kinetically limited competitive grain growth in TM nitrides is shown in Figure 9.5. Initially both grain orientations nucleate. An adatom on a (111) surface has three backbonds to N-surface atoms and consequently a low potential energy and small surface mobility. In contrast, adatoms on (001) surfaces, with only one N-backbond, diffuse at much higher rates and may cross a grain boundary to a 111-oriented grain. Thus, adatoms that are stochastically deposited near grain boundaries and, through surface diffusion, sample sites on both sides of the boundary have a higher probability of becoming incorporated at the low-diffusivity (111) surface that provides the more stable, lower potential energy sites. Conversely, adatoms on high diffusivity planes have larger mean free paths with correspondingly higher probabilities to move off the plane and become trapped on adjacent grains. The consequence is a net adatom flux from the 002- to the 111-oriented grains, leading to the 111 grains overgrowing the 002 that die out, as illustrated in Figure 9.5b.

The simple schematic in Figure 9.5 is consistent with detailed *ab initio* density functional studies of adsorption and binding energies of adatoms on TiN(001) and TiN(111).[21] Reported adsorption energies for Ti adatoms are 3.30 and 10.09 eV, with activation barriers for surface diffusion of 0.35 and 1.74 eV on TiN(001) and TiN(111), respectively. Thus, Ti

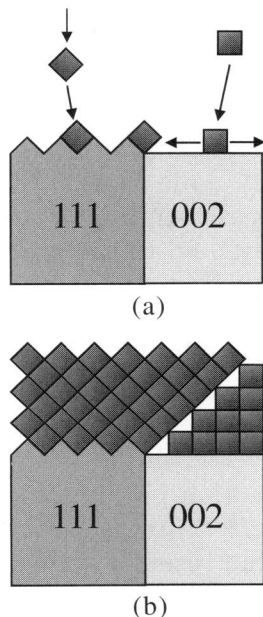

Figure 9.5 Schematic for kinetically limited competitive columnar growth in transition metal nitride layers. (a) The lower adatom mobility and potential energy on the 111 versus the 002 oriented grain result in (b) the overgrowth of the 002 oriented grain.

adatoms are much more stable and have a much lower surface mobility on 111- versus 002-oriented grains.

In summary, the texture of transition-metal nitride layers gradually evolves toward 111, due to a combination of a large cation surface diffusion length on the (001) surface combined with a considerably lower adatom potential energy on the (111) vs. the (001) surface, leading to a net flux from 002- to 001-oriented grains. However, this effect can be totally reversed by high-flux ion-irradiation during growth, as shown in the following section.

9.2.2 Role of N_2^+ Ion Irradiation

Magnetically unbalanced magnetron sputter deposition methods have been developed in order to investigate plasma–surface interactions during film growth.[22] In such experiments, an external magnetic field is used to shape the plasma near the substrate, resulting in an increase by up to two orders of magnitude in the ion flux incident at the growing film with essentially no effect on the sputtered atom flux. The primary advantage of this technique is that it allows independent control over ion energy, determined by the substrate bias, and ion flux. It has been applied to the growth of polycrystalline and single-crystal transition-metal nitride layers with dramatic effects on microstructural evolution and texture.[13,14,18,19,23,24] For example, TiN layers grown at 350°C with low ion-to-Ti flux ratios, $J_{N_2^+} / J_{Ti} \leq 1$, and low ion energies E_i = 20 eV exhibit 111 texture. However, increasing $J_{N_2^+} / J_{Ti}$ to $\geq$ 5, with all other deposition parameters remaining constant, results in 001 texture,[13,14] a total reversal.

TaN, a material that recently gained considerable interest due to its potential as a diffusion barrier in microelectronic devices based on Cu technology, exhibits a similar trend. Figure 9.6 shows the results of an x-ray diffraction study on the effect of ion-irradiation on the texture of δ-TaN.[19] The δ-phase refers to the rock-salt structure (see Figure 9.1), and is specified here because TaN exhibits a wide range of alternative stoichiometric and off-stoichiometric phases.[25] The ω-2θ scans in Figure 9.6a are obtained from 500-nm-thick TaN layers grown on SiO_2 at T_s = 350°C in mixed Ar+15%N_2 discharges with an ion-energy E_i = 20 eV and ion-to-Ta flux ratios J_i/J_{Ta} ranging from 1.3 to 10.7. Films grown with high flux ratios ($J_i/J_{Ta} \geq 7.4$) exhibit a complete 002 texture while those grown with $J_i/J_{Ta} \leq 6.3$ have a strong 111 preferred orientation with a small volume fraction of 002, 022, and 113 grains, as indicated in Figure 9.6b, a plot of the normalized hkl XRD peak intensities ($I_{hkl}/[I_{111} + I_{002} + I_{022} + I_{113}]$) as a function of J_i/J_{Ta}. These results clearly demonstrate that the incident ion-to-metal flux ratio can be used to selectively and controllably vary the preferred orientation of δ-TaN films from predominantly 111 to 002.

The energy E_i of the ions impinging on the surface during growth of transition-metal nitrides strongly determines

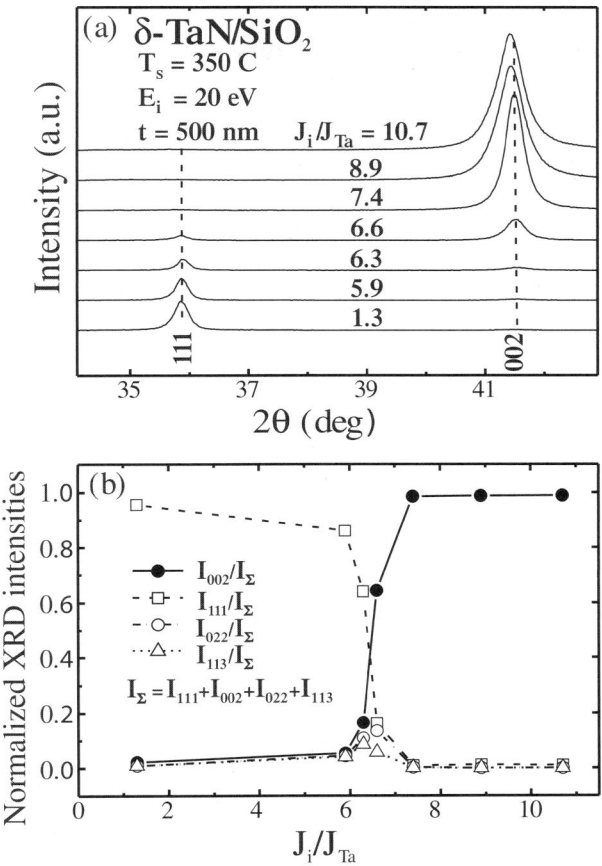

Figure 9.6 (a) XRD ω-2θ scans from 500-nm-thick δ-TaN layers deposited on amorphous SiO₂ at 350°C as a function of J_i/J_{Ta} with E_i = 20 eV. (b) Normalized intensities of the (111), (002), (022), and (113) XRD peaks in (a) as a function of J_i/J_{Ta}. (C.-S. Shin, D. Gall, Y.-W. Kim, N. Hellgren, I. Petrov, and J.E. Greene, *J. Appl. Phys.* 92, 5084, 2002. With permission.)

their effect. Low E_i values (≤ 15 eV) typically have no noticeable effect, E_i ≈ 20 eV provides sufficient momentum transfer to alter surface kinetics, while larger energies, E_i ≥ 30 eV, cause bulk defects and compressive stress. In the case of TaN, 8.5 eV is insufficient to observe an effect on texture evolution while E_i = 30 eV results in nucleation of a secondary hexagonal phase ε-TaN.[19] For ScN, growth with E_i = 13 eV yields layers with mixed grain orientation, texture is controlled using 20 eV ions, and E_i = 50 eV causes high compressive stress due to the inclusion of N_2 gas bubbles in the ScN matrix.[23] The onset for ion-irradiation-induced bulk defect formation in TiN is reported to occur at 43 eV with T_s = 650°C.[6] Considerable amounts of N_2 bubbles form in TiN at E_i > 300 eV.[26]

Several models have been proposed to explain the effect of ion irradiation on texture evolution in cubic transition-metal nitrides. Pelleg et al.[27] and later, Oh and Je[15,16] proposed that the orientation of polycrystalline TiN films should initially be 001, corresponding to the lowest energy surface.[26] However, ion irradiation causes compressive stress, leading to preferred 111 texture, since the elastic modulus is lower in the (111) direction. The change in orientation in this model is driven by the film/substrate system minimizing the total free energy.

However, investigations by Greene et al.,[13,14] mentioned above, showed that low-energy (20 eV) N_2^+ ions impinging on the layer surface during reactive TiN deposition have a dramatic effect on texture, without introducing strain. They therefore proposed a model where texture evolution is determined by competition between thermodynamic and kinetic driving forces. The formation of 111 texture under low-flux conditions was ascribed to limited adatom mobilities on 111 surfaces, leading to larger cation residence times, and consequently, higher cation incorporation probabilities on 111- versus 002-oriented grains. In contrast, high-flux ion irradiation gives rise to increased surface diffusivity due to ion-adatom momentum transfer and therefore has an effect similar to raising the growth temperature. Thus, under high ion-flux conditions, texture evolves toward the lowest surface energy, which is the 001.[26]

More recently, a detailed atomistic model has been proposed based on results from density functional calculations.[21] Texture evolution toward 111-orientation has been ascribed primarily to the lower cation adatom chemical potential on the (111) surface, also related to the low surface mobility on 111 grains, similar to the argument of Greene et al.[13,14] and the simple schematic shown in Figure 9.5. However, it was found that the impingement of high-flux low-energy (~ 20 eV) N_2^+ ions on the layer surface results in the presence of excess atomic N on the (001) surface, due to collisionally induced N_2^+ dissociation. The atomic N reduces the Ti diffusion length, enhances the surface island nucleation rate, and lowers the Ti chemical potential on the (001) surface, leading to preferential growth of 002-oriented grains due to ion irradiation.

Therefore, low-energy ion irradiation alters adatom mobilities, surface island nucleation, layer texture, and ultimately layer microstructures. Figure 9.7 shows TEM and XTEM micrographs from TaN layers grown with low J_i/J_{Ta} = 1.3 and high J_i/J_{Ta} = 10.7 ion-to-Ta flux. The plan-view image in Figure 9.7a of the low-flux sample, which exhibits 111-preferred orientation, reveals an underdense microstructure containing open grain boundaries. The cross section of the same sample (Figure 9.7b) shows a columnar structure with both inter- and intracolumnar voids and a surface that is very irregular with a root-mean-square (rms) roughness of 4.1 nm. In contrast, the 002-oriented layer grown with J_i/J_{Ta} = 10.7 is fully dense, as observed by both plan-view and cross-sectional micrographs in Figure 9.7c and d. A dense microstructure is in general preferred for applications such as hard corrosion-resistant coatings or diffusion barriers. Dark-field imaging investigations indicate that both layers presented in Figure 9.7 initially contain both 111- and 002-oriented grains.[19] However, one orientation emerges as the preferred orientation while the other grains die out within a thickness $\leq$ 200 nm. The underdense microstructure of the 111-oriented layer is due to the limited adatom mobility on (111) surfaces, resulting in increased roughness and, in turn, atomic shadowing. In contrast, the higher cation mobility on 002-oriented grains results in smoother surfaces and a dense microstructure of layers grown at high J_i/J_{Ta} values.

In summary, this section presented the current level of understanding of atomistic processes during the growth of transition-metal nitride layers, deposited primarily by reac-

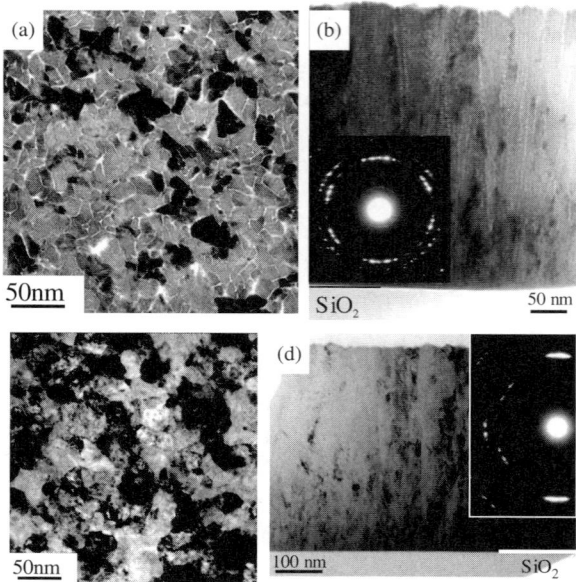

Figure 9.7 Bright-field TEM images from 500-nm-thick δ-TaN layers deposited on amorphous SiO$_2$ at 350°C with E$_i$ = 20 eV: (a) plan view and (b) cross section with J$_i$/J$_{Ta}$ = 1.3; (c) plan view and (d) cross section with J$_i$/J$_{Ta}$ = 10.7. (C.-S. Shin, D. Gall, Y.-W. Kim, N. Hellgren, I. Petrov, and J.E. Greene, *J. Appl. Phys.* 92, 5084, 2002. With permission.)

tive magnetron sputtering. The technologically relevant issue of texture evolution shows that there are primarily two surfaces of importance, (111) and (001), which exhibit distinctively different growth kinetics. The (111) surfaces are under typical growth conditions fully N terminated,[21] leading to low cation adatom potential energies and mobilities. In contrast, cations on (001) surfaces diffuse fast until they reach a step edge, surface island, grain boundary, or form a metal–N admolecule with a N adatom. The surface coverage of the latter can be altered by changing the flux of low-energy N$_2^+$ ion irradiation, which in turn allows control over growth kinetics and microstructural evolution. This anisotropy between (111) and (001) surfaces, together with kinetic surface roughening and atomic shadowing, is essential to the growth of nanopipes as described in the following section.

9.3 NANOPIPES

Nanopipes were first observed in single crystal ScN(001) layers,[23,28] and then also in CrN,[24] TaN,[29] and $Ti_{1-x}W_xN$.[30] Figure 9.8 shows narrow regions of typical XRD reflection and transmission ω-2θ scans for a ScN layer grown at T_s = 800°C on a TiN(001) buffer layer on MgO(001) with layer thicknesses t_{ScN} = 62 nm and t_{TiN} = 220 nm.[28] The scans, offset by one decade to avoid overlap, contain only $(00l)$ peaks in reflection and $(h00)$ peaks in transmission, indicating that both the NaCl-structure ScN overlayer and the TiN buffer layer grow with a cube-on-cube orientational relationship to the substrate: $(001)_{ScN} | 2 | (001)_{TiN} | 2 | (001)_{MgO}$ and $[100]_{ScN} | 2 | [100]_{TiN} | 2 | [100]_{MgO}$. This is in contrast to results presented in Figure 9.4, showing that growth of ScN on MgO(001) without TiN buffer layer leads to a polycrystalline film exhibiting a mixture of two grain orientations. The MgO, TiN, and ScN 002 reflection peaks at 2θ = 19.39, 19.23, and 18.06° correspond to lattice constants along the film growth direction $a_\perp$ of 0.4213, 0.4247, and 0.4519 ± 0.0005 nm, respectively. Both $a_{\perp,TiN}$ and $a_{\perp,ScN}$ are larger than relaxed values, a_{TiN} = 0.4242 nm[31] and a_{ScN} = 0.4501 nm,[28] indicating that the layers are in a state of mild in-plane compressive stress, which is attributed to differential thermal contraction during cooling from the film growth temperature, 800°C, to room temperature, assuming fully relaxed layers during growth and no relaxation during the cooling process. The thermal expansion coefficients for ScN, TiN, and MgO are 4 × 10^{-6} K^{-1} (Ref. 32), 9 × 10^{-6} K^{-1} (Ref. 33), and 1.3 × 10^{-5} K^{-1} (Ref. 34), respectively. The in-plane compressive strain is measured from the ScN 200 peak position at 2θ = 18.28°, obtained in transmission geometry,[28] yielding an in-plane lattice constant $a_{\|,ScN}$ = 0.4465 ± 0.0005 nm, 0.8% smaller than the relaxed value. The Gaussian fit used to determine the 200 peak position is shown in the inset of Figure 9.8.

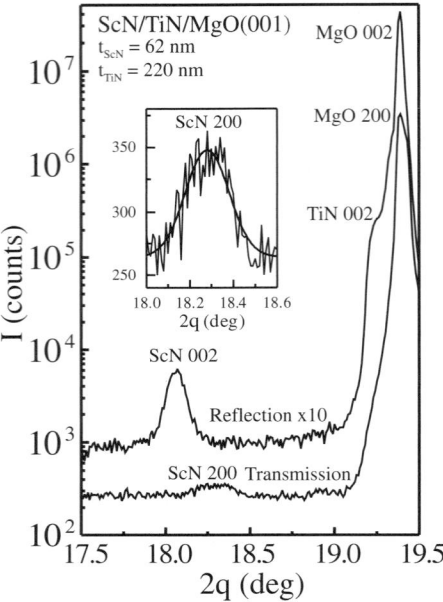

Figure 9.8 XRD ω-2θ scans in transmission and reflection from an epitaxial 62-nm-thick ScN layer grown on TiN/MgO(001). The inset shows a Gaussian fit of the ScN 200 peak. (D. Gall, I. Petrov, P. Desjardins, and J.E. Greene, *J. Appl. Phys.* 86, 5524, 1999. With permission.)

A typical XTEM image with a corresponding selected area electron diffraction (SAED) pattern, in this case from an epitaxial 82-nm-thick ScN film deposited on a TiN(001) buffer layer, is shown in Figure 9.9. The 100 (Figure 9.9a) zone axis diffraction pattern, obtained with a 0.4 μm aperture centered at the ScN/TiN interface, exhibits two sets of symmetric reflections characteristic of single crystals and separated by a distance consistent with the lattice-constant mismatch between ScN and TiN. XTEM tilting experiments reveal an abrupt ScN/TiN interface, with misfit dislocations visible, at the position indicated in Figure 9.9b. The ScN layer appears defect-free (other than the presence of dislocations) up to a thickness t_c of ≈ 15 nm. Above the defect-free region, nan-

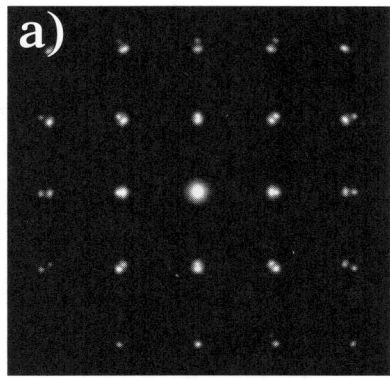

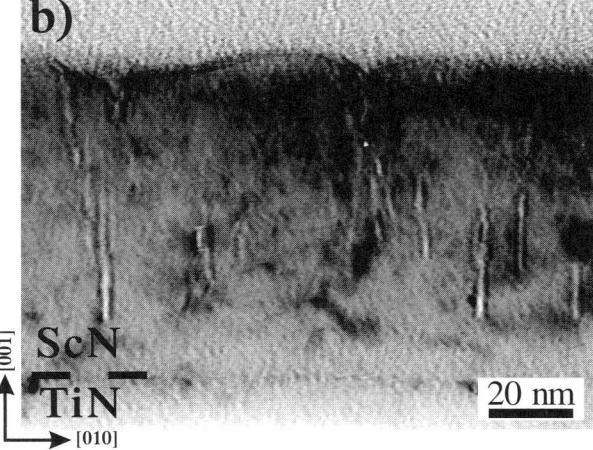

Figure 9.9 (a) Selected-area electron diffraction pattern and (b) corresponding bright-field XTEM micrograph from an epitaxial ScN film grown on TiN/MgO(001). (D. Gall, I. Petrov, P. Desjardins, and J.E. Greene, *J. Appl. Phys.* 86, 5524, 1999. With permission.)

opipes with a width of ~ 1 nm extend along the growth direction and continue to the free surface where they terminate at cusps. The ScN surface is rough with an average mound height of ≈ 7 nm. The apparent termination of some nanopipes below the surface in Figure 9.9b is due to a gradient in the thickness of the XTEM specimen that is thinnest at the TiN buffer layer and thickest at the ScN free surface.

The nanopipes exhibit contrast reversal in under- and over-focus imaging.

Ex situ XPS depth profile measurements from uncapped samples also provide evidence for the presence of nanopipes. Figure 9.10 shows a typical depth profile from a 70-nm-thick ScN layer on TiN/MgO(001). The relative intensity of the Sc 2p, N 1s, O 1s, and Ti 2p peaks are plotted versus sample depth. The near-surface oxygen concentration is high due to sample air exposure, atomic mixing during ion etching, and surface roughness. The O 1s intensity I_{O1s} falls to a steady-state non-zero value over a distance, 5–10 nm, that is consistent with the measured average peak-to-valley surface roughness. I_{O1s} then remains constant for a film thickness t = 45–50 nm before falling to zero. This region of the sample contains the nanopipes whose internal surfaces adsorb oxygen upon air exposure. The remaining ≈ 15 nm of ScN, prior to the rise of the Ti and the fall of the Sc 2p intensities, is fully dense with no nanopipes and thus exhibits no detectable oxygen signal. In contrast to these results, XPS depth profiles of samples that were capped *in situ* with 20-nm-thick W over-layers prior to air exposure showed that the entire ScN layers were oxygen free.[28] This confirms that the oxygen detected in uncapped films enters the nanopipes from the free surface.

The combination of the experimental evidence presented in Figures 9.8–9.10 shows that (a) nanopipes are not due to tensile stress, as could be imagined, since the presented layers exhibit even a slight compressive stress, (b) they do not form at the substrate/layer interface but develop after a certain critical thickness, (c) they terminate at surface cusps, and (d) they provide a continuously open path trough the entire layer with exception of the nanopipe-free bottom. Figure 9.11 illustrates a model for the formation of nanopipes that is consistent with these experimental observations. Initially the surface roughens kinetically to form a periodic mound structure as shown in Figure 9.11a. A high-magnification cross-sectional micrograph in Figure 9.11e from the surface of a 345-nm-thick ScN layer grown with J_i/J_{Sc} = 14 and E_i = 50 eV illustrates the periodic surface roughness.[23] Similar mound structures have also been observed in semiconductor[35–39] and

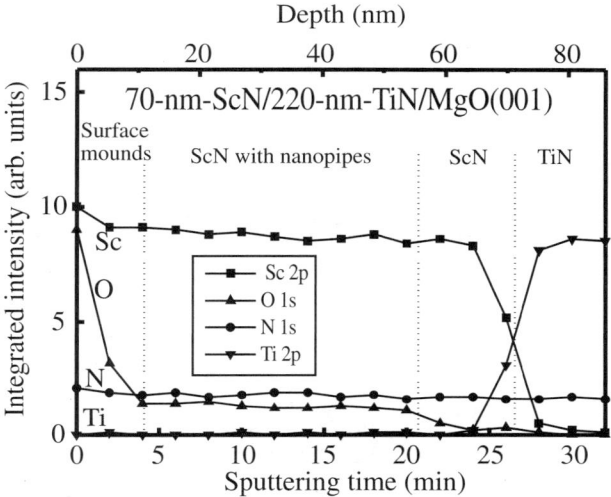

Figure 9.10 XPS profiles from a 70-nm-thick epitaxial ScN layer grown on TiN/MgO(001) showing the Sc 2p, O 1s, N 1s, and Ti 2p intensities as a function of sample depth. (D. Gall, I. Petrov, P. Desjardins, and J.E. Greene, *J. Appl. Phys.* 86, 5524, 1999. With permission.)

metal[40] films grown at temperatures that are low with respect to the melting point. The primary origin of kinetic roughening is the presence of Ehrlich–Schwöbel barriers[41] to the migration of adatoms over down-steps and/or deep traps at step edges on growing surfaces. This leads to a divergence in adatom flux and, hence, increased nucleation on terraces. Adatoms have a higher probability to be incorporated at ascending compared with descending step edges, which results in a net adatom flux toward surface mounds, giving rise to surface roughening and faceting during film growth. When the mounds reach a certain height (Figure 9.11b), roughening is exacerbated due to atomic shadowing from surface mounds. Sputter deposition typically results in a cosine distribution of the incoming atom flux with the highest flux at an azimuthal angle of 45°.[20] Atoms reaching the surface at non-normal angles have a higher probability to impinge on mounds than

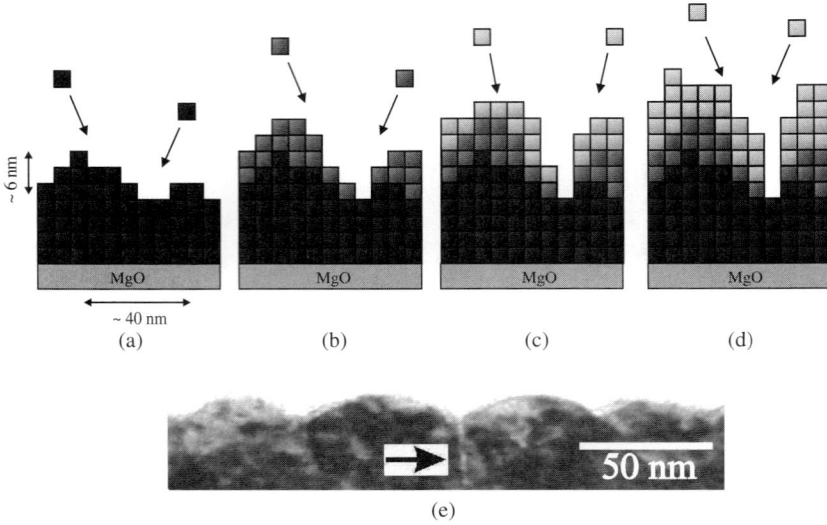

Figure 9.11 (a–d) Schematic of nanopipe formation in single crystal transition metal nitride layers. (e) TEM micrograph of the periodic surface mound structure from a 345-nm-thick ScN layer grown on MgO(001). (D. Gall, I. Petrov, N. Hellgren, L. Hultman, J.-E. Sundgren, and J.E. Greene, *J. Appl. Phys.* 84, 6034, 1998. With permission.)

in valleys, leading to a decrease in the local deposition rate in the valleys, resulting in deep surface cusps (Figure 9.11c). The surface around a cusp contains a high density of step edges or may even form (111) facets. Consequently, the adatom mobility around surface cusps is relatively low (see also comparison of diffusion on (001) and (111) surfaces in the previous section). Thus, surface diffusion is insufficient for filling of the trenches, leading to the formation of nanopipes as shown in Figure 9.11d.

The observation that nanopipes form only above a critical film thickness ($t \simeq t_c \approx 15$ nm) indicates that surface roughening occurs gradually during growth and a critical mound height, which depends on both the angular incident flux distribution and the adatom mobility, is required to maintain continuously open nanopipe structures. Nanopipes elongate along the growth direction because the average direction of the incoming atom flux is normal to the layer surface. This directionality is drastically enhanced by the high adatom mobility on {100} surfaces, which causes nanopipes to be preferentially bound by {100} planes, as shown in the following using plan-view transmission electron microscopy.

Figure 9.12 shows an image taken from a ScN(001)/MgO(001) layer grown at 750°C with $J_i/J_{Sc} = 14$ and $E_i = 20$ eV.[23] This plan-view sample was prepared by step-wise ion milling through the film from both sides. Thus, the micrograph is a plan-view from the center of the 345-nm-thick layer, showing a cross section through the nanopipes that are elongated along the growth direction perpendicular to the image-plane. The nanopipes are rectangular in cross section with a width of ~ 1 nm by 1–15 nm along orthogonal ⟨100⟩ directions. They are bound by high-adatom-mobility {100} planes. Their rectangular elongated shape is explained by their formation; approximately square-shaped surface mounds, with edges along low-energy ⟨100⟩ directions (i.e., along Sc-N bonding directions), adjoin along ⟨100⟩ with the development of valleys, which are consequently also elongated along ⟨100⟩. These transform, with increasing layer thickness, to deep cusps and ultimately nanopipes as shown in Figure 9.11. The square-shaped surface mounds, which have an average mound-to-mound spacing of 30 ± 5 nm, are centered between the nanopipes and can easily be imagined when looking at Figure 9.12. Since the nanopipes are directly linked to the periodic surface roughness, they form a self-organized periodic array with, however, only very limited long-range order. Long-range order and complex nanostructures may, however, be obtained by initial patterning (see Section 9.3.6) and controlled surface morphological evolution.

9.3.1 Surface Morphological Evolution

Nanopipes form due to atomic shadowing from surface mounds under low adatom mobility conditions and highly anisotropic surfaces. Developing quantitative models for surface morphological evolution is therefore essential in order to

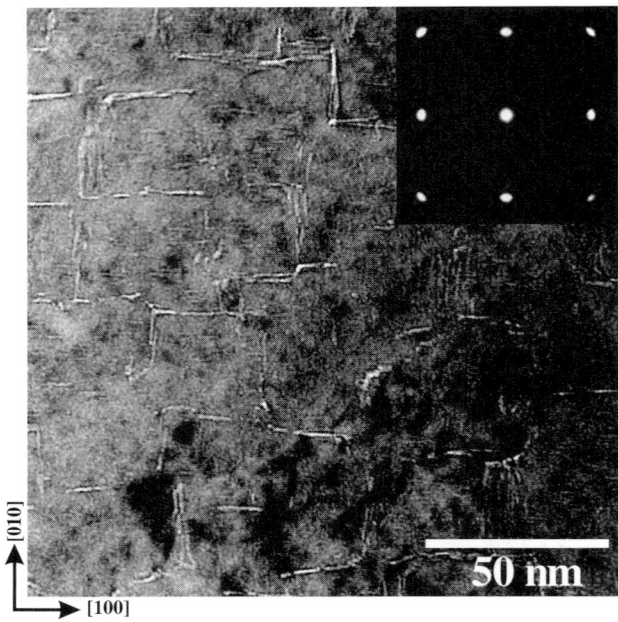

Figure 9.12   Plan-view TEM mircrograph and corresponding diffraction pattern from a ScN(001)/MgO(001) layer showing the arrays of rectangular shaped nanopipes aligned along orthogonal ⟨100⟩ directions. (D. Gall, I. Petrov, N. Hellgren, L. Hultman, J.-E. Sundgren, and J.E. Greene, *J. Appl. Phys.* 84, 6034, 1998. With permission.)

control nanopipe formation. Surface diffusion, surface island growth and decay kinetics, and surface mound formation have been investigated for single crystal TiN(001) layers using *in situ* scanning tunneling microscopy (STM) studies.[6,42–47]

Karr et al.[42] were the first to report *in situ* STM results from TiN(001) surfaces. The images shown in Figure 9.13 are from TiN(001) layers with thicknesses t = 25, 50, 100, and 230 nm. The layers were grown on MgO(001) by reactive magnetron sputter deposition at 750°C. In all images, atomic height steps (0.21 nm) are clearly observable. The surface morphology on a large scale is controlled by a regular array of growth mounds, ~ 1 nm high, that are separated by a

Figure 9.13 Scanning tunneling microscopy images of TiN(001) films grown on MgO(001) by reactive magnetron sputter deposition at T = 750°C as a function of film thickness t = 25, 50, 100, and 230 nm. (B.W. Karr, I. Petrov, D.G. Cahill, and J.E. Greene, *Appl. Phys. Lett.* 70, 1703, 1997. With permission.)

characteristic length d ~ 100 nm. The sequence of images indicates that both d and the vertical amplitude of the surface roughness slowly increase with film thickness. This can be quantified using the height difference correlation function $G(\rho) = \left\langle \left(h_i - h_j \right)^2 \right\rangle$ where h_i and h_j are the heights of the surface at two locations labeled i and j separated by a distance ρ, and the brackets signify an average over pairs of points, i, j.[48] $G^{1/2}(\rho)$ increases with ρ < d and approaches a constant value for ρ > d. The value for d can, therefore, be directly determined from the $G^{1/2}(\rho)$ curve. However, for the case of the images in Figure 9.13, d was more accurately obtained by taking the square root of the surface mound density.[42] The amplitude of the surface roughness $A = G^{1/2}(d/2)$ is approximately equal to $\sqrt{2}W$, where W is the rms surface roughness measured on a length scale comparable to d.[6]

The in-plane length scale d and the amplitude for surface roughness A are the fundamental length scales that characterize the surface morphology and are plotted in Figure 9.14 as a function of layer thickness. The open diamonds represent the data obtained from the samples presented in Figure 9.13 and grown at 750°C. Circles and triangles are from TiN(001) grown at 650°C from Ref. 6. The latter are grown under high-flux ion irradiation with J_i/J_{Ti} = 25 and ion energies E_i of 3 and 43 eV, as indicated in the figure. Both d and A increase with increasing layer thickness for all samples, indicating growth of the surface mounds in both lateral and vertical directions. The formation of the mounds is due to the asymmetry in the attachment of adatoms at ascending versus descending steps, which destabilizes growth on low miscut surfaces, and has been reported to lead to mound formation on both metal and semiconductor surfaces.[49–51] While this process is well established, the reported roughening rates vary widely with exponents ranging between 0.16 and ≈ 1.[42] The three datasets in Figure 9.14 exhibit a slope that follows a power law with an exponent of 0.25, for both d and A, as indicated by the dashed lines that are obtained using $d \propto h^{0.25}$ and $A \propto h^{0.25}$. Therefore, the roughening and coarsening rates for TiN(001) layer growth are within the range of rates reported for other materials.

Figure 9.14 also provides insight into the effects of growth temperature and ion irradiation on surface morphological evolution. With no anode bias, i.e., E_i = 3 eV, an increase in growth temperature from 650 to 750°C increases d slightly and decreases A more significantly.[6] Thus, the

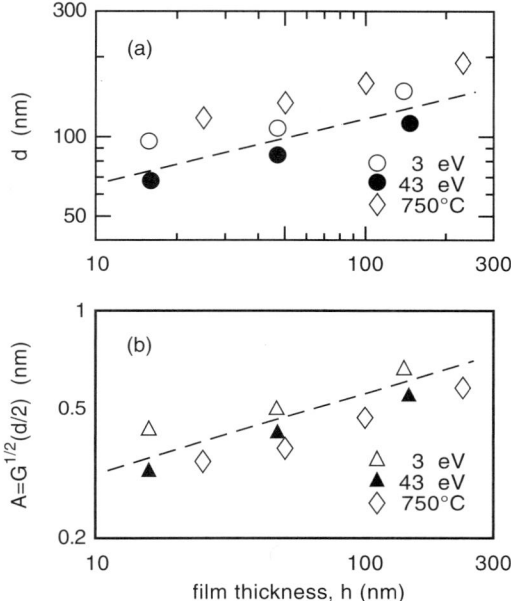

Figure 9.14 Summary of (a) the in-plane length scale d and (b) the roughness amplitude A plotted as a function of layer thickness h. The different symbols indicate growth at 650°C with ion energy of 3 and 43 eV, and growth at 750°C. The dashed line shows a power law dependence of $h^{1/4}$ for comparison. (B.W. Karr, I. Petrov, D.G. Cahill, and J.E. Greene, *Phys. Rev.* B 61, 16137, 2000. With permission.)

surface mounds become wider and less high at higher temperature, due to larger adatom diffusion lengths. A similar temperature dependence of d and A has also been reported for low-temperature growth of Ge(001).[52] In contrast, an increase in E_i from 3 to 43 eV at 650°C causes d to decrease slightly while A also decreases. Thus, ion irradiation has little effect on A/d, the average aspect ratio of the surface. Ion irradiation reduces the effective strength of the growth instability that drives the formation of growth mounds.[6] It also reduces, due to the presence of atomic N on the surface as discussed in Section 9.2.2 and Ref. 21, the diffusion length of Ti adatoms and consequently d.

A study by Wall et al.[47] focuses on surface diffusion and nucleation kinetics in order to develop a quantitative understanding of the relevant processes determining the surface morphological evolution of TiN(001). Large, atomically smooth TiN(001) terraces were obtained using a series of preparation steps including the anneal of MgO(001) substrates at 1400°C for 4 h followed by the successive growth of TiN buffer layers at 650, 750, and 1050°C with thicknesses t = 20, 40, and 8 nm, respectively. A typical *in situ* STM image of a resulting TiN(001) surface is shown in Figure 9.15a. The steps in the image are single atomic layers, 2.1 Å high, and the surface has large terraces, ≈ 60 nm wide, indicating layer-by-layer growth.

The nucleation of surface islands on such large terraces is then investigated by an additional deposition of 0.3 monolayers. Figure 9.15b–d shows STM images after this additional deposition at 510, 600, and 800°C. The islands, which are one atomic step high, are dendritically shaped and increase in size with increasing temperature (note the decreasing magnification for the images b through d). The nucleation length L_n, which is defined as the characteristic separation between islands, is calculated using $L_n = 1/\sqrt{n}$, where n is the average areal density of islands, excluding areas close to step edges. L_n increases from 6 nm at 510°C to 9 nm at 600°C to 17 nm at 800°C.

Alternatively, the nucleation of surface islands can also be studied for the case of nucleation on growing two-dimensional islands, which corresponds more closely to the process during continuous thin film growth. Figure 9.16 shows STM images[47] from 8-nm-thick TiN(001) layers deposited at 650, 750, and 920°C on TiN(001) buffer layers comparable to that shown in Figure 9.15a. At the lowest temperature $T_s = 650°C$, the surface morphology is dominated by growth mounds formed by kinetic roughening. For higher T_s, the spacing between the growth mounds increases, and the surface width decreases. At $T_s \geq 870°C$ (not shown) the surface shows only two or three layers, indicating a transition to layer-by-layer growth.[47] The straight steps in Figure 9.16b are slip-steps from dislocations that form during cooling due to the

Figure 9.15 Scanning tunneling micrographs from (a) a TiN(001) buffer layer grown at 1050°C and from 0.3 TiN monolayers grown at (b) 510°C, (c) 600°C, and (d) 800°C on a buffer layer comparable to that shown in (a). (M.A. Wall, D.G. Cahill, I. Petrov, D. Gall, and J.E. Greene, *Phys. Rev.* B 70, 035413, 2004. With permission.)

unmatched coefficients of thermal expansion for MgO and TiN.[42] The island shape is isotropic at the lower temperatures but shows fourfold symmetry, corresponding to the crystal symmetry (Figure 9.1), at higher temperatures. The critical island radius R_c necessary to nucleate a new layer is determined by fitting an ellipse within the rough boundary of the island. R_c is then one-half the length of the minor axis.[47]

According to derivations by Venables[53] and Tersoff,[54] L_n and R_c follow an exponential temperature dependence:

Figure 9.16 STM images from 8-nm-thick epitaxial TiN(001) layers grown at (a) 650°C, (b) 750°C, and (c) 920°C. (M.A. Wall, D.G. Cahill, I. Petrov, D. Gall, and J.E. Greene, *Phys. Rev.* B 70, 035413, 2004. With permission.)

$$L_n \approx \frac{1}{\sqrt{\eta_i}} \left(\frac{\upsilon a^2}{F} \exp\left(\frac{-E_s}{kT}\right) \right)^{1/6} \qquad (9.1)$$

$$R_c \approx \left(\frac{192}{\pi}\right)^{1/6} \left(\frac{\upsilon a^2}{F} \exp\left(\frac{-E_s}{kT}\right) \right)^{1/6} \qquad (9.2)$$

where υ is the attempt frequency, a^2 is the area of a surface unit cell, F is the deposition flux, E_s is the activation energy for diffusion, and η_i is a slowly varying function that relates the coverage and critical cluster size to the saturated nucleation length; typically, $\eta_i \approx 0.2$. Equations (9.1) and (9.2) assume that two adatoms that meet on the surface stay sufficiently long together for other adatoms to join and nucleate a cluster. Thus, the largest unstable cluster of adatoms contains one adatom (i = 1). This is, as shown below, only true for $T_s < 865°C$.

Figure 9.17 shows the L_n and R_c values measured by Wall et al.[47] from STM images comparable to those shown in Figures 9.15 and 9.16 as a function of $1/T_s$. Both L_n and R_c decrease exponentially with $1/T_s$ for $T_s < 865°C$, as expected from Equations (9.1) and (9.2). The fit provides values for the activation energy for surface diffusion $E_s = 1.4 \pm 0.1$ eV and attempt frequency $\upsilon = 10^{14 \pm 1}$ s^{-1}.

At $T_s > 865°C$, the critical radius decreases exponentially with a greater slope compared with the lower temperature regime. This is because two adatoms are no longer a stable cluster, i.e., i > 1. Venables[53] has shown for the i > 1 regime that the slope of L_n as a function of $1/T_s$ is a function of the activation barrier for diffusion plus the binding energy of the largest unstable cluster. The fit of the high-temperature data in Figure 9.17 provides, assuming the limiting case of i = ∞, a value for the sum of binding energy per adatom plus activation energy for surface diffusion of 2.8 eV.

An alternative approach to study surface diffusion and island kinetics was reported by Kodambaka et al.[45] They deposited 0.2 monolayers of TiN on atomically smooth TiN(001) surfaces. This was followed by annealing experiments to examine island decay kinetics. Figure 9.18 shows

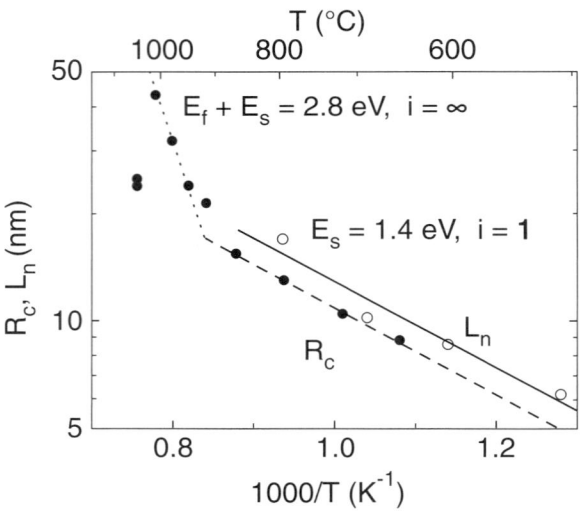

Figure 9.17   The critical radius of the island necessary to nucleate a second layer cluster, R_c, and the nucleation length on a large terrace, L_n, as a function of temperature. The solid and dashed lines correspond to $E_s = 1.4$ eV, $\nu = 10^{14}$ s^{-1}, and $i = 1$. The dotted line shows R_c with an activation-energy sum of 2.8 eV, $i = \infty$ and $\nu = 10^{14}$ s^{-1}. (M.A. Wall, D.G. Cahill, I. Petrov, D. Gall, and J.E. Greene, *Phys. Rev.* B 70, 035413, 2004. With permission.)

representative STM images obtained at times $t_a = 0$, 74, and 171 min during a 3-h anneal at $T_a = 890°C$. The islands in Figure 9.18a, obtained after thermal equilibration of the STM tip, are roughly square-shaped, corresponding to the equilibrium shape of the islands. Most of the smaller islands, including islands 1 through 4 in Figure 9.18a, have disappeared by $t_a = 171$ min. Note that a denuded zone is formed around island 5 (Figure 9.18c), which coarsens at the expense of the smaller neighbors. The radial extent of this region corresponds to the quasi-equilibrium adatom diffusion length at this temperature.,

The decay of islands in the presence of larger islands is due to coarsening (Ostwald ripening), as described by the Gibbs–Thompson equation showing that the equilibrium ada-

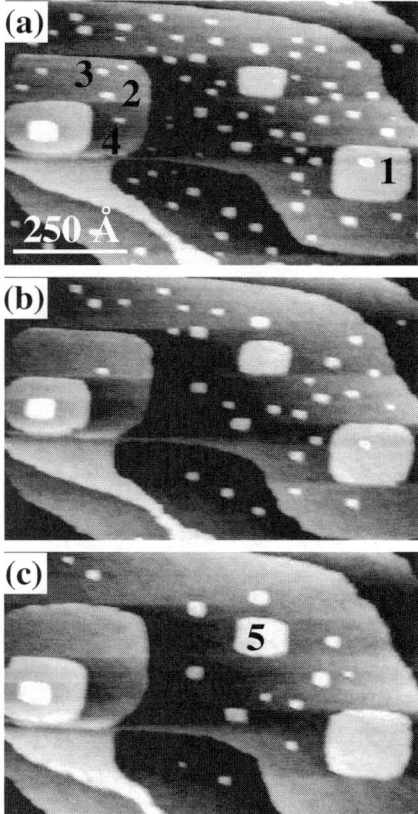

Figure 9.18 STM images from a TiN(001) surface covered by 0.2 monolayers of TiN. Images were obtained during annealing at 890°C for times (a) t_a = 0, (b) 74, and (c) 171 min. (S. Kodambaka, V. Petrova, P. Desjardins, I. Petrov, D.G. Cahill, and J.E. Greene, *Thin Solid Films* 392, 164, 2001. With permission.)

tom concentration around an island increases with decreasing radius r. Small islands have a higher curvature and hence a higher two-dimensional spreading pressure than larger islands, resulting in the decay of the smaller and growth of the larger islands. Modeling island decay, adatom diffusion, and adatom density distributions using STM data from these annealing experiments provides values for line tension and for

Figure 9.19 Atomic force microscopy images (3×3 μm^2 scans) from 500-nm-thick epitaxial CrN/MgO(001) layers grown at $T_s = 600°C$ with (a) $J_{N_2^+} / J_{Cr} = 1.7$ and (b) $J_{N_2^+} / J_{Cr} = 14$. The black-to-white scale corresponds to 14.6 and 7.1 nm for (a) and (b), respectively. (D. Gall, C.-S. Shin, T. Spila, M. Odén, M.J.H. Senna, J.E. Greene, and I. Petrov, *J. Appl. Phys.* 91, 3589, 2002. With permission.)

the product of surface diffusivity times the adatom concentration associated with a straight step. Kodambaka et al.[45] performed such an analysis using STM data in the temperature range between 750 and 950°C, and extracted an activation energy for adatom detachment plus diffusion of 3.4 ± 0.3 eV.

9.3.2 Ion-Irradiation Effects on Surface Morphology and Nanopipe Development

All cubic transition metal nitrides exhibit surface morphologies similar to the most commonly studied TiN, presented in the previous section. Figure 9.19 shows atomic force microscopy (AFM) images from 0.5-μm-thick epitaxial CrN(001) layers grown at T_s = 600°C by ultra-high vacuum magnetron sputter deposition on MgO(001) substrates.[24] Deposition was done at constant ion energy E_i of 12 eV. However, the ion-to-neutral ratio was varied from $J_{N_2^+} / J_{Cr}$ = 1.7, corresponding to typical conditions for standard magnetron sputter deposition systems, to $J_{N_2^+} / J_{Cr}$ = 14, representing a high ion flux. Under standard deposition conditions (Figure 9.19a), the surface morphology exhibits primarily square-shaped growth mounds with edges predominantly aligned along low-energy ⟨100⟩ directions. Lateral mound sizes range from 130 to 300 nm with a surface width ⟨w⟩ of 1.80 ± 0.04 nm and an average peak-to-valley mound height ⟨h⟩ = 5.09 ± 0.11 nm. The in-plane correlation length ⟨d⟩ is 208 ± 20 nm. Thus, the mound aspect ratio ⟨h⟩/⟨d⟩ is 0.024.

Some mounds in Figure 9.19a (see example marked with white arrow) have corners elongated along ⟨110⟩ directions. The formation of these mounds, which have the shape of a distorted four-cornered star, is attributed to the incorporation of small mounds into larger ones through coalescence at adjacent corners along ⟨110⟩ directions.[24] Preferential corner coalescence, which begins to occur, as discussed below, at layer thicknesses of > 30 nm, can be explained as follows. Growth mounds form due to kinetic surface roughening during deposition under conditions of low adatom mobility in the presence of Ehrlich barriers to adatom migration over descending step edges.[41] Mound formation is exacerbated during sputter deposition due to the large component of non-normal deposition flux,[20] which results in atomic shadowing and, hence, a decrease in the flux reaching the valleys between the mounds. Shadowing is strongest along ⟨100⟩ valleys and less pronounced at corners. Consequently, ⟨100⟩-oriented valleys are deeper than the dips between mound corners along ⟨110⟩ directions. This, in turn, favors coalescence along ⟨110⟩ directions.[24]

The AFM image in Figure 9.19b is obtained from a CrN(001) layer grown with a high ion-to-metal flux ratio, $J_{N_2^+} / J_{Cr} = 14$. The surface of this sample exhibits a strikingly different morphology than that shown in Figure 9.19a, which is characteristic of sputter deposition under low ion-to-metal flux ratios. The growth mounds in Figure 9.19b are much larger and elongated along ⟨110⟩ while dendritic in the orthogonal direction. They have an average length-to-width ratio of 4 ± 1. The elongation occurs along the ⟨110⟩ direction that forms the largest angle with the steps of the substrate miscut, i.e., surface mounds tend to be elongated perpendicular to the edges of the substrate miscut. Growth along ⟨110⟩ is due to the following process. Under sufficiently large adatom mobility conditions, deposited atoms primarily incorporate into the growing layer by diffusion on the terrace and attachment at ascending steps. There is consequently a gradient in the adatom concentration on the terrace that drives the diffusion flux toward the ascending step. The surface steps, which are, on a large scale, oriented at an arbitrary polar angle determined by the substrate miscut, are made up of low-energy step edges along [100] and [010] and therefore exhibit a zigzag shape. The step edge close to corners that stick into the lower terrace experience, due to the above-described adatom gradient, grows faster than the step edge near the concave corner farther away from the lower terrace. Due to geometrical arguments, the edges on both sides of the corners ([100] and [010]) grow at the same rate, resulting in the growth along ⟨110⟩ directions. In summary, mound growth along ⟨110⟩ occurs when the adatom mobility is sufficiently high so that the nucleation length is larger or comparable to the terrace size. This is consistent with STM results in Figure 9.16, showing islands with dendritic fingers along ⟨110⟩ at a high growth temperature ($T_s = 920°C$), with large nucleation length, but isotropic islands at lower temperatures (T_s 750°C). The transition from square-shaped to ⟨110⟩ elongated mounds in Figure 9.19 is therefore a clear indication that the increased ion-irradiation flux results in an enhanced adatom mobility.

The in-plane areas A of growth mounds on CrN(001) layers deposited with $J_{N_2^+} / J_{Cr} = 14$ range from 0.25 to 1.5 μm^2, approximately an order of magnitude larger than A-values from samples grown with $J_{N_2^+} / J_{Cr} = 1.7$ (Figure 9.19a). In addition, the average mound height, $\langle h \rangle = 2.52 \pm 0.03$ nm, decreases by a factor of two while the in-plane correlation length, $\langle d \rangle = 520 \pm 100$ nm, is 2.5 times larger. This results in an average aspect ratio $\langle h \rangle / \langle d \rangle = 0.005$ for the CrN(001) layer grown with $J_{N_2^+} / J_{Cr} = 14$, a factor of 5 lower than that obtained for the $J_{N_2^+} / J_{Cr} = 1.7$ CrN(001) layer.

The above results from epitaxial CrN(001) layers show that increasing the incident ion-to-metal flux ratio during growth, using low N_2^+ ion energies, results in smoother surfaces with both smaller mound heights and decreased mound aspect ratios.[24] High-flux, low-energy ion irradiation leads to higher adatom mean surface diffusion lengths and, consequently, larger surface features. In addition to larger adatom diffusion lengths on terraces, momentum transfer from low-energy ion irradiation increases the probability of adatoms crossing descending step edges. This contributes to the filling of trenches between growth mounds, thereby reducing atomic shadowing. The surface smoothening obtained at higher ion-to-metal flux ratios has important consequences on the formation of nanopipes, as shown in the following.

Figure 9.20 shows cross-sectional transmission electron micrographs from epitaxial CrN/MgO(001) layers, corresponding to the AFM images in Figure 9.19, as reported in Ref. 24. The layers were grown with ion-to-metal flux ratios $J_{N_2^+} / J_{Cr}$ of 1.7, 6, and 10. The selected area electron diffraction (SAED) pattern in Figure 9.20c is typical for all epitaxial CrN(001) layers. It exhibits symmetric single-crystal reflections revealing a cube-on-cube epitaxial relationship: $(001)_{CrN} | | (001)_{MgO}$ with $[100]_{CrN} | | [100]_{MgO}$.

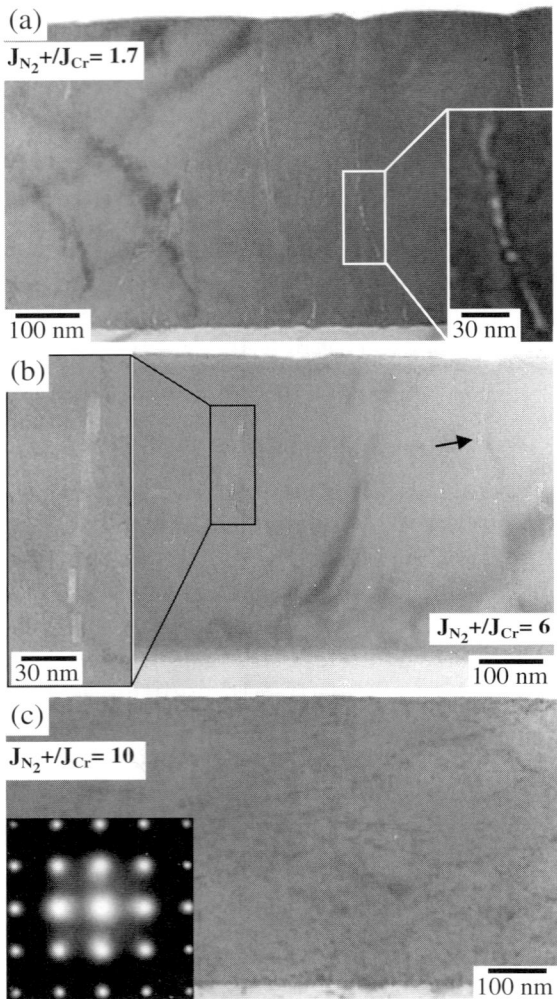

Figure 9.20　XTEM micrographs from 500-nm-thick epitaxial CrN/MgO(001) layers grown at $T_s = 600°C$ with (a) $J_{N_2^+} / J_{Cr} = 1.7$, (b) $J_{N_2^+} / J_{Cr} = 6$, and (c) $J_{N_2^+} / J_{Cr} = 10$. The insets in (a) and (b) show higher resolution images of typical nanopipes, while the inset in (c) is a corresponding SAED pattern. (D. Gall, C.-S. Shin, T. Spila, M. Odén, M.J.H. Senna, J.E. Greene, and I. Petrov, *J. Appl. Phys.* 91, 3589, 2002. With permission.)

The most prominent features in the micrograph from the layer grown with $J_{N_2^+} / J_{Cr} = 1.7$ (Figure 9.20a) are 1–5 nm wide nanopipes along the [001] growth direction. A higher-resolution image of a short section of a typical nanopipe is shown in the inset in Figure 9.20. The nanopipes are underdense regions that appear bright when the image is underfocused, as is the case for all micrographs in Figure 9.20, but appear dark when viewed in overfocused conditions.

The XTEM image in Figure 9.20a reveals 10 nanopipes in the field of view, all originating from the CrN/MgO(001) interface. The average lateral separation $\langle s \rangle$ between nanopipes is 90 ± 20 nm. At film thickness $t \cong 30$ nm, corresponding to the onset of mound coalescence, half of the nanopipes terminate. That is, the nanopipe density decreases by a factor of two while $\langle s \rangle$ increases to 180 ± 40 nm, approximately equal to the growth mound in-plane correlation length $\langle d \rangle = 208 \pm 20$ nm determined from the AFM image of this sample in Figure 9.19a. The remaining five nanopipes continue to the free surface where they terminate at cusps between growth mounds. This suggests that the formation of nanopipes is directly related to the presence of the surface cusps between the periodic growth mounds that cause strong atomic shadowing, as discussed above (see Figure 9.11). The nanopipe density, determined from large-area plan-view TEM micrographs, is ≈ 870 μm^{-2}.[24] This number is larger than the surface mound density, 48 μm^{-2}, since each valley between adjacent mounds contains several nanopipes. The surface exhibits a periodic mound structure with an average mound height $\langle h \rangle = 5$ nm and a mound-to-mound spacing of 220 ± 70 nm, in good agreement with the AFM result in Figure 9.19a.

The XTEM micrograph in Figure 9.20b was obtained from a CrN(001) layer grown with $J_{N_2^+} / J_{Cr} = 6$. The layer appears defect-free (no nanopipes) over a layer thickness $t^* > 250$ nm with a low nanopipe density at $t > t^*$. Two nanopipes are observed in the field of view in Figure 9.20b; one is shown at higher magnification while the other is indicated by a black arrow. The nanopipe density determined from low-magnification plan-view micrographs is ≈ 270 μm^{-2}, more than a factor of three less than in CrN(001) layers grown with $J_{N_2^+}/J_{Cr} = 1.7$. The nanopipes, as is the case for the layer grown at the low ion-to-metal ratio, have widths ranging from 1 to 5 nm. However, they are not continuous and appear and disappear along the growth direction (see higher resolution view). While the nanopipe on the left side of Figure 9.20b terminates well below the surface, the nanopipe on the right reaches the surface at a cusp between two mounds. The average mound-to-mound spacing determined from XTEM is 280 ± 80 nm and the mound height ≈ 4 nm. Thus, as $J_{N_2^+}/J_{Cr}$ is increased from 1.7 to 6, the in-plane mound size increases while the height decreases, confirming the trend observed by AFM in Figure 9.19.

Figure 9.20c is a micrograph from a CrN(001) sample grown with $J_{N_2^+}/J_{Cr} = 10$. The surface appears flat to within the resolution of the microscope. No nanopipes are observed. A detailed analysis of multiple XTEM images from many layers grown with $J_{N_2^+}/J_{Cr} = 10$ and 14 reveals, however, the presence of nanopipes above a film thickness $t^* \cong 400$ nm, but with a very low number density. Large-area plan-view micrographs provide an upper limit of 20 μm^{-2} for the nanopipe number density in layers grown with $J_{N_2^+}/J_{Cr}$ 10, more than forty times less than for CrN(001) layers grown with $J_{N_2^+}/J_{Cr} = 1.7$.

Thus, the strong effects of low-energy N_2^+ ion irradiation during deposition on the surface morphology of CrN(001) layers, as discussed above, also controls the formation rate of nanopipes. Layers grown with low ion-to-metal ratios ($J_{N_2^+}/J_{Cr} = 1.7$) exhibit rough surfaces composed of growth mounds with relatively high aspect ratios, 0.024, and short in-plane coherence lengths, <d>, = 208 nm, which, within the experimental uncertainty, are equal to the measured average separations between nanopipes. The higher nanopipe density observed at t < 30 nm for layers grown with $J_{N_2^+} / J_{Cr} = 1.7$ results from the fact that mound sizes are smaller (~ 90 nm) during the initial stages of CrN(001) growth prior to mound coalescence.

For CrN(001) layers grown with medium ion-to-metal ratios, $J_{N_2^+} / J_{Cr} = 6$, the in-plane correlation length is larger, corresponding to a lower mound number density. This in turn results in a reduction in the nanopipe number density by more than a factor of three compared with CrN(001) layers grown with $J_{N_2^+} / J_{Cr} = 1.7$. In addition, nanopipes form only above a critical film thickness (t* ≥ 250 nm) and are found to be discontinuous along the growth direction. The rate of surface roughening occurs more gradually during growth with $J_{N_2^+} / J_{Cr} = 6$. Thus, at t < t*, the mound aspect ratio is too small to provide sufficient atomic shadowing to form cusps and, hence, nanopipes. Even at t > t*, the surface roughness is not adequate to sustain continuous nanopipe formation, leading to discontinuous nanopipes. At high ion-to-metal flux ratios, $J_{N_2^+} / J_{Cr} ≥ 10$, layer surfaces are relatively smooth, <w>, = 0.89 nm, such that shadowing is negligible and essentially no nanopipes are formed.

9.3.3 Atomic Shadowing Theory

Atomic shadowing is, as indicated in Figure 9.11, key to the development of nanopipes and also affects surface morphological evolution prior to pipe formation, as discussed in the previous section. Atomic shadowing (called also self-shadowing) in the context of thin film growth refers to the effect that the surface morphology plays in the capture probability of incident atoms from the gas phase. This is primarily due to surface mounds having a higher probability than surface valleys to capture incident atoms, leading to an inequality in layer growth rates as a function of location on the surface. The development of surface morphology in the presence of atomic shadowing, has been modeled by a variety of theoretical approaches.[55-63]

Karunasiri et al.[55] proposed a growth model for sputter deposition of thin amorphous films, based on a one-dimensional solid-on-solid formalism. They set up an equation for the local growth rate at point x and accounted for non-local atomic shadowing using the exposure angle, which itself depends on the surface morphology surrounding point x. The model also accounts for surface diffusion and contains a Gaussian white-noise term. They found that for small values of the diffusion constant, the growing layer surface develops a self-similar mountain landscape. At larger diffusion values, the growth of compact flat films is possible up to a critical height above which surface roughening starts to occur.

Bales and Zangwil[56,57] improved the above-mentioned model for amorphous layers by treating the deposition term more completely. In particular, they accounted for the fact that the local growth rate occurs in the direction perpendicular to the local surface, which is in general not equal to the macroscopic growth direction. Their two-dimensional model explicitly accounts for both surface diffusion and atomic shadowing, and assumes equal flux from all angular directions that are not shadowed by other parts of the surface, as illustrated in Figure 9.21. This assumption is correct for growth at relatively high gas pressures, where the mean free path of the deposited species is much smaller than the source-substrate distance, leading to a fully randomized direction of the deposition flux. They show that for the case of negligible surface diffusion, the maxima (i.e. surface mounds) flatten as growth proceeds while, in contrast, the minima steadily evolve toward groves. Also, in the absence of diffusion, no characteristic length scale is favored, leading to self-similar starting surfaces resulting in self-similar final morphology, consistent with "cauliflower"-like morphologies observed experimentally.[64] However, when surface diffusion is present, i.e., at growth temperatures that are a significant fraction of the melting point, smoothing of the surface occurs with a characteristic length scale, determined by the diffusion length. This smoothing is in direct competition to the destabilizing shad-

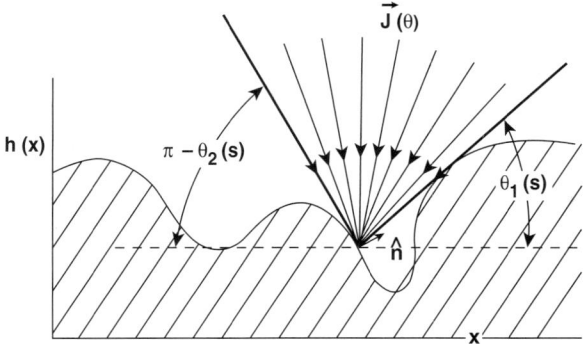

Figure 9.21 Deposition flux according to the model of Bales and Zangwill.[57] Each point on the surface receives a fraction of the total incident flux as defined by the shadow angles θ_1 and θ_2. Growth occurs perpendicular to the local surface orientation. (G.S. Bales and A. Zangwill, *J. Vac. Sci. Technol.* A 9, 145, 1991. With permission.)

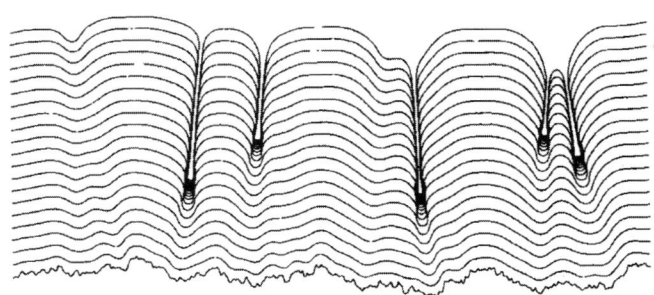

Figure 9.22 Simulated surface morphological evolution. (G.S. Bales and A. Zangwill, *J. Vac. Sci. Technol.* A 9, 145, 1991. With permission.)

owing term that causes groove formation. Figure 9.22 shows the results of a numerical simulation by the same authors, illustrating the surface morphological evolution predicted by this model. The morphology is consistent with the columnar growth observed experimentally. It shows wide, relatively flat columns that are separated by narrow grooves. The average column width is directly related to the diffusion length. Figure

9.22 resembles the images of layers containing nanopipes shown above. However, nanopipes in transition-metal nitrides form with a more uniform inter-nanopipe spacing as well as direction and width. This is due to, in addition to shadowing effects, a combination of highly anisotropic surface diffusion, faceting, and kinetic surface roughening as discussed above.

Guo and co-workers[58–60] applied a discrete model and direct simulations on a square lattice to study the shadowing effect on surface morphological evolution. Their simulations included a parameter θ_{max}, which defines the angle spread of the incoming deposition flux. For low θ_{max} values ($< 5°$), shadowing effects are minor and the surface simply roughens. However, larger θ_{max} results in columnar structures with deep grooves, consistent with the prediction of Bales and Zangwil.[57] Column coalescence and competition was also observed in their simulations. Some smaller columns were overshadowed and evolved into grooves while wide column tops were susceptible to tip splitting. The column-coarsening exponent for this model was determined to be 0.56 ± 0.02 and the power spectrum for the interface shape was found to exhibit dynamic scaling. Guo and co-workers also discuss the transition from a rough to a columnar microstructure (as a function of temperature and θ_{max}) in terms of a non-equilibrium phase transition. This is comparable to the transition, discussed above, for the growth of epitaxial CrN layers, where a high density of nanopipes is observed at low ion fluxes but high ion irradiation leads to higher surface diffusion, lower surface roughness, the absence of shadowing, and consequently no nanopipes.

Krug and Meakin[61] employed a similar simulation approach to study the roughening exponent under growth condition where significant shadowing is present. They found that the exponent is considerably affected by the angular distribution of the incoming flux, i.e., the exposure angle θ, as well as the initial conditions, i.e., the morphology of the starting surface. However, there is also a stable regime for the case of large angular distribution, where the roughening exponent is constant and equal to 1. The roughening exponent of 1 was also obtained by Drotar et al.[62] who reported results from numerical calculations and Monte Carlo simulations in 2 + 1 dimensions. Their simulated microstructure exhibited tall columns with correlation lengths that increase with deposition time, following a power law.

Karpenko et al.[63] simulated thin film growth focusing on the combined effect of atomic shadowing and crystallographic growth anisotropy, which is due to anisotropic surface free energies and/or surface diffusion. They found that neither of the two effects by itself can explain the development of in-plane texture during polycrystalline thin film growth. However, if both are present, grains that have the preferred crystallographic direction aligned with the preferred geometric direction (determined by the flux geometry) will overgrow misoriented grains and, consequently, texture will develop. This is consistent with the results presented in Section 9.2.1, where anisotropic surface diffusion in transition-metal nitrides leads to texture development.

9.3.4 Glancing-Angle Deposition Modeling

The previous section focuses on the general understanding of shadowing effects primarily for the case of a randomly oriented incoming deposition flux. However, a much bigger technological interest exists for deposition techniques where the flux impinges on the substrate at a well-defined angle, allowing controlled shadowing and the creation of novel nanostructures.

Figure 9.23 illustrates how non-normal deposition leads to a layer that exhibits a columnar microstructure where the columns are tilted by an angle β toward the direction of the deposition flux, which occurs at an angle $\alpha \geq \beta$ from the surface normal. Several theoretical approaches[65–72] have been used to address the relationship of α and β, which likely depends on other deposition and materials parameters including the surface diffusion length and the initial surface morphology.[66] Nieuwenhuizen and Haanstra[73] developed, based on a wide range of experimental results, the empirical "tangent rule" that relates β and α by the following expression:

$$\tan \beta = \frac{1}{2} \tan \alpha \qquad (9.3)$$

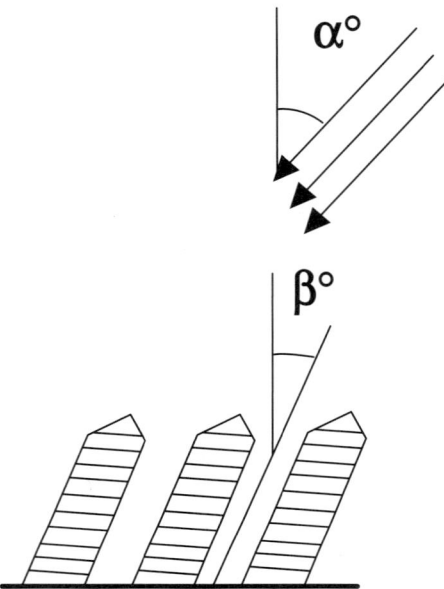

Figure 9.23 Schematic of oblique deposition geometry. α is the angle between surface normal and deposition flux. β denotes the mean column angle. (F. Paritosh and D.J. Srolovitz, *J. Appl. Phys.* 91, 1963, 2002. With permission.)

Early atomistic simulations and growth experiments by Dirks and Leamy[65,74] confirmed the tangent-rule (Equation (9.3)). They showed that $\beta < \alpha$ and that the microstructure, consisting of a voided network surrounded by an array of parallel uniform-sized rods of higher density, is understood based upon shadowing effects.

Lichter and Chen[66] developed a more complex relationship between α and β, based on an analytical investigation that incorporates also surface diffusion. They found that the tangent rule is an approximation to a more general relationship that also depends on the ratio of the deposition flux divided by the surface diffusion coefficient. Their model predicts that columns occur only at sufficiently low temperatures and that the column width is determined by the mean-square diffusion length.

Krug and Meakin[67,68] employed analytical as well as numerical simulation methods to study microstructural evolution during oblique deposition. Their analytical expression for column growth angle β shows an approximately linear relationship with the flux angle α for small α, in good agreement with their simulations. Characteristic in-plane and out-of-plane length scales increase with layer thickness according to power laws. They also find that faceting of the column tips stabilizes the columnar morphology.

Tait et al.[69] used computer simulations to study the α–β relationship and the layer density obtained by oblique deposition. They developed an empirical relationship that deviates from the tangent-rule. For the case of limited surface diffusion, they found $\beta \approx \alpha$ for small α (i.e., a larger β than predicted by the tangent rule). In contrast, for α approaching 90°, they found β to converge to a constant value < 90°, which means that columns cannot be tilted by more than a critical angle β_{max}.

Paritosh and Srolovitz[70] report yet another α–β relationship based on their model using growth of faceted films from randomly oriented nuclei. Their results show a strong dependence on crystal structure. Four-fold symmetric crystallites follow $\beta \approx 1/2\alpha$, in agreement with the tangent rule at low α. However, higher symmetries result in a more complex α–β relationship. They also provide a functional form for the void-fraction $V_f = \tan^2(\alpha/2)$, which is constant with film thickness t above an initial transient range. While V_f is constant with t, both void and column width increase with t.

An incremental growth model was employed by Hodgkinson et al.,[71] to investigate dielectric films with spatially modulated columnar nanostructures. They used the tangent rule, however, with a material-dependent coefficient, and found good agreement with experimental results on complex structured layers.

Smy et al.[72] also reported the simulation of complex nanostructures obtained using glancing angle deposition. A three-dimensional ballistic deposition simulator showed that complex substrate motion and three-dimensional shadowing leads to underdense layers consisting of isolated microcolumns with a variety of engineered forms.

The theoretical results summarized in this section provide insight into the processes that determine the columnar microstructure of layers grown from oblique angles. There is a general consensus that deposition angle and surface diffusion are key parameters. However, the various analytical and numerical approaches do not agree on the exact form for the α versus β relationship and also indicate that material parameters like the crystal structure may be important. Therefore, there is still more theoretical work needed until predictive tools are developed that accurately predict columnar microstructures for specific materials.

9.3.5 Glancing-Angle Deposition

The growth of thin films from oblique angles to obtain layers with anisotropic microstructure and therefore anisotropic physical properties, including optical activity, is a technique that has been known for several decades,[75,76] but has gained considerable attention in more recent years due to the wide range of achievable nanostructures.[77–100] Atomic shadowing under limited adatom mobility conditions results, if the growth occurs at glancing angles ($\alpha \geq 70°$), in a microstructure that is dominated by well-separated columns that themselves can exhibit complex shapes by virtue of substrate rotation during deposition.[79] The term columnar thin film (CTF) refers to such layers consisting of separated columns while sculptured thin films (STFs) are layers where the columns themselves have an engineered complex structure. Figure 9.24 shows scanning electron micrographs of exemplary STFs reported by Robbie and Brett,[81] obtained using the glancing-angle deposition (GLAD) technique. Figure 9.24a is obtained using a deposition flux from the right side at an angle of 85°. The layer exhibits a columnar structure with the columns tilted toward the deposition flux. The layer imaged in Figure 9.24b was obtained by alternating deposition from both sides, resulting in zigzag-shaped columns. Helices are obtained (Figure 9.24c) by a continuous substrate rotation during growth. The porosity of the CTFs as well as the column tilt angle β increase with increasing deposition angle α. However, the

(a)

(b)

(c)

Figure 9.24 Typical microstructures obtained by the glancing angle deposition technique. Oblique deposition at 85° from one side (a) results in tilted columns, from alternating directions (b) leads to zigzag towers, and from rotating substrate motion (c) produces helices. (K. Robbie and M.J. Brett, *J. Vac. Sci. Technol.* A 15, 1460, 1997. With permission.)

porosity and β can be varied independently using controlled substrate rotation.[81] For example, if substrate rotation is fast relative to the time needed to grow a thickness comparable to the column width, columns grow straight (perpendicular to the substrate surface) although the deposition flux is at glancing angle.[82]

A major application for STFs are optically active coatings.[83] Layers with tilted columns exhibit a strong dependence of optical transmission on the angle of incidence and polarization.[84] Biaxial films with large birefringence as well as thin film wave plates have been achieved using the GLAD technique.[85,86] Layers containing helices act as helicoidal bianisotropic medium (HBM).[87,88] They directly affect circular polarized light. For example, a circular-polarization spectral-hole filter can be realized by two structurally left-handed chiral STFs with a 90° twist between them.[89,90] Regular arrays of columns are obtained using initial patterning of the substrate followed by GLAD. Such layers have potential applications as photonic crystals.[82,91,92]

Columnar thin films exhibit also exceptional magnetic properties. Magnetic domains in sputtered Co films are strongly affected by the growth angle, showing transitions from elongated domains in the column tilt direction to stripe domains to circular and elliptic magnetic bubbles.[93] Strong magnetic anisotropy was also observed in Co layers on top of obliquely sputtered Ta underlayers.[94] GLAD was employed to synthesize Co and Co–Cr magnetic pillars in periodic and random arrangements with possible applications in magnetic storage.[95,96]

Shadowing effects during deposition at oblique angle was also applied to grow Fe nanowire arrays on NaCl(110) templates,[98] amorphous Si square-spirals on colloid substrates,[99] and nanoflowers using a periodic varying substrate rotation speed.[100]

In summary, deposition at oblique angles is a powerful technique to grow nanostructured thin films with a variety of column sizes, shapes, and arrangements, which can be built using a wide selection of materials. The deposition angle is used in two ways: (1) The azimuthal angle, i.e., the angle between substrate surface normal and incoming flux, determines the degree of shadowing and thus the level of porosity and the column tilt angle; (2) the polar angle controls the growth direction of the columns. Varying both angles in a controlled way during thin film growth provides tremendous possibilities for nanostructure shaping. Additional simultaneous use of directional ion irradiation will further enhance the accessible range of nanostructures by the GLAD technique.

9.3.6 Future Directions for Nanopipe Design

Nanopipes in transition metal nitrides form due to atomic shadowing from a surface mound structure under limited adatom mobility conditions, as described in detail above. Nanopipe density, average spacing, and length can be controlled by thin film deposition parameters, including growth temperature, N_2 partial pressure, and ion-irradiation energy and flux. This is due to changes in the surface morphology, which is primarily determined by surface diffusion and reaction processes on both (001) and (111). Figure 9.25a is a schematic showing nanopipe distribution and surface morphology corresponding to ScN growth under high-flux ion-irradiation in pure N_2 at 750°C. The surface mounds are roughly square-shaped with edges along perpendicular $\langle 100 \rangle$ directions. The cusps between the surface mounds lead to the formation of nanopipes that are aligned along the [001] growth direction. The nanopipes are, in this case, rectangular in cross section with a width of $1 \times (10$–$40)$ nm^2. The 1 nm width is determined by the angular distribution of the deposition flux, thus the degree of shadowing, while the orthogonal elongation, in this case up to 40 nm, is defined by the width of the surface mounds. The average inter-nanopipe separation is equal to the in-plane correlation length and is therefore fully determined by the surface morphology.

While deposition parameters control the average distance between nanopipes by virtue of affecting surface diffusion processes and thus surface morphology, no long-range order of nanopipes has been achieved yet. Figure 9.25b illus-

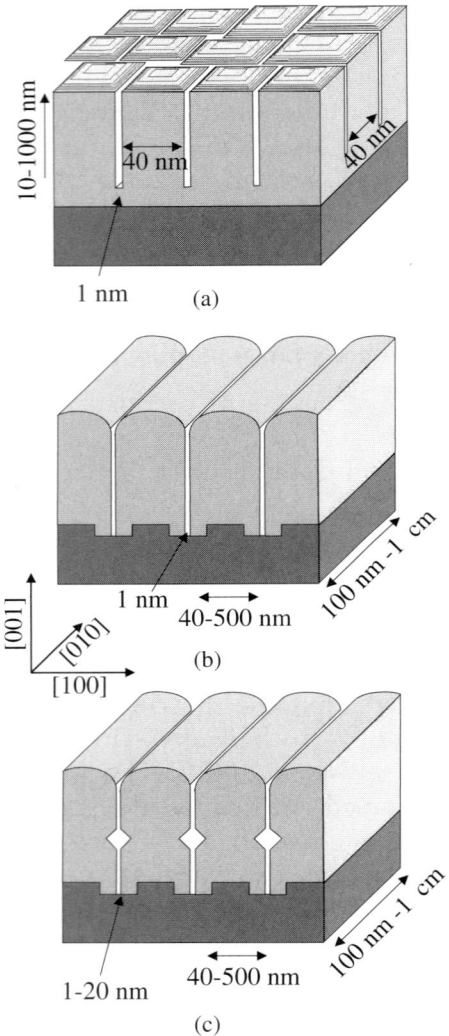

Figure 9.25 Schematic illustrating nanopipe arrangement and surface morphology of (a) self-organized nanopipes typical for epitaxial transition metal nitrides, (b) nanoslots/nanosheets obtained by growth on patterned substrates, and (c) nanochannels fabricated using glancing angle deposition during a selected period in the deposition process.

trates how this may be achieved. The substrate is initially patterned to obtain a periodic or other desired surface mound structure. Subsequently, an epitaxial transition metal nitride layer is grown under conditions that lead to nanopipe formation. The nanopipes will nucleate at the bottom of the valleys of the initial patterned surface structure and will, due to the self-limiting shadowing processes described above, exhibit a width of ~ 1 nm. The schematic in Figure 9.25b shows the case for a substrate that was patterned to form an array of one-dimensional ridges and corresponding valleys. The nanopipes are consequently 1 nm wide along [100] but elongated along [010], forming planar nanoslot structures in (100) planes. The strength of this technique is due to the combination of self-organized bottom-up processes for the formation of nanopipes and the patterning on a much larger length scale, which allow the controlled arrangement of nanopipes into desired structures.

Further development of this technique could be envisioned. In particular, it should be possible to control the nanopipe width by glancing-angle deposition techniques that directly affect the degree of atomic shadowing as described above. In that way, arrays of nanopipes can be synthesized that exhibit a desired width. The width could even be varied as a function of layer thickness by simply changing the incidence angle of the deposition flux. Figure 9.25c illustrates a possible structure obtained using such a technique. Growth is performed as for Figure 9.25b. However, after about half the deposition time, the substrate is tilted so that deposition occurs from a glancing angle. This leads to an increase in atomic shadowing and consequently a larger width of the nanopipes. Finally, thin film growth is continued at normal orientation resulting in closure of the nanopipe width to ~ 1 nm. The obtained structure in Figure 9.25c exhibits nanochannels. Their arrangement is controlled by the initial substrate patterning while their width is determined by growth conditions and deposition angles.

These future nanostructures presented in Figure 9.25 have potential applications as templates for tunneling devices, sensors, and two-dimensional conductors. The nanochannels may be used for molecular transport and single molecule chemistry. Periodic arrays will be useful in optoelectronics and photonic materials. Other potential applications include magnetic storage, large surface-area devices, thermal barriers, and optical coatings.

REFERENCES

1. L.E. Toth, *Transition Metal Carbides and Nitrides*, Academic Press, New York, 1971.

2. V.A. Gubanov, A.L. Ivanovsky, and V.P. Zhukov, *Electronic Structure of Refractory Carbides and Nitrides*, Cambridge University Press, Cambridge, 1994.

3. K. Schwarz, *Crit. Rev. Solid State Mater. Sci.*, 13, 211 (1987).

4. D. Gall, M. Städele, K. Järrendahl, I. Petrov, P. Desjardins, R.T. Haasch, T.-Y. Lee, and J.E. Greene, *Phys. Rev.* B 63, 125119 (2001).

5. D. Gall, I. Petrov, and J.E. Greene, *J. Appl. Phys.* 89, 401 (2001).

6. B.W. Karr, I. Petrov, D.G. Cahill, and J.E. Greene, *Phys. Rev.* B 61, 16137 (2000).

7. Y. Tsuchiya, K. Kosuge, Y. Ikeda, T. Shigematsu, S. Yamaguchi, and N. Nakayama, *Mater. Trans.* JIM 37, 121 (1996).

8. D. Gall, C.-S. Shin, R.T. Haasch, I. Petrov, and J.E. Greene, *J. Appl. Phys.* 91, 5882 (2002).

9. J.-S. Chun, J.R.A. Carlsson, P. Desjardins, D.B. Bergstrom, I. Petrov, J.E. Greene, C. Lavoie, C. Cabral, Jr., and L. Hultman, *J. Vac. Sci. Technol.* A 19, 182 (2001).

10. J.-S. Chun, I. Petrov, and J.E. Greene, *J. Appl. Phys.* 86, 3633 (1999).

11. J.-S. Chun, P. Desjardins, C. Lavoie, I. Petrov, C. Cabral, Jr., and J.E. Greene, *J. Vac. Sci. Technol.* A 19, 2207 (2001).

12. J.-S. Chun, P. Desjardins, C. Lavoie, C.-S. Chin, C. Cabral, Jr., I. Petrov, and J.E. Greene, *J. Vac. Sci. Technol.* A 19, 2207 (2001).

13. J.E. Greene, J.-E. Sundgren, L. Hultman, I. Petrov, and D.B. Bergstrom, *Appl. Phys. Lett.* 67, 2928 (1995).

14. L. Hultman, J.-E. Sundgren, J.E. Greene, D.B. Bergstrom, and I. Petrov, *J. Appl. Phys.* 78, 5395 (1995).

15. U.C. Oh and J.H. Je, *J. Appl. Phys.* 74, 1692 (1993).

16. J.H. Je, D.Y. Noh, H.K. Kim, and K.S. Liang, *J. Appl. Phys.* 81, 6126 (1997).

17. D. Gall, I. Petrov, L.D. Madsen, J.-E. Sundgren, and J.E. Greene, *J. Vac. Sci. Technol.* A 16, 2411 (1998).

18. F. Adibi, I. Petrov, J.E. Greene, L. Hultman, and J.-E. Sundgren, *J. Appl. Phys.* 73, 8580 (1993).

19. C.-S. Shin, D. Gall, Y.-W. Kim, N. Hellgren, I. Petrov, and J.E. Greene, *J. Appl. Phys.* 92, 5084 (2002).

20. H. Huang, G.H. Gilmer, and T.D. de la Rubia, *J. Appl. Phys.* 84, 3636 (1998).

21. D. Gall, S. Kodambaka, M.A. Wall, I. Petrov, and J.E. Greene, *J. Appl. Phys.* 93, 9086 (2003).

22. I. Petrov, F. Adibi, J.E. Greene, W.D. Sproul, and W.-D. Münz, *J. Vac. Sci. Technol.* A 10, 3283 (1992).

23. D. Gall, I. Petrov, N. Hellgren, L. Hultman, J.-E. Sundgren, and J.E. Greene, *J. Appl. Phys.* 84, 6034 (1998).

24. D. Gall, C.-S. Shin, T. Spila, M. Odén, M.J.H. Senna, J.E. Greene, and I. Petrov, *J. Appl. Phys.* 91, 3589 (2002).

25. C.-S. Shin, Y.-W. Kim, D. Gall, J.E. Greene, and I. Petrov, *Thin Solid Films* 402, 172 (2001).

26. L. Hultman, J.-E. Sundgren, and J.E. Greene, *J. Appl. Phys.* 66, 536 (1989).

27. J. Pelleg, L.Z. Zevin, and S. Lungo, *Thin Solid Films* 197, 117 (1991).

28. D. Gall, I. Petrov, P. Desjardins, and J.E. Greene, *J. Appl. Phys.* 86, 5524 (1999).

29. C.-S. Shin, Y.-W. Kim, N. Hellgren, D. Gall, I. Petrov, and J.E. Greene, *J. Vac. Sci. Technol.* A 20, 2007 (2002).

30. F. Tian, J. D'Arcy-Gall, T.-Y. Lee, M. Sardela, D. Gall, I. Petrov, and J.E. Greene, *J. Vac. Sci. Technol.* 21, 140 (2003).

31. Inorganic Index to Powder Diffraction File (Joint Committee on Powder Diffraction Standards, International Center for Powder Diffraction Data, Swarthmore, PA): MgO (Card number 04-0820), TiN (Card number 38-1420).

32. B. Hajek, V. Brozek, and H. Duvigneaud, *J. Less-Common Met.* 33, 385 (1973).

33. K. Aigner, W. Lengauer, D. Rafaja, and P. Ettmayer, *J. Alloys and Compounds* 215, 121 (1994).

34. H. Landolt and R. Börnstein, *Numerical Data and Functional Relationships in Science and Technlolgy, Group III*, vol. 7, Pt. b1, Springer, Berlin, 1975, p. 27.

35. N.-E. Lee, D.G. Cahill, and J.E. Greene, *J. Appl. Phys.* 80, 2199 (1996).

36. G. Xue, H.Z. Xiao, M.-A. Hasan, J.E. Greene, and H.K. Birnbaum, *J. Appl. Phys.* 74, 2512 (1993).

37. N.-E. Lee, G.A. Tomasch, and J.E. Greene, *Appl. Phys. Lett.* 65, 3236 (1994).

38. N.-E. Lee, G. Xue, and J.E. Greene, *J. Appl. Phys.* 80, 2199 (1996).

39. D.J. Eaglesham, H.-J. Gossmann, and M. Cerullo, *Phys. Rev. Lett.* 65, 1227 (1990).

40. J.A. Stroscio, D.T. Piercee, M.D. Stiles, A. Zangwill, and L.M. Sander, *Phys. Rev. Lett.* 75, 4246 (1995).

41. S.C. Wang and G. Ehrlich, *Phys. Rev. Lett.* 70, 41 (1993) and 71, 4177 (1993); G. Ehrlich, *Surf. Sci.* 331/333, 865 (1995); A. Gölzhäuser and G. Ehrlich, *Phys. Rev. Lett.* 77, 1334 (1996).

42. B.W. Karr, I. Petrov, D.G. Cahill, and J.E. Greene, *Appl. Phys. Lett.* 70, 1703 (1997).

43. B.W. Karr, I. Petrov, P. Desjardins, D.G. Cahill, and J.E. Greene, *Surf. Coat. Technol.* 94-95, 403 (1997).

44. S. Kodambaka, V. Petrova, A. Vailionis, P. Desjardins, I. Petrov, D.G. Cahill, and J.E. Greene, *Surf. Rev. Lett.* 7, 589 (2000).

45. S. Kodambaka, V. Petrova, S.V. Khare, D. Gall, A. Rockett, I. Petrov, and J.E. Greene, *Phys. Rev. Lett.* 89, 176102 (2002).

46. S. Kodambaka, V. Petrova, P. Desjardins, I. Petrov, D.G. Cahill, and J.E. Greene, *Thin Solid Films* 392, 164 (2001).

47. M.A. Wall, D.G. Cahill, I. Petrov, D. Gall, and J.E. Greene, *Phys. Rev.* B 70, 035413 (2004).

48. J. Lapujoulade, *Surf. Sci. Rep.* 20, 191 (1994).

49. J.E. Nostrand, S. Jay Chey, M.-A. Hasan, D.G. Cahill, and J.E. Greene, *Phys. Rev. Lett.* 74, 1127 (1995).

50. H.-J. Ernst, F. Fabre, R. Folkerts, and J. Lapujoulade, *Phys. Rev. Lett.* 72, 112 (1994).

51. D. Johnson, C. Orme, A.W. Hunt, D. Graff, J. Sudijono, L.M. Sander, and B.G. Orr, *Phys. Rev. Lett.* 75, 116 (1994).

52. J.E. Nostrand, S. Jay Chey, and D.G. Cahill, *Phys. Rev.* B 57, 12536 (1999).

53. J.A. Venables, *Phil. Mag.* 27, 693-738 (1973).

54. J. Tersoff, A.W. Denier van der Gon, and R.M. Tromp, *Phys. Rev. Lett.* 72, 266-269 (1994).

55. R.P.U. Karunasiri, R. Bruinsma, and J. Rudnick, *Phys. Rev. Lett.* 62, 788 (1989).

56. G.S. Bales and A. Zangwill, *Phys. Rev. Lett.* 63, 692 (1989).

57. G.S. Bales and A. Zangwill, *J. Vac. Sci. Technol.* A 9, 145 (1991).

58. C. Roland and H. Guo, *Phys. Rev. Lett.* 16, 2104 (1991).

59. J.H. Yao, C. Roland, and H. Guo, *Phys. Rev.* A 45, 3903 (1992).

60. J.H. Yao and H. Guo, *Phys. Rev.* E, 47 1007 (1993).

61. J. Krug and P. Meakin, *Phys. Rev.* E 47, 17 (1993).

62. J.T. Drotar, Y.-P. Zhao, T.M. Lu, and G.-C. Wang, *Phys. Rev.* B 62, 2118 (2000).

63. O.P. Karpenko, J.C. Bilello, and S.M. Yalisove, *J. Appl. Phys.* 82, 1397 (1997).

64. R. Messier and J.E. Yehoda, *J. Appl. Phys.* 58, 3739 (1985).

65. A.G. Dirks and H.J. Leamy, *Thin Solid Films* 47, 219 (1977).

66. S. Lichter and J. Chen, *Phys. Rev. Lett.* 56, 1396 (1986).

67. P. Meakin and J. Krug, *Phys. Rev.* A 46, 3390 (1992).

68. J. Krug and P. Meakin, *Phys. Rev.* A 43, 900 (1991).

69. R.N. Tait, T. Smy, and M.J. Brett, *Thin Solid Films* 226, 196 (1993).

70. F. Paritosh and D.J. Srolovitz, *J. Appl. Phys.* 91, 1963 (2002).

71. I. Hodgkinson, Q.H. Wu, and A. McPhun, *J. Vac. Sci. Technol.* B 16, 2811 (1998).

72. T. Smy, D. Vick, M.J. Brett, S.K. Dew, A.T. Wu, J.C. Sit, and K.D. Harris, *J. Vac. Sci. Technol.* A 18, 2507 (2000).

73. J.M. Nieuwenhuizen and H.B. Haanstra, *Phillips Tech. Rev.* 27, 87 (1966).

74. H.J. Leamy and A.G. Dirks, *J. Appl. Phys.* 49, 3430 (1978).

75. N.O. Young and J. Kowal, *Nature* 183, 104 (1959).

76. D.O. Smith, M.S. Cohen, and G.P. Weiss, *J. Appl. Phys.* 31, 1755 (1960).

77. K. Robbie, M.J. Brett, A. Lakhtakia, *Nature* 384, 616 (1996).

78. K. Robbie, D.J. Broer, and M.J. Brett, *Nature* 399, 764 (1999).

79. R. Messier, V.C. Venugopal, and P.D. Sunal, *J. Vac. Sci. Technol.* A 18, 1528 (2000).

80. K. Robbie and M.J. Brett, *J. Vac. Sci. Technol.* A 15, 1460 (1997).

81. K. Robbie, J.C. Sit, and M.J. Brett, *J. Vac. Sci. Technol.* B 16, 1115 (1998).

82. M. Malac and R.F. Egerton, *J. Vac. Sci. Technol.* A 19, 158 (2001).

83. A. Lakhtakia, *Materials Sci. Eng.* C 19, 427 (2002).

84. M. Suzuki and Y. Taga, *J. Appl. Phys.* 71 2848 (1992).

85. I.J. Hodgkinson and Q.H. Wu, *Appl. Optics* 38, 3621 (1999).

86. I.J. Hodgkinson and Q.H. Wu, *Optical Eng.* 37, 2630 (1998).

87. K. Robbie, M.J. Brett, A. Lakhtakia, *J. Vac. Sci. Technol.* A 13, 2991 (1995).

88. R.M.A. Azzam, *Appl. Phys. Lett.* 61, 3118 (1992).

89. A. Lakhtakia and R. Messier, *Opt. Photon. News* Sept. 01, 27 (2001).

90. I.J. Hodgkinson, Q.H. Wu, K.E. Thorn, A. Lakhtakia, and M.W. McCall, *Opt. Commun.* 184, 57 (2000).

91. M. Malac, R.F. Egerton, M.J. Brett, and B. Dick, *J. Vac. Sci. Technol.* B 17, 2671 (1999).

92. M. Malac and R.F. Egerton, *Nanotechnology* 12, 11 (2001).

93. A. Lisfi and J.C. Lodder, *Phys. Rev.* B 63, 174441 (2001).

94. R.D. McMichael, C.G. Lee, J.E. Bonevich, P.J. Chen, W. Miller, and W.F. Egelhoff, Jr., *J. Appl. Phys.* 88, 3561 (2000).

95. B. Dick, M.J. Brett, T.J. Smy, M.R. Freeman, M. Malac, and R.F. Egerton, *J. Vac. Sci. Technol.* A 18, 1838 (2000).

96. A. Hagemeyer, H.J. Richter, H. Hibst, V. Maier, and L. Marosi, *Thin Solid Films* 230, 199 (1993).

97. K. Starbova, J. Dikova, N. Starbov, *J. Non-Cryst. Solids* 210, 261 (1997).

98. A. Sugawara, T. Coyle, G.G. Hembree, and M.R. Scheinfein, *Appl. Phys. Lett.* 70, 1043 (1997).

99. Y.-P. Zhao, D.-X. Ye, P.-I. Wang, G.-C. Wang, and T.-M. Lu, *Int. J. Nanosci.* 1, 87 (2002).

100. Y.-P. Zhao, D.-X. Ye, G.-C. Wang, and T.-M. Lu, *Nano. Lett.* 2, 351 (2002).

Chapter 10

Nuclear Tracks and Nanostructures

ROBERT L. FLEISCHER

CONTENTS

ABSTRACT

This chapter focuses on a special fine-scale effect, track production — the result of irradiation with ions. Tracks typically
are elongated, needle-shaped regions that are densely populated with atomic defects. But if tracks are short enough (at
low deposited energy), they are less well-defined, low-density
globs of damage, rather than being needle-like.

10.1 INTRODUCTION

Ion beams can alter thin films in many ways, for example by
displacing individual atoms, which (if given enough energy)
might in turn displace other nearby atoms or groups of atoms;
by re-arranging atoms to produce new crystal structures or
amorphous material; by altering the local composition by
either adding new atoms or sputtering away atoms of the
target; and by creating stresses that alter physical properties.
The focus in this chapter is on another possible result of
irradiation with ions, the production of tracks. These features
typically are elongated, needle-shaped regions that are
densely populated with atomic defects. As we will also see, if
tracks are short enough (at low deposited energy), instead of
being needle-like, they are less well-defined globs of damage.

There is a vast literature on the nature, production, and
uses of ion tracks (nuclear tracks in solids), including books
that survey tracks broadly (Fleischer et al. 1975, Durrani and
Bull 1987, Fleischer 1998) and one that concentrates on technological uses (Spohr 1990). Although many of the early observations utilized transmission electron microscopy to image
tracks in thin sheets of material (Silk and Barnes 1959, Price
and Walker 1962a), the most widely used way to observe

tracks is by using preferential chemical etching (Young 1958, Price and Walker 1962b, Fleischer and Price 1963a,b) to produce channels or pits at the sites of damage tracks. Figure 10.1 shows tracks from fission of ^{235}U in muscovite mica, unetched and etched, viewed via electron microscopy. The damage and its associated strain fields that display the unetched tracks are transformed into cylindrical tunnels by etching.

10.1.1 Organization

The plan here is first to distinguish two major categories of tracks, then describe their properties, list the types of materials in which they are found, and note what ions produce tracks in different substances (the *registration* properties of materials). Next the geometry of tracks is described — the dimensions of tracks and of etched tracks, and how these dimensions have been measured. Etched tracks are pores, which can have their own special usefulness. The chapter concludes with a discussion of the present understanding of how tracks form in different types of solids.

10.2 TYPES OF ION TRACKS

High-energy and low-energy tracks originate in very different ways. The low-energy ones — from particles with velocities corresponding to at most a few keV/atomic mass unit (amu) — have short ranges, typically tens of nanometers. The particles lose their energy primarily by billiard-ball-like collisions with other atoms (Lindhard and Scharff 1961, Lindhard and Thomsen 1962) and produce damaged regions that do not have a simple, clean geometry. They result in localized groups of atomic defects. As will be pointed out shortly in more detail, the defects are seldom seen directly, but can be inferred by general disordering of crystals and by accelerated localized preferential leaching of assorted materials. In the laboratory, particles of such energies are provided by ion-implantation devices and in nature by recoiling nuclei from alpha decay and by solar wind nuclei, ions that are boiled off from the

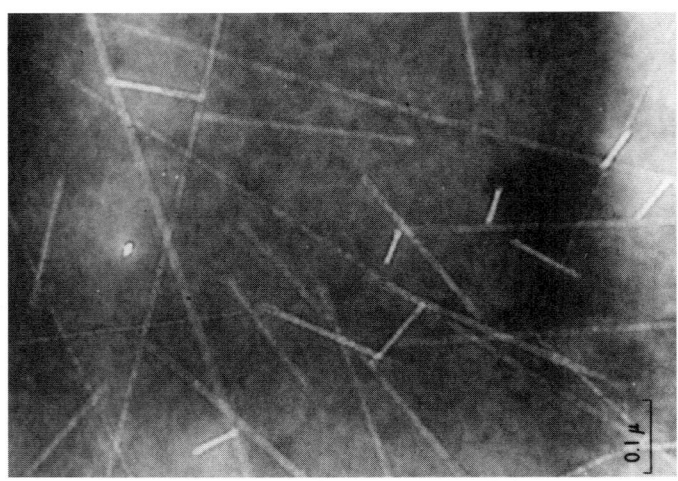

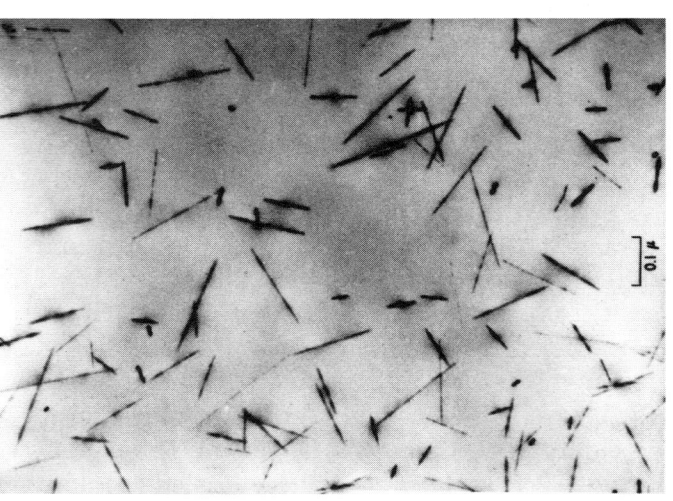

Figure 10.1 Transmission electron microscopic images of tracks in a synthetic fluor-phlogopite mica. (a) The periodic variations in the darkness of the tracks are from diffraction contrast, not from track widths. (b) The light lines are holes etched in the mica and imaged by thickness contrast. They show the holes at their earliest visible stage and thus are the true widths of the severe track damage. (Courtesy of the University of California Press.)

sun. When man first brought lunar samples to earth, the (low-energy) solar wind nuclei were found to have transformed the top 50 nm of exposed lunar crystals into amorphous layers (Bibring et al. 1972).

High-energy particles (a few MeV/amu or more) produce truly track-shaped damage — highly elongated cylinders that are only a few nanometers in radius, but may extend for millimeters or more depending on the energy of the particles (Fleischer et al. 1975, Fleischer 1998) — even crossing through space helmets, and astronauts' heads (Comstock et al. 1971). The damage results not from atomic collisions but from localized ionization, i.e., by a very different mechanism, as will shortly be described. As noted just above, such tracks can be seen easily using preferential chemical etching and an ordinary optical microscope (see Figure 10.2). Accelerators are now able to supply such ions from all of the chemical species. On Earth such particles in nature result from spontaneous fission of ^{238}U (the tracks from which allow dating of minerals) and above the Earth's atmosphere heavy ions are major constituents of the cosmic radiation.

The low-energy damage can be produced in almost any solid. High-energy track production is mostly a phenomenon of dielectric solids — insulating crystals, glasses, and polymers (Fleischer et al. 1975). But, under special circumstances they are produced in good conductors — metals, intermetallic compounds, and oxide superconductors (Provost et al. 1995, Barbu et al. 1995, Fleischer 1995).

10.2.1 Why the Difference?

What physical factor leads to the profound difference between the radiation damage at low and at high energy? The answer is the ionization state of the projectile. A fast-moving ion loses some of its electrons — more electrons the faster it moves — and thus it becomes more highly charged and smaller. The loss of an orbital electron on an atom at rest is governed by the speed of the ion relative to that of the electron circling the atom (Knipp and Teller 1941, Betz 1972). When the speed of the ion is much greater than that of the electron,

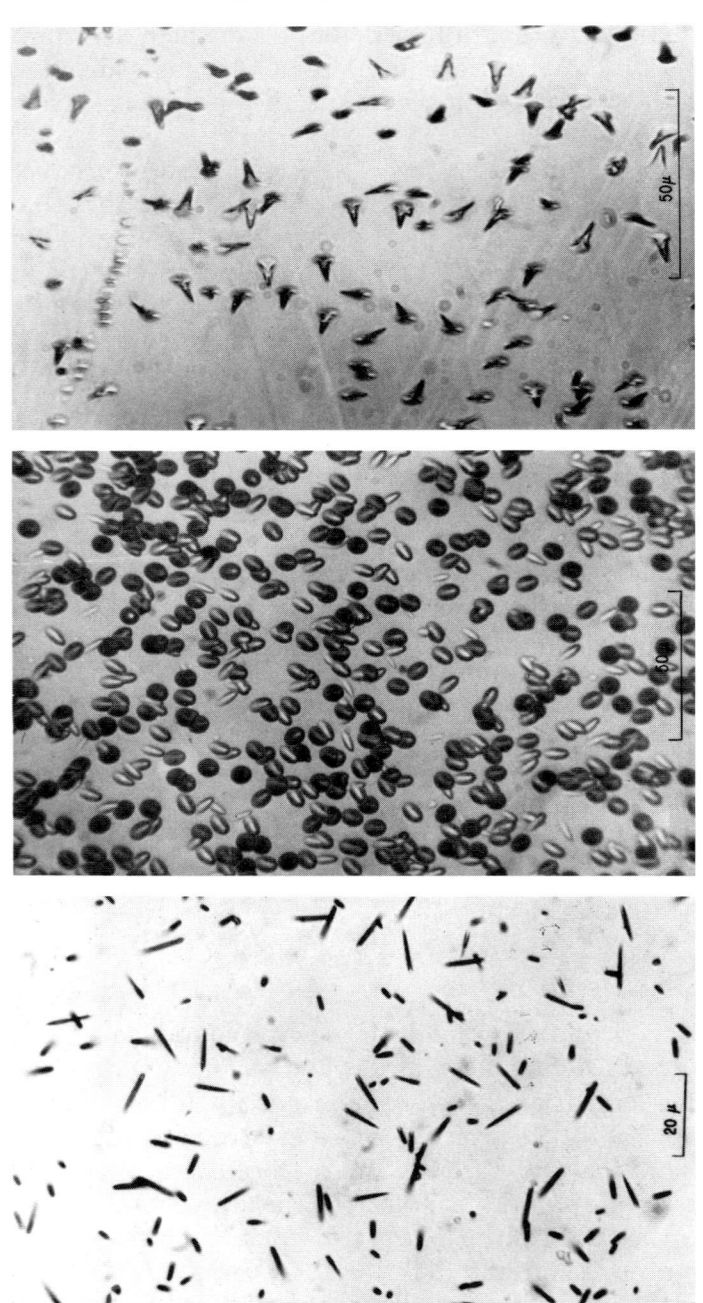

Figure 10.2 Etched tracks in three different classes of dielectric materials. Left: A plastic, Lexan polycarbonate; middle: a silicate glass; right: a crystal — the feldspar mineral orthoclase. The shapes are the result of preferential etching along tracks and simultaneous general, bulk etching away from them.

the ion interacts as though the electron were at rest; and the ion can eject the electron if the Coulomb force is great enough. Where the crossover occurs from atomic collisions being dominant to ionization being more important is a complicated function of the atomic number and atomic mass of both the projectile and the stopping medium (Lindhard and Scharff 1961, Lindhard and Thomsen 1962). In some typical cases (Powers et al. 1968) it lies around 10 keV/amu.

Thus the high charge on a fast ion causes ionization, yanking electrons from atoms along the trajectory; and, because the ion is far smaller than the atom, direct knock-on effects are not very important. Later in this chapter we will seek to understand how ionization by fast ions causes atomic disorder and thus track formation.

Very slow atoms retain all, or the vast majority, of their electrons and therefore these projectiles cause no, or very little, ionization. Because they are of atomic size, they are massive objects relative to the open spaces in solids and therefore they will displace atoms repeatedly before they come to rest.

10.3 SIZES OF DAMAGED REGIONS

The dimensions of tracks and track-like damage are important in at least two regards. The first aspect has a practical aim. If the track volumes are well established, their size tells how many tracks one needs in order to fill a desired volume fraction of an irradiated sample — for example the whole volume if a phase change is desired, or a critical fraction if mechanical or magnetic strengthening is the objective. For certain magnetic and superconducting properties, particular directions for tracks relative to either the sample axes or special crystallographic directions may optimize properties by placing barriers to motion of magnetic flux lines in strategically useful orientations.

The second consideration is of scientific importance. Knowing the dimensions may be diagnostic to identifying the track-formation mechanisms, or at least eliminating some possibilities. For example, in many minerals the etchable range of heavy-ion tracks is somewhat less than the distance traveled by the track-producing particle. The result is a *range deficit* of 1 to 4 µm in different minerals between the low-energy termination of the etched track to where the particle came to rest (Price et al. 1968). Since the track damage is missing just where atomic collisions are dominant, such collisions are not the cause of the high-energy tracks.

10.3.1 Sizes of High-Energy Tracks

The tracks of fast ions are relatively easy to measure because in most cases they are continuous and change only slowly in diameter along their length. However, particles that are close to the threshold for detection may form tracks in which statistical fluctuations cause the damage to be discontinuous and the diameters therefore irregular (Meftah et al. 1993). Similarly, continuous tracks that are heated until they are on the verge of disappearing can become discontinuous, as has been inferred to occur in muscovite mica (Fleischer et al. 1964a).

10.3.1.1 Electron Microscopy Measurements

Figure 10.1 shows tracks in mica that are visibly continuous. Each line displays damage produced by the passage of a fission fragment (a heavy particle that is approximately half of a uranium nucleus that was split as a result of the energy release from gobbling up a neutron). One might think that the photo at the top of unetched tracks would give the most reliable measure of the diameter, but that is not the case. The usual transmission microscopy for direct observation of tracks in crystals uses reflection from a particular crystal plane, and the tracks show up because they distort the crystal and thus locally tilt the plane that is imaged. Also, the spatial extent of the local tilt of the crystal is a result of the changed volume (usually enlarged) of the region of damage, and of the elastic properties of both the damage and the surrounding undamaged matrix — possibly related to, but not likely to be identical with, the diameter of the damage.

    Etched tracks such as those in the lower portion of Figure 10.1 are more directly useful in displaying track size (Price and Walker 1962b). For short etching times in mica only the damage is removed. Extended etching will dissolve more material from the sides of the cylindrical holes but for this type of mica, this is a far slower process than the etching of the damage. From the scale in the figure, a track diameter of about 7 nm can be inferred, i.e., 25 to 30 atomic diameters.

Only a limited fraction of the materials that display tracks by etching can yield track diameters this way. Where general etching of the material is not far less rapid than the preferential rate along tracks, the etched shape is conical or dagger-like, as shown by glass and orthoclase in Figure 10.2. Only for cylindrical, or very nearly cylindrical, tracks (such as Lexan polycarbonate in Figure 10.2) can the diameter of the damage be measured simply.

More recently, thinned cross sections of head-on tracks have been imaged in transmission microscopy with *atomic resolution*, allowing the disorder to be seen. Such a picture is given in Figure 10.3 for a Bi-based superconducting material (Provost et al. 1995). Here the diameter is 13 nm, nearly twice that seen in mica; but the ion is Pb at 0.3 GeV, a particle that is about twice as massive as a typical fission fragment and far more heavily ionizing. When this discussion turns to track-production mechanisms, it will become obvious that the ionization level should influence the diameter of the disorder.

10.3.1.2 Electrical Conductivity

A clever, novel method for measuring the interiors of tracks during etching was devised by Bean (1975). This technique has given detailed, quantitative measurements of tracks as a function of time. By way of background, if you shoot a large number of fission fragments across a thin sheet of plastic and then etch, you produce perforations that are all of a single size, and that size can be controlled by how long the etching is continued (Fleischer et al. 1964b). Such membranes (called *nuclepore* filters) were early practical outputs of nuclear-track etching. In 1995 Kuznetzov et al. estimated nuclear-track filters to be a US$30 million/year business. Bean's new tech-

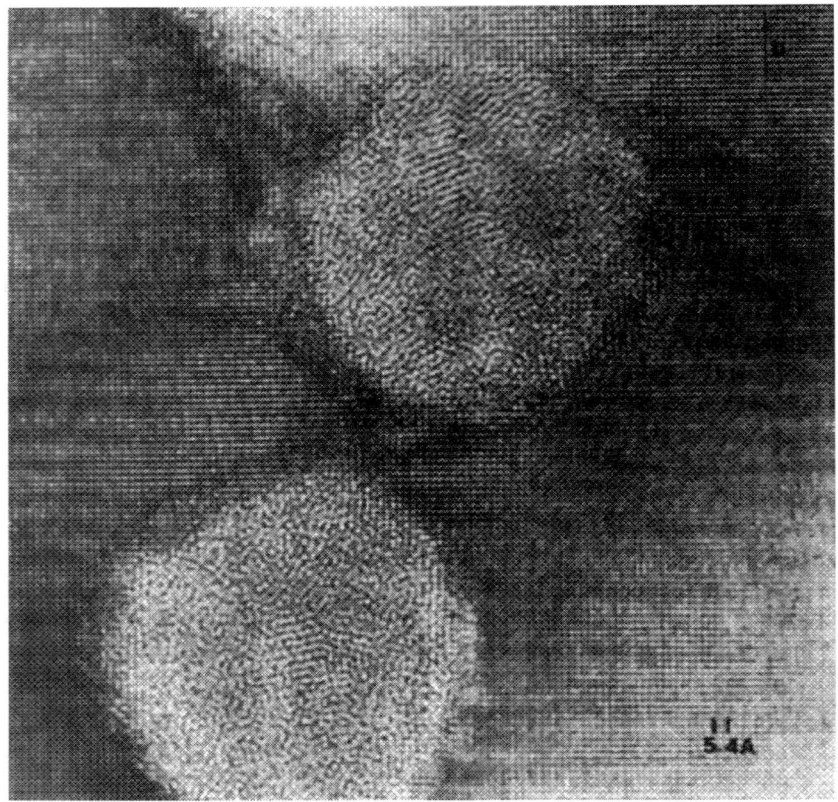

Figure 10.3　High-resolution electron microscopic images of head-on tracks of 300 MeV Pb ions in the superconducting oxide Bi-2212. The core is amorphous with a diameter of 13 nm (Provost et al. 1995). (From Provost, J., Simon, Ch., Hervieu, M., Groult, D., Hardy, V., Studer, F., and Toulemonde, M. (1995) *MRS Bull.* 20(12), 24. Courtesy of the Materials Research Society.)

nique arose, at least in part, because of the need for quality control in commercial production of filters.

The method consists of using the material that is crossed by tracks as a membrane that separates the two electrodes of an electrochemical cell. The resistance across the cell is high until holes etch through; it then drops continuously as the holes enlarge radially. As Figure 10.4 shows (Bean et al.

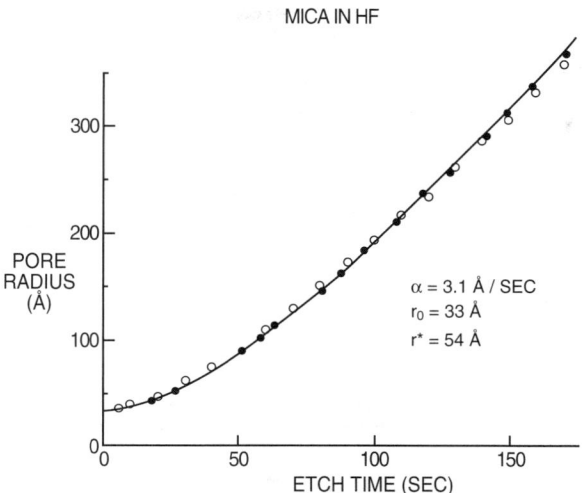

Figure 10.4 Growth of track-produced holes, as measured by electrical conductivity through pores. The inferred radius as a function of time is given for two irradiated samples of muscovite mica — each of thickness 4.2 μm — etched in 34% HF at 25°C. The solid curve is a theoretical result that includes the effect of surface energy in slowing the initial radial etching. The derived original diameter when the etchant first punches through is 6.6 nm (Bean, 1975). (Courtesy of the University of California Press.)

1970), the diameters of the holes in muscovite mica are a sensitive function of the duration of etching. In this case, the initial holes measured 6.6 nm in diameter. For similar experiments in Lexan polycarbonate (DeSorbo 1979) the knee of the resistance curve is taken as the radius where most of the fission fragment tracks have etched through. The radius is 4.5 to 5.0 nm, or a diameter of 9 to 10 nm. Since Lexan is a more sensitive detector than mica, the observation is reasonable that the track damage is more extensive there.

A variation of the Bean technique (DeBlois and Bean 1970) uses small, electrically insulating particles as probes of etched tracks, as depicted in Figure 10.5a. The method is economical of irradiation requirements, since it is best done using a membrane with just a single hole. As a particle

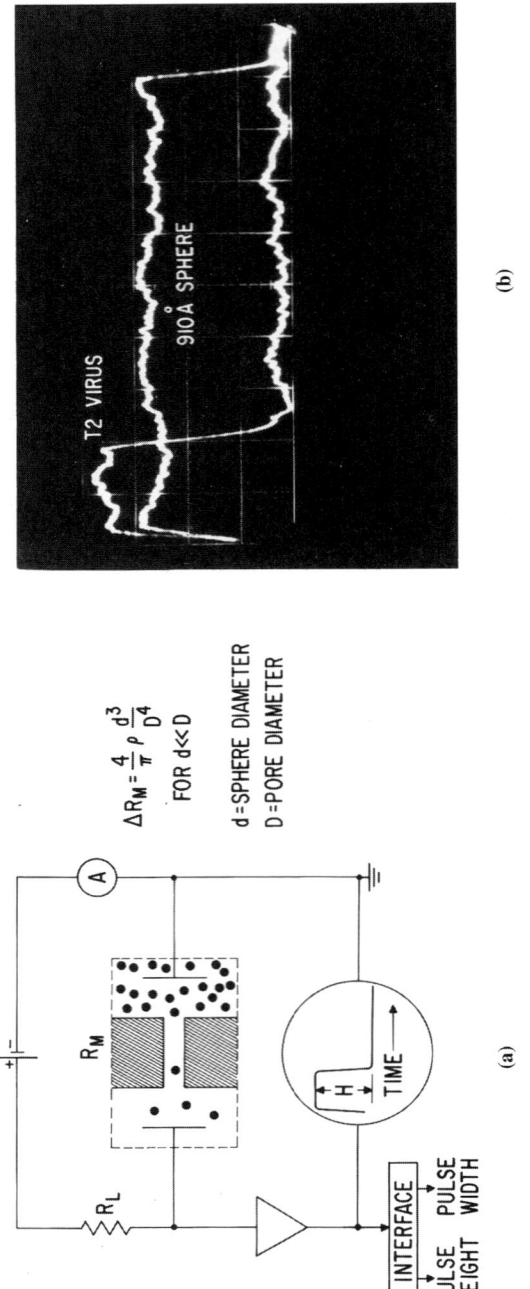

Figure 10.5   Principle of the DeBlois-Bean counter (DeBlois and Bean, 1970). (a) Schematic of the method. (b) Oscilloscope resistive-pulse traces from the passage of standard 91 nm polystyrene spheres and of T-2 viruses through the hole. The pulse heights give sizes and the transit times the charge on the particles. The virus is concluded to have a diameter of 98 nm. (Courtesy of the University of California Press.)

transits a pore, it crowds the current lines and therefore increases the resistance. The height of the voltage pulse is proportional to the volume of the particle and varies as the inverse 4th power of the diameter of the pore. Thus, the voltage can be used to characterize a particle, or a particle of known character can be used to document the size of a hole and variations in diameter along a pore. As shown in Figure 10.5b, the pulse from a T-2 virus is 25% higher than that from a standard 91 nm sphere, allowing the diameter of the virus to be measured as 98 nm. Furthermore, as shown in Figure 10.6, a sea-urchin sperm produced a pulse from the head, followed by a much lower pulse from the narrow tail. The second signal is from a sperm that, unexpectedly, was dragged in tail first.

10.3.1.3   Calculated Track Sizes

It is also of interest to see whether calculation can give reasonable track sizes. Models are limited in the number of atoms that can be used computationally to represent a crystal and in the details of the physical assumptions that can be accommodated. The most advanced such model is that of Young (2002), who modeled a 12,800-atom KCl crystal and used an ion-spike model (Fleischer et al. 1965), which will be described under the discussion on track mechanisms. The results of the calculation gave (in his Figure 3) an atomic disorder that is much like that shown in Figure 10.3. The diameter was 7.6 nm, i.e., within the range of the experimental results.

10.3.2 Sizes of Low-Energy Tracks

The low-energy "tracks" deserve quotation marks because their shapes are not well determined and most of the natural low-energy defects are probably close to being equi-axial, low-intensity blobs of damage. Consider the most common low-energy, natural particles on Earth — alpha-recoil nuclei. Using Lindhard and Scharff (1961) and Lindhard and Thomsen (1962), the recoiling nucleus from, for example, alpha decay of ^{235}U has computed ranges in the materials quartz, muscovite, orthoclase, glass, apatite, zircon, silicon, and epoxy resin of 24,

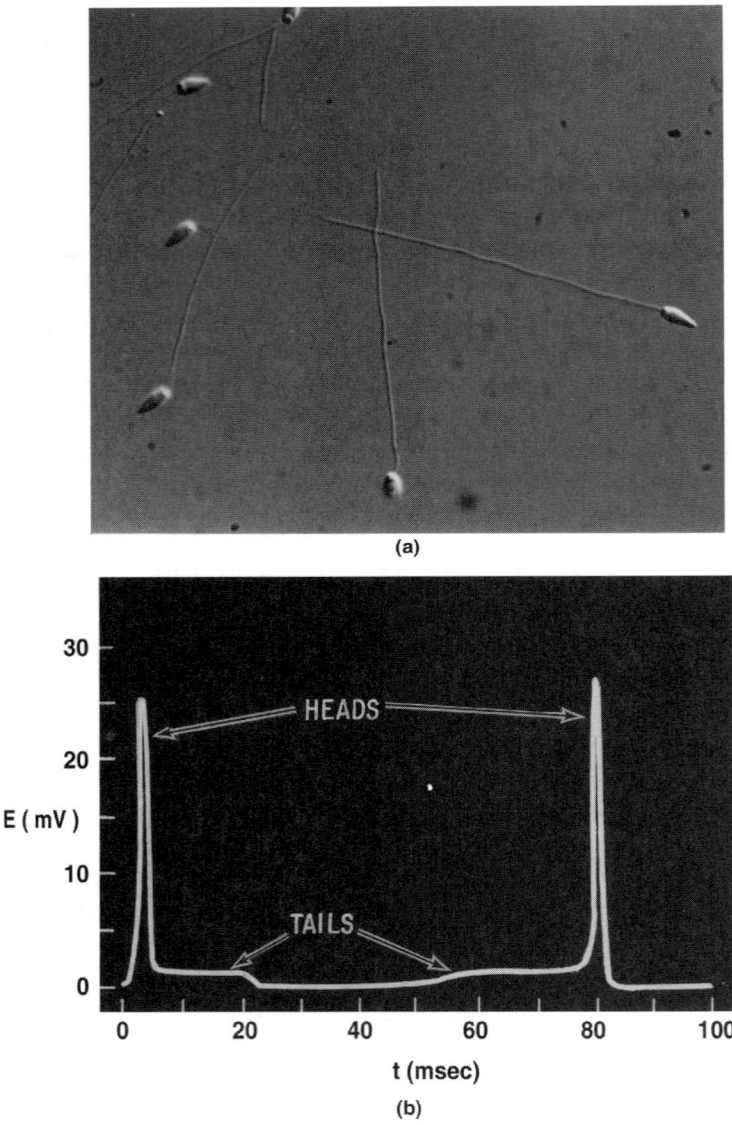

(a)

(b)

Figure 10.6 As sea-urchin sperm (top) pass through a hole in a membrane, the pulses (bottom) show one entering head-first and the second (presumably kicking and screaming) tail-first. (Photo from C.P. Bean, pulses from R.W. DeBlois. From Fleischer, R.L. (1995) *MRS Bull.* 20 (12), 36.)

22, 25, 26, 21, 16, 27, and 53 nm, respectively. For comparison, Huang and Walker (1967) calculated 20 nm for muscovite. Since the atomic collisions that are involved have wide fluctuations in location and direction, the diameters that include most of the displaced atoms are expected to be larger than those measured for fast-ion tracks, probably 10 to 15 nm or more. Thus, if high-energy tracks are needles, then the (low-energy) alpha-recoil tracks will be footballs. Direct observations of low-energy tracks has proved elusive; so we turn to less direct information.

10.3.2.1   Observations of Low-Energy Tracks by Etching

Alpha-recoil tracks were first seen in etched muscovite mica by Huang and Walker (1967) as shallow etched pits. Although the tracks can be seen using ordinary illumination in an optical microscope, phase-contrast microscopy is far more revealing. Curiously, after recognizing that phase contrast was highly effective in revealing the etched tracks, Huang and Walker found that in many earlier optical photos of fission tracks the alpha-recoil tracks were clearly visible in the background. Things are far easier to recognize once you know what you are looking for.

Using a new tool for viewing tracks, the atomic-force microscope, Snowden-Ifft et al. (1993) measured depths of such pits of about 30 nm, an example being the etch pit shown in Figure 10.7. This value is an upper limit for a single recoil track, since natural tracks may be extended in length by multiple decays in the uranium or thorium decay series. Turkowsky (1969) etched albite, shadow-cast replicas of the surface, and measured about 20 nm for the depth of tracks from single alpha decays. Since albite is a feldspar mineral, like orthoclase, the 20 nm can be compared with the calculated value of 25 nm noted for orthoclase in a previous paragraph.

10.3.2.2   Size from Dose to Disorder Crystals

One noteworthy effect of low-energy particles on crystalline materials is disordering to a state that is referred to in the

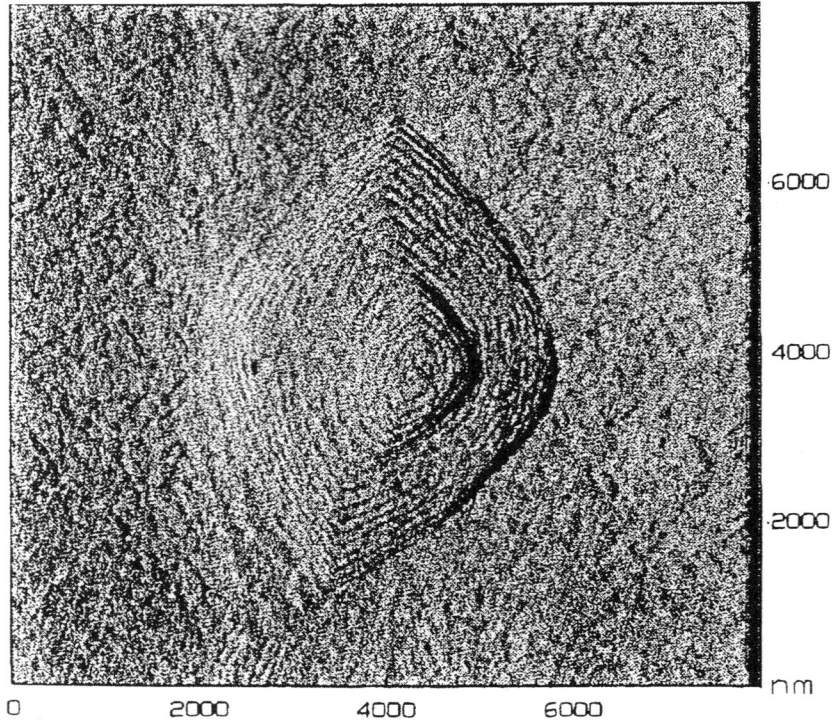

Figure 10.7 Atomic-force micrograph of an etched alpha-recoil track in muscovite mica. The step heights are 2 nm. The depth of the track is 30 nm (Snowden-Ifft et al. 1993). (Courtesy of the American Physical Society.)

geological literature as *metamict* (a subject surveyed by Ewing 1994). We may think of this process as an accumulation of many disordered regions until what was a crystal is completely disordered. The number of alpha decays per unit volume then tells the effective size of the individual regions of disorder. As an example, for the most studied mineral, zircon ($ZrSiO_4$), disorder saturates at 10^{19} recoils/g, or 4.7 × 10^{19} recoils/cm^3. Allowing for tracks that overlap tracks, this number implies that about 3 × 10^{19}/cm^3 would suffice to fill all of the volume if there was no overlap. Since there are about 6×10^{22} atoms/cm^3 in zircon, about 2000 atoms are disordered per recoil event. This number of atoms is sufficient to fill a sphere 4 nm in diameter. This size is about 2% of that inferred in the preceding paragraph, and this fact is of profound importance (Fleischer, 2003): the displaced atoms are not in a single, dense clump. This description will be returned to later.

10.3.2.3 Information from Leaching of Recoil Atoms

Recoils that enter a solid from a surface can be removed by exposure to aqueous solutions (Fleischer 1980, 1982a,b) — presumably by preferential solubility of damage that was created by the recoiling atoms. In a series of experiments with a variety of materials, recoiling ^{235}U from alpha decay of thin ^{239}Pu sources was injected into the surfaces. After separate treatments of separate samples with multiple solutions, the samples were tested for the fraction of the injected uranium that remained by using neutrons to induce fission and recording the fission tracks. For quartz, muscovite mica, diopside, orthoclase, apatite, sphene, obsidian, silicon, soda glass, and epoxy resin fractions of down to 0.6, 0.55, 0.6, 0.4, 0.8, 0.8, 0.55, 0.3, 0.3, and 0.2 remained after 24-hour treatments, and for the mica as little as 15% after a week-long treatment. In short, a considerable fraction was leached out for all the materials, but total removal did not happen for any. Therefore, a continuous path for etching was not provided to many of the recoil atoms.

The leaching of recoils from PuO_2 revealed that tiny particles of PuO_2 (containing from 60 to 10,000 Pu atoms) were going into solution (Fleischer and Raabe 1977). At the time two alternative explanations were offered. The preferred one was that atoms recoiling from the surface were blasting off little chunks of the material into the water. A later experiment (Fleischer and Raabe 1978) showed that only the single recoil atom emerged, leaving the alternative: The damage making up the tracks consists of discrete blocks that can be loosened by the leachant.

10.4 MECHANISMS OF TRACK PRODUCTION I

10.4.1 Low-Energy Damage

Since the nature of the process of atomic collisions is long established by Lindhard and co-workers (Lindhard and Scharff 1961, Lindhard and Thomsen 1962), the mechanism of damage is not in question, but the distribution of disorder in a "track" needs clarification. The information is specific to alpha-recoil damage, but it applies generally to low-energy events. From above, a typical recoil track size is 10 by 10 by 20 nm, but only 2000 atoms are displaced per event, and only 2% of those in the track region. It follows that the track is a region of diffuse damage that contains many atomic defects, but only a low atomic fraction of them. Such a picture explains why transmission electron microscopy has not been very revealing in imaging recoil events and why leaching is not fully effective in releasing track-forming particles from the ends of tracks.

Nevertheless, the region is damaged enough that preferential etching can occur (Huang and Walker 1967,Turkowsky 1969). Snowden-Ifft et al. (1993) show a shallow etched recoil track in muscovite mica that is 30 nm deep and 3500 nm across (Figure 10.7). In short, it is strikingly different from an etched fission track, which is deep and narrow (the second part of Figure 10.1). In general, the shapes of etched recoil tracks are conical and result from a competition of preferential etching at a rate v_T and general removal of surrounding material at a lower rate v_G (Fleischer and Price 1964, Fleischer et al. 1975). The half cone angle ϑ is given by $\sin\vartheta = v_G/v_T$. For the usual fission track in muscovite mica, ϑ is very close to zero; but for the recoil track such as is shown in Figure 10.7, it is 89°. Since v_T is $v_G/\sin^{-1}\vartheta$, v_T is far less for the recoil track, in agreement with the weak, diffuse nature of the inferred damage. Fleischer (2003) discusses low-energy "tracks" in more detail.

10.5 MECHANISMS OF TRACK PRODUCTION II

10.5.1 High-Energy Damage

Tracks from ionization have the remarkable characteristic that, although they are created by disturbance of the electrons around atoms, the actual damage consists of atomic defects — atoms out of place and vacant sites where atoms normally sit (Fleischer et al. 1965). Figure 10.8 illustrates the situation graphically. For the particular glass $P_2O_5 \cdot 5V_2O_5$ the activation energy for motion of electronic charges is known to be 0.23 eV, and the value for healing tracks is distinctly different — more than 5 times larger, 1.2 eV. The latter is at the level expected for the migration of atomic defects, and typical track erasure activation energies are even higher — a few eV, up to 10 eV (Tables 2–4 in Fleischer et al. 1975), i.e., also in the range of values expected for atomic diffusion.

10.5.1.1 Detection Thresholds

Extensive data fit the model shown in Figure 10.9, in which the intense burst of ionization close to the path of an ion leaves behind an unstable, positively charged region whose order self-destructs through the mutual Coulomb repulsions of nearby ions (Fleischer et al. 1965, 1975, Fleischer 1981, Young 1958, 2002). This description agrees with the qualitative observations that tracks can be made in virtually all insulators, but not normally in good conductors, where electronic conduction is thought to dissipate the localized net charge in a potential track before the ions can move into a lattice.

Figure 10.10 is a remarkable illustration of the ionization spike in LiF (Knorr 1964). For an ion moving just in the plane of a free surface, the mutually repelling ions are unconfined, so that they are sprayed outward to produce grooves that in this case are about 10 nm in width. Where the ions have passed just below the surface, a bulge reveals the volume expansion where the heavy ion passed much like the effect of a mole burrowing under sod. The semi-quantitative model of track formation given by Fleischer et al. (1965) correlates the

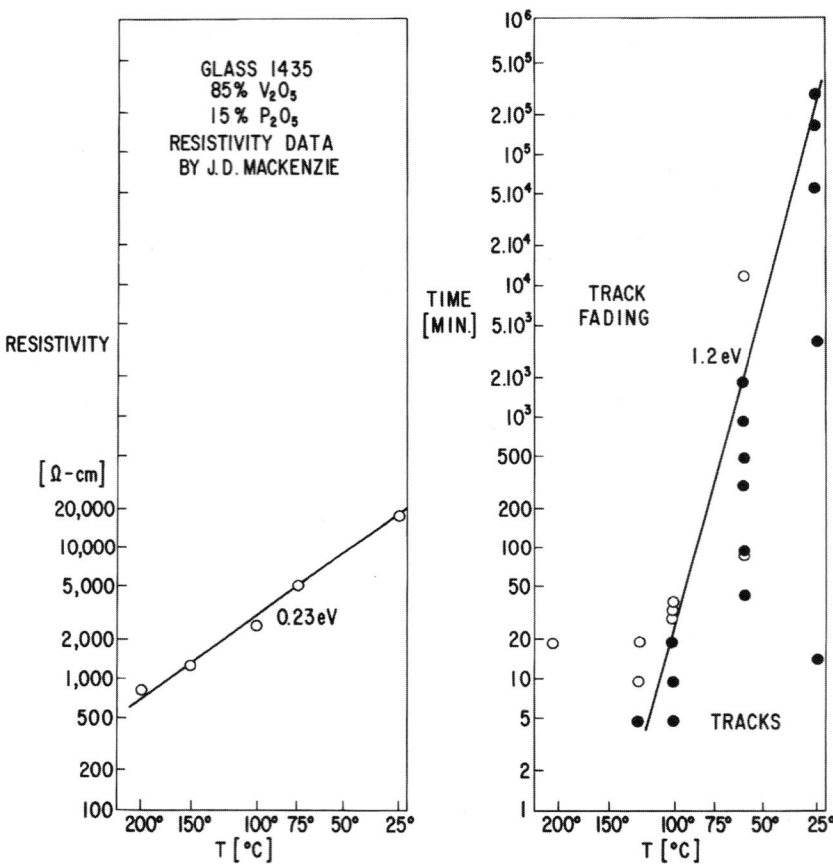

Figure 10.8 Variation with temperature of the conductivity (a measure of the motion of electronic defects) and track annealing time (a measure of the mobility of the defects composing the tracks) for a $P_2O_5 \cdot 5V_2O_5$ glass. The widely different slopes indicate that the track does not consist of electronic defects, and they suggest that atomic defects are what are moving. (Courtesy of the University of California Press.)

relative ease of track production with the elastic strength of the solid (via the modulus E), the tightness of bonding (through the atomic spacing b), and with the forces between charges (as mediated by the dielectric constant ε). The sequence of sensitivities that are predicted — according to

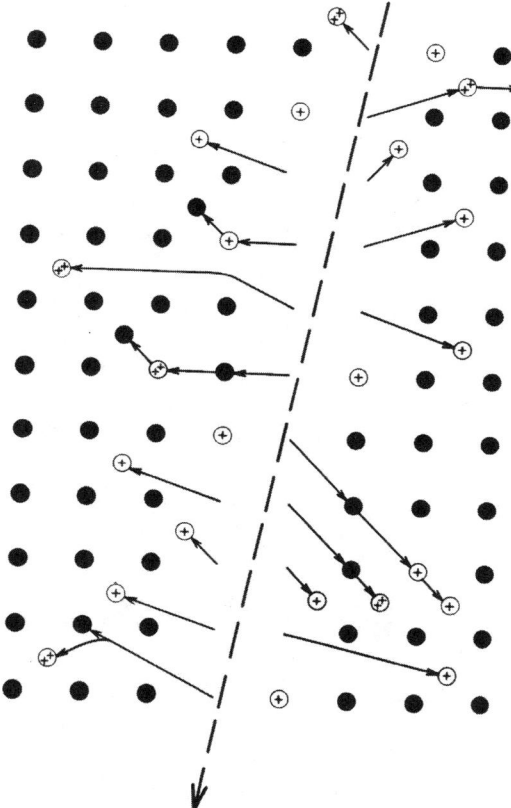

Figure 10.9 Ion-explosion-spike mechanism of track formation. Electrons are ejected by a moving charged particle. The remaining positively charged region is unstable, and the Coulomb forces produce atomic disorder (Fleischer et al. 1965). (Courtesy of the American Institute of Physics.)

εEb^4 — put most dielectric detectors into groups in the correct sequence.

An expectation from the ionization model is that the ionizing particles should have a threshold for track formation that is set by what is called primary ionization — the ionization caused directly by the fast particle close to the core of the track. Figure 10.11 is the best-documented example of this behavior. For a plastic, Lexan polycarbonate, data extend

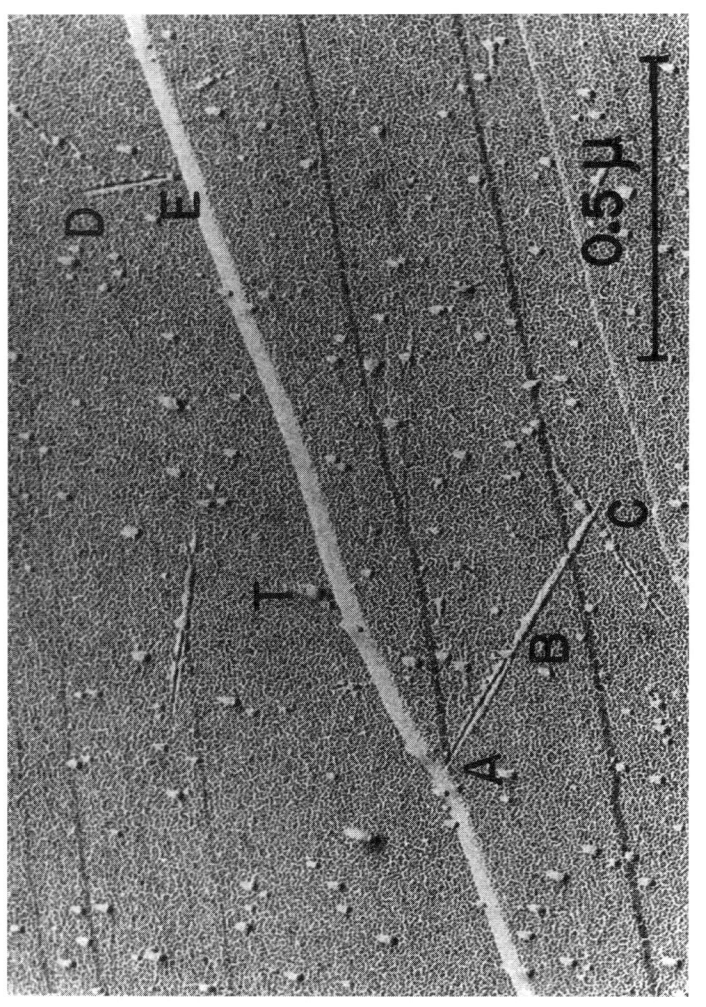

Figure 10.10 Nuclear-track grooves and bulges in a lithium fluoride crystal with a step. A fission fragment at glancing-angle incidence caused ion repulsion (as in Figure 10.9) — creating a groove by ejecting ions from the surface, and then a bulge where the disordered track-region is contained. Coming out at the edge of a step the particle approached the somewhat lower surface and repeated the process (Knorr, 1964). (Courtesy of the Materials Research Society.)

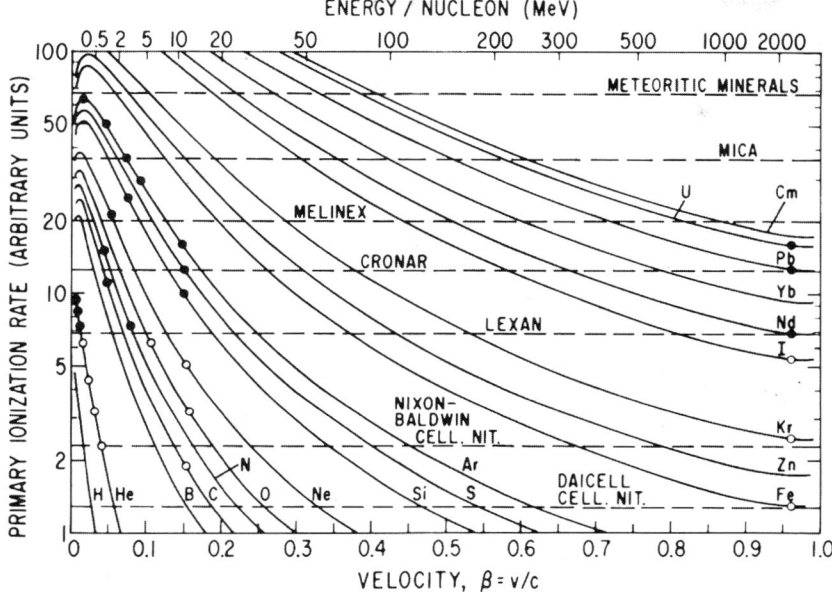

Figure 10.11 Ionization versus velocity for various charged nuclei. Each track detector has an ionization level below which no tracks are observed and above which all can be revealed. Data are given for Lexan polycarbonate (for filled circles, tracks are seen; for open circles, none are seen). Thresholds are given for other detectors. (Courtesy of the University of California Press.)

from low velocities up to relativistic ones. The open circles (no tracks) are cleanly separated from the filled circles (100% registration). Several other criteria that have been proposed, but the data have been tested and shown to be inconsistent with total energy loss (Fleischer et al. 1967), restricted energy loss (Fleischer 1981), and energy deposited by the ejected electrons (Fleischer 1981). As a first guess, we proposed total energy loss, since it fit our early, limited data (Fleischer et al. 1964c). Our later work (Fleischer et al. 1967), however, ruled that idea out.

Other materials, as indicated on the diagram, have a variety of different thresholds. In addition, the track-etching rate v_T increases monotonically for ionization rates above the threshold — a phenomenon that allows individual track-forming particles to be identified (Price et al. 1967, Fleischer et al. 1970).

From time to time the thermal pulse associated with track formation has been suggested as *the* cause of track formation, most recently by Szenes (1995). In a process that is as complex in detail as track formation, it is unlikely that any simple model will fit the behaviors of all solids. Szenes' model fits a selection of magnetic insulators, but he does not show whether other models, including the ion spike, do also. At present, thermal-spike models have not supplied many detailed predictions that can be compared with observations. However, one expectation is that the heating, and therefore track formation, would depend on the total energy deposition along a particle path. The test noted above by Fleischer (1981) contradicts that idea.

10.5.1.2   Further Details and Adjustments

The immense variety of solids that exist allows and requires other factors in track formation to be recognized. For example, most polymeric materials are more sensitive than inorganic track detectors and, if no other factor entered, the thresholds would imply that ionization occurs more readily in plastics than in the inorganic materials. However, what is thought to be the case is that excitation of electrons, followed by bonds that are broken by de-excitation, plays a major role in the plastics. This result comes about because bonds can be broken with less energy transferred (1 or 2 eV) than is required for ionization (around 10 eV or more). Bovey (1958) noted that the excited electrons along a polymer chain can then decay and break bonds, converting the initial chain into two or more shorter ones, and shorter chains are dissolved by chemical etching more rapidly than longer ones. Thus, the track before etching can be regarded as being a low-molecular-weight region in the sense of having a high density of chain ends (Figure 10.12), as well as containing the usual number of ionized atoms from the process shown in Figure 10.9.

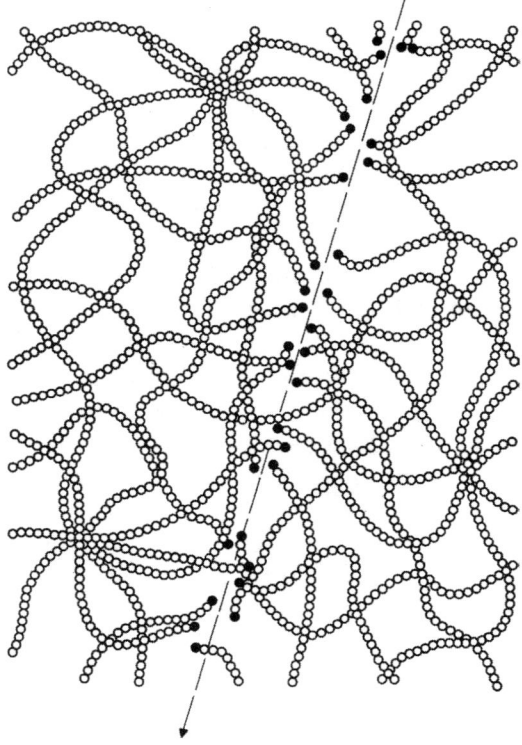

Figure 10.12 The atomic character of a particle track in a high polymer. New chain ends and other chemically reactive sites are formed.

10.5.1.3 Unexpected Track-Detecting Materials

Earlier, it was noted that under special circumstances tracks form in a limited selection of good conductors — metals, inter-metallic compounds, and oxide superconductors (Provost et al. 1995, Barbu et al. 1995, Fleischer 1995). For references see my review of the observations and possible explanations (Fleischer 2002). What is special about the actual track-detecting materials among the many that do not record tracks in each category?

The *metal* in which tracks of fast heavy ions (U or Pb) have been seen is titanium. However, there are others (Fe and Zr) that show tracks from accelerated molecules of fullerene (C_{60}), a much heavier projectile; and Bi, Co, and Zr show anomalously high defect production under irradiation by very heavy ions (shown by increases in resistivity), implying that they are close to allowing track formation. In contrast, Ag, Al, Au, Cu, Nb, Pd, Pt, W, and Zn do not record tracks.

The common aspect of most of the metals in which tracks occur or damage is large (except Bi) is that they have allotropic forms. In short, the fact that a second, or third phase is available with slight difference in free energy suggests a route to track production. In titanium the change of phase within tracks has been seen. Tracks consist of a different phase from the usual room-temperature hexagonal (hP2) alpha structure. Interestingly that phase is not the common high-temperature cubic (cI2) structure, but the hexagonal (hP3) ω, high-pressure form (Dammak et al. 1993, 1996). The exception to having an allotropic form, Bi, is doubly unusual in different ways: (1) it has a complicated rhombohedral (hP2) structure, and (2) its density is decreased strikingly by disordering.

Intermetallic compounds and *oxide superconductors* that show tracks mostly have two common characteristics. First, the materials have complicated crystal structures (so that if disordered they would find it hard to rearrange to the normal structure). Among the intermetallics the importance of this consideration is bolstered by the fact that the particular intermetallics that show tracks are complicated — for example, $NiZr_2$ (tetragonal with 12 atoms per unit cell) and BNi_3, which is orthorhombic with 16 atoms per cell. In contrast, the structures $AuCu_3$ and $AlZr_3$ have cubic symmetry and do not show tracks. NiTi does record tracks and is a simple structure, but it is complicated in a different way: as a shape-memory alloy (Wayman and Inoue 1995), it changes structure readily under modest stress at room temperature. In this way it is similar to the metals that do record tracks.

The other characteristic that all three of the unusual classes of track recorders share is that they are less sensitive than the least sensitive of the insulators (which almost universally register tracks). Fission fragments suffice to create tracks in usual track detectors, but ions that are about twice as heavily ionizing, or more, are required for the special cases.

10.6 CONCLUDING REMARKS

Earlier, I noted that the low-energy "tracks" were diffuse and that the actual displaced atoms only occupy about 2% of the apparent track volume. This open space implies that a many-fold overlap of tracks is required to disorder a material fully by alpha-recoil events. This expectation is consistent with what Weber et al. (1997) inferred from the non-linear, roughly parabolic increase in amorphous fraction f with alpha dose; they concluded that two hits must overlap. The algebra in the model that they used to calculate f would also fit when extended to multiple overlaps. These defects can be used in many materials to produce amorphous structures or, after leaching, to roughen a surface.

High-energy tracks are (usually) continuous regions of densely packed disorder with reasonably distinct diameters (See Figure 10.3). The tracks have many uses on a macro-scopic scale (Fleischer 1998), but also a number on a nanos-cale. Fine filters are one example (Fleischer et al. 1964b). The use of a single hole to count and characterize minute particles in a liquid is another (DeBlois and Bean 1970). The use of tracks as obstacles to the motion of lines of magnetic flux, to improve superconducting properties, is a third (Bourgault et al. 1989, Fleischer et al. 1989).

REFERENCES

Barbu, A., Dammak, H., Dunlop, A., and Lesueur, D. (1995) Ion tracks in metals and intermetallic compounds. *MRS Bull.* 20 (12), 29.

Bean, C.P. (1975) Electrical measurements of track diameter, in Fleischer, R.L., Price, P.B., and Walker, R.M. (Eds.), *Nuclear Tracks in Solids*, University of California Press, Berkeley, 1975, pp. 11–17.

Bean, C.P., Doyle, M.V., and Entine, G. (1970) Etching of submicron pores in irradiated mica. *J. Appl. Phys.* 41, 1454–1459.

Betz, H.-D. (1972) Charge states and charge-changing cross sections of fast heavy ions penetrating through gaseous and solid media. *Rev. Mod. Phys.* 44, 465–539.

Bibring, J.P., Duraud, J.P., Durrieu, L., Jouret, C., Maurette, M., and Meunier, R. (1972) Ultra-thin amorphous coatings on lunar dust grains. *Science* 175, 753–757.

Bourgault, D., Bouffard, S., Toulemonde, M., Groult, D., Provost, J., Studer, F., Nguyen, N., and Raveau, B. (1989) Modifications of the physical properties of the high-T_c superconductors $YBa_2Cu_3O_7$ by 3.5-GeV xenon ion bombardment. *Phys. Rev. B* 39, 6549–6554.

Bovey, F.A. (1958) *The Effects of Ionizing Radiation on Natural and Synthetic High Polymers,* Wiley Interscience, New York.

Comstock, G.M., Fleischer, R.L., Giard, W.R., Hart, H.R. Jr., Nichols, G.E., and Price, P.B. (1971) Cosmic ray tracks in plastics: the Apollo helmet dosimetry experiment. *Science* 172, 154-157.

Dammak, H., Barbu, A., Dunlop, A., Lesueur, D., and Lorenzelli, N. (1993) Alpha-to-omega phase transformation induced in titanium during ion irradiations in the electronic slowing-down regime. *Phil. Mag. Lett.* 67, 253–259.

Dammak, H., Dunlop, A., and Lesueur, D. (1996) Phase transformation induced by swift heavy ion irradiation of pure metals. *Nucl. Instrum. Methods B* 107, 204–211.

DeBlois R.W. and Bean C.P. (1970) Counting and sizing of submicron particles by the resistive pulse technique. *Rev. Sci. Instrum.* 41, 909–916.

DeSorbo, W. (1979) Ultraviolet effects and aging effects on etching characteristics of fission tracks in polycarbonate films. *Nucl. Tracks* 3, 13–32.

Durrani, S. and Bull, R.K. (1987) *Solid State Nuclear Track Detection*, Pergamon, Oxford.

Ewing, R.C. (1994) The metamict state: 1993 — the centennial. *Nucl. Instrum. Methods in Phys. Research* B91, 22–29.

Fleischer, R.L. (1980) Isotopic disequilibrium of uranium: alpha-recoil damage and preferential solution effects. *Science* 207, 979–981.

Fleischer, R.L. (1981) Nuclear track production in solids, in *Chalmers Anniversary Volume* (J.W. Christian, P. Haasen, and T.B. Massalski, Eds.) *Progress in Materials Science* 97–123.

Fleischer, R.L. (1982a) Alpha-recoil damage and solution effects in minerals: uranium isotopic disequilibrium and radon release. *Geochim. Cosmochim. Acta* 46, 2191–2201.

Fleischer, R.L. (1982b) Nature of alpha-recoil damage: evidence from preferential solution effects. *Nuclear Tracks* 6, 35–42.

Fleischer, R.L. (1995) Ion tracks in solids — from science to technology, to diverse applications, *MRS. Bull.* 20 (12) 17–19.

Fleischer, R.L. (1998) *Tracks to Innovation — Nuclear Tracks in Science and Technology,* Springer-Verlag, New York.

Fleischer, R.L. (2002) Ion tracks in intermetallic compounds, in J.H. Westbrook and R.L. Fleischer (Eds.), *Intermetallic Compounds — Principles and Practice*, Vol. 3, Wiley, Chichester, UK, Chapter 14, pp. 263–273.

Fleischer, R.L. (2003) Recoil tracks in solids. *Geochim. Cosmochim. Acta* 67, 4769–4774.

Fleischer, R.L. and Price, P.B. (1963a) Tracks of charged particles in high polymers. *Science* 140, 1221–1222.

Fleischer, R.L. and Price, P.B. (1963b) Charged particle tracks in glass. *J. Appl. Phys.* 34, 2903–2904.

Fleischer, R.L. and Price, P.B. (1964) Glass dating by fission fragment tracks. *J. Geophys. Res.* 69, 331–339.

Fleischer, R.L. and Raabe, O.G. (1977) Fragmentation of respirable PuO2 particles in water by alpha decay — a mode of "dissolution". *Health Phys.* 32, 253-257.

Fleischer, R.L. and Raabe, O.G. (1978) On the mechanism of fragmentation in liquids of PuO2 by alpha decay. *Health Phys.* 35, 545-549.

Fleischer R.L., Price P.B., Symes, E.M., and Miller, D.S. (1964a) Fission track ages and track-annealing behavior of some micas. *Science* 143, 349–351.

Fleischer, R.L. Price, P.B., and Symes, E.M. (1964b) A novel filter for biological studies, *Science* 143, 249-250.

Fleischer, R.L., Price, P.B., Walker, R.M., and Hubbard, E.L. (1964c) Track registration in various solid state nuclear track detectors. *Phys. Rev.* 133A, 1443–1449.

Fleischer, R.L., Price, P.B., and Walker, R.M. (1965) The ion explosion spike mechanism for formation of charged particle tracks in solids. *J. Appl. Phys.* 36, 3645–3652.

Fleischer, R.L., Price, P.B., Walker, R.M., and Hubbard, E.L. (1967) Criterion for registration in dielectric track detectors. *Phys. Rev.* 156, 353–355.

Fleischer, R.L., Hart, H.R., Jr., and Giard, W.R. (1970) Particle track identification: application of a new technique to Apollo helmets, *Science* 170, 1189-1191.

Fleischer, R.L., Price, P.B., and Walker, R.M. (Eds.) (1975) *Nuclear Tracks in Solids*. University of California Press, Berkeley.

Fleischer, R.L, Hart, H.R., Jr., Lay, K.W., and Luborsky, F.E. (1989) Increased flux pinning upon thermal-neutron irradiation of uranium-doped Y-Ba-Cu-O. *Phys. Rev. B* 40, 2163-2169.

Huang, W.H. and Walker, R.M. (1967) Fossil alpha-particle recoil tracks: a new method of age determination. *Science* 155, 1103–1106.

Knipp, J. and Teller, E. (1941) On the energy loss of heavy ions. *Phys. Rev.* 59, 659–669.

Knorr, T.G. (1964) Fission-fragment tracks and directional effects in the surface of LiF crystals. *J. Appl. Phys.* 35, 2753–2760.

Kuznetksov, V.I., Kuznetksov, L.V., and Shestakov, V.D. (1995) Track membranes of the third generation. (TMG-3): their properties and industrial application. *Nucl. Instrum. Methods B* 105, 250–253.

Lindhard, J. and Scharff, M. (1961) Energy dissipation by ions in the keV region. *Phys. Rev.* 124, 128–130.

Lindhard, J. and Thomsen, P.V. (1962) Sharing of energy dissipation between electronic and atomic motion. In *Radiation Damage in Solids* 1, IAEA, Vienna, pp. 65–75.

Meftah, A., Brisard, F., Costantini, J.M., Hage-Ali, M., Soquert, J.P., Studer, F., and Toulemonde, M. (1993) Swift heavy ions in magnetic insulators: a damage-cross-section effect. *Phys. Rev. B* 48, 920–925.

Powers, D., Chu, W.K., and Bourland, P.D. (1968) Range of Ar, Kr, and Xe ions in solids in the 500-keV to 2-MeV energy region. *Phys. Rev.* 165, 376–387.

Price, P.B. and Walker, R.M. (1962a) Observation of charged-particle tracks in solids. *J. Appl. Phys.* 33, 3400–3406.

Price, P.B. and Walker, R.M. (1962b) Chemical etching of charged-particle tracks in solids. *J. Appl. Phys.* 33, 3407–3412.

Price, P.B., Fleischer, R.L., Peterson, D.D., O'Ceallaigh, C., O'Sullivan, D., and Thompson, A. (1967) Identification of isotopes of energetic particles with dielectric track detectors, *Phys. Rev.* 164, 1618-1620.

Price, P.B., Fleischer, R.L., and Moak, C.D (1968) The identification of very heavy cosmic ray tracks in meteorites, *Phys. Rev.* 167, 277-282.

Provost, J., Simon, Ch., Hervieu, M., Groult, D., Hardy, V., Studer, F., and Toulemonde, M. (1995) Swift, heavy ions in insulating and conducting oxides: tracks and physical properties. *MRS Bull.* 20(12), 22–28.

Silk, E.C.H. and Barnes, R.S. (1959) Examination of fission fragment tracks with an electron microscope. *Phil. Mag.* 4, 970–971.

Snowden-Ifft, D., Price, P.B., Nagahara, L.A., and Fujishima, A. (1993) Atomic-force-microscopic observations of dissolution of mica at sites penetrated by keV/nucleon ions. *Phys. Rev. Lett.* 70, 2348–2351.

Spohr R. (1990) *Ion Tracks and Microtechnology*. Vieweg, Braunschweig.

Szenes, G. (1995) General features of latent track formation in magnetic insulators irradiated with swift heavy ions. *Phys. Rev. B* 51, 8026.

Turkowsky, C. (1969) Electron-microscopic observation of artificially produced alpha-recoil tracks in albite. *Earth Planet. Sci. Lett.* 5, 492–496.

Wayman, C.M. and Inoue, H.R.P. (1995) Crystallographic transformations, in J.H. Westbrook and R.L. Fleischer (Eds.), *Intermetallic Compounds — Principles and Practice*, vol. 1, Wiley, Chichester, pp. 827–847.

Weber, W.J., Ewing, R.C., and Meldrum, A. (1997) The kinetics of alpha-decay-induced amorphization in zircon and apatite containing weapons-grade plutonium or other actinides. *J. Nuclear Mater.* 250, 147–155.

Young, D.A. (1958) Etching of radiation damage in lithium fluoride. *Nature* 182, 375–377.

Young, D.A. (2002) Evolution of a model ion explosion spike in potassium chloride by molecular dynamics. *Europhys. Lett.* 59, 540–545.

Chapter 11

Forensic Applications of Ion-Beam Mixing and Surface Spectroscopy of Latent Fingerprints

CHARLES H. KOCH

CONTENTS

ABSTRACT

Ion-beam enhancement and detection of latent fingerprints using Auger and SIMS analysis have potentially significant forensic applications. The ion implantation process is used to ion-beam mix the materials of latent fingerprints into a substrate, such that the atoms that form the latent fingerprints become an integrated part of the substrate material. This now permanent record of the fingerprint can be imaged optically or with a scanning electron microscope. In addition, surface analysis techniques such as secondary ion mass spectrometry, particle induced x-ray emission or Auger electron spectroscopy can be used to identify the chemical composition of the fingerprint material. Once identified, the elements, molecular fragments and molecules unique to the fingerprint can be mapped using computer-assigned intensities to their relative abundance. The result is a computer-aided map of the latent fingerprint drawn with elements, molecules or molecular fragments. These can be of human origin or residue from the person leaving the fingerprint. A further benefit of the ion implantation process is that the fingerprint is permanently fixed in the substrate and can serve as a long-term record. This work involved ion-beam mixing fingerprints into selected classes of substrates and used Auger spectrometry and computer-aided Auger mapping to image fingerprints with low concentrations of the fingerprint materials. This combination of techniques produces images of fingerprints with surface analysis techniques that have not been previously applied. The application of these surface analysis techniques is made possible by the ion-beam mixing process. Applications are now possible utilizing techniques that are more sensitive to lower concentrations of fingerprint material than those currently being used by the forensic community. Using modern surface spectroscopy methods it is possible to detect and map fingerprints at very low concentrations.

11.1 INTRODUCTION

Methods of ion-beam mixing and surface analysis can be applied for forensic purposes when materials associated with a latent fingerprint are considered as thin films. Most latent fingerprints are organic in origin but may also contain residual materials transferred from the handling of other products. Ion implantation is used to ion-beam mix materials associated with latent fingerprints into a substrate. The atoms, molecules and molecular fragments that form the latent fingerprints become an integrated part of the substrate material. Once ion-beam mixed, the fingerprint becomes more permanent and durable. The ion-beam mixed fingerprint can be imaged with scanning electron microscopy (SEM); importantly, the fingerprint is now more resistant to the effects of the energetic electron beam of the microscope. At a time when interpretation of macro features by human experts has been called into question, a treated fingerprint can now be greatly magnified to the point where unique micro-features can be identified with sophisticated computer algorithms. These micro-features may prove to be the definitive feature needed for positive identification.

Once treated, surface analysis techniques such as secondary ion mass spectrometry (SIMS), particle induced x-ray emission (PIXE) or Auger electron spectroscopy (AES) can be used to identify the chemicals associated with the fingerprint material. The elements unique to the fingerprint can be mapped using computer-assigned intensities to abundance. These results can be used to more positively link the latent fingerprints with the person leaving them. An additional advantage of this process to the forensic community is that fingerprints can now be mapped from very low concentrations of a single element, possibly revealing fingerprints that would not be obtainable by any other method. A further benefit of the ion-implantation process is that the fingerprints are permanently affixed in the substrate and can be recalled years later as original evidence.

11.2 EXPERIMENTAL PROCEDURE

Ion implantation is a technology that uses an accelerator, or a small self-contained ion source to create a beam of charged atoms, or ions.[1] The ion beam is shaped and directed toward a target surface, embedding ions into the material. When the target surface contains other material on the surface, such as a fingerprint, the effect of the incident ions is to drive that material into the target. This process is referred to as "ion-beam mixing." A schematic cross-sectional view of an ion-beam mixed fingerprint is provided in Figure 11.1. For ion-beam mixing to be successful, the mass and energy of the incident ion must be appropriate to the substrate. If the incident ion has too high a mass with too low an energy it will sputter the first surface layer (the fingerprint) away, without any mixing taking place. If the incident ion is too low in mass,

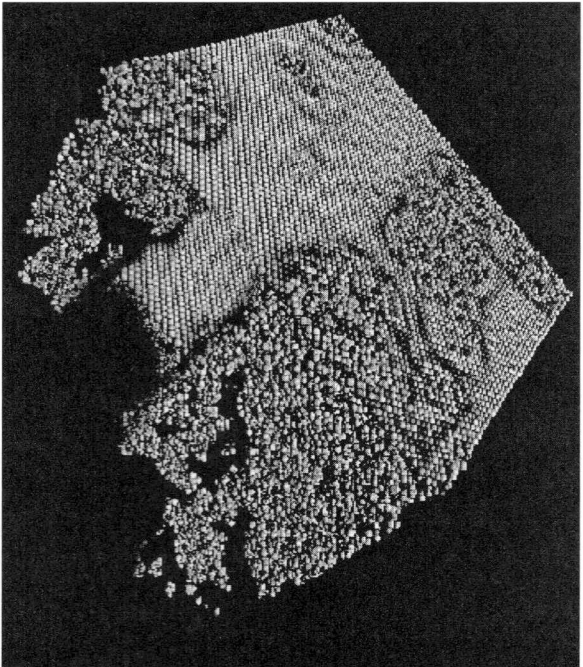

Figure 11.1   Representation of an ion beam mixed fingerprint. The raised areas represent the fingerprint material. After ion implantation they go into the body of the material itself. The process has been shown to work on metals, ceramics, plastics, paper and other materials.

and too high in energy, it will pass through the fingerprint layer and deposit its energy deep in the second layer (the fingerprint substrate). For optimal results the ion should have the correct mass and energy to come to rest in the area of the interface between the fingerprint and the substrate material. The ions will come to rest at a depth described by a roughly Gaussian distribution, and some of the atoms from the fingerprint will be projected deep into the substrate material and remain there.

The ion implantation process takes volatile and fragile fingerprint material and imbeds those atoms and molecules into the substrate, making them more durable and permanent. The finger material is now detectable by normal optical methods and by sophisticated material analysis techniques that are sensitive to very low concentrations. Following ion implantation, the fingerprint is no longer only on the surface, but is a permanent part of the substrate material.

Experiments were performed on a variety of substrates to test the effectiveness of the different classes of ions chosen for the ion-beam mixing process. Other variables considered were dose (i.e., ions implanted per unit area) and the ion beam energy, which affects the depth and profile of the implanted ions. These variables could have different effects on the subsequent analysis techniques. A complex series of experiments was performed to find optimum combinations of ion species, dose and beam energy for different classes of substrates. Results from experiments on varied substrate types and their analysis are discussed below.

11.3 SAMPLE RESULTS AND ANALYSIS

A fingerprint was lightly placed on a cleaned glass slide; it was almost invisible before implantation (see Figure 11.2). The fingerprint material was ion-beam mixed into the glass substrate. After the implantation, the fingerprint turned a dark violet, and could be rubbed or scratched with tweezers without smearing. On the very top of the slide, there appeared a fragment of another fingerprint that was revealed only after the implantation process. The slide was placed in a scanning

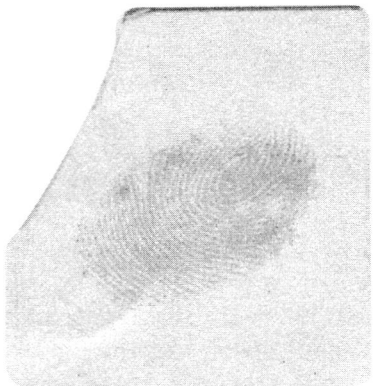 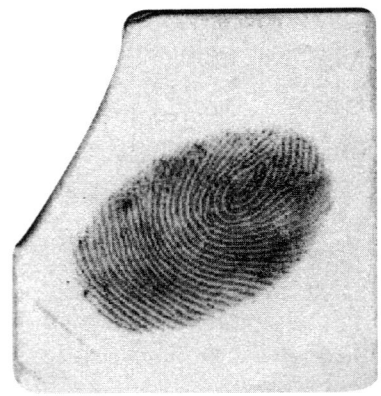

Figure 11.2 Fingerprint lightly placed on a cleaned glass slide, nearly invisible before ion implantation (left). The fingerprint material was ion-beam mixed into the glass substrate. After implantation, it turned a dark violet, and could be rubbed or scratched with tweezers without smearing (right). On the very top of the slide, there appeared a fragment of another fingerprint that was revealed only after the implantation process.

Auger analyzer, for imaging with secondary electrons. An image of one area was maintained for hours without any apparent loss or degradation. This would not have been possible before the ion-beam mixing process. This allows for analysis of micro-features of the fingerprint. The secondary electron image can be saved digitally and used for identification, as a whole fingerprint, or with sophisticated computer algorithms that could compare micro-features found in a fingerprint or fingerprint fragment.

The second part of the ion-beam enhancement and detection of latent fingerprints consists of analyzing ion-beam mixed fingerprints with computer-aided, topological atomic/chemical mapping surface analysis techniques. This unique combination of processes allows the detection of the atoms, molecules and molecular fragments left by the fingerprints. These analysis techniques use energetic ion or electron beams that would remove or degrade untreated fingerprint atoms and/or molecules if they were not first ion-beam mixed into the substrate. Some of the surface analysis techniques that can be used are SEM, AES, SIMS and PIXE.

When an impinging electron beam ionizes a core level of a surface atom, the atom can decay to a lower energy state through the emission of an Auger electron, which will have a kinetic energy characteristic of the parent atom. When the Auger electrons are created near the surface (about 1–3 nm), many reach the vacuum without loss of energy and have the characteristic energy distribution of the atom emitting the electrons. The energy and shape of these Auger peaks can be used to positively identify the composition of the solid surface. Using these Auger electrons the fingerprints will be chemically mapped and imaged. Auger is not very sensitive in general, having a detection limit of ~ 1%, so it will only give part of the desired information.[2] To enhance sensitivity time-of-flight secondary ion mass spectrometry (TOF-SIMS) can be applied.

A scanning energetic primary ion beam sputters a sample surface. Secondary ions formed in this sputtering process are extracted from the sample and analyzed in a mass spectrometer.[3] The SIMS technique is capable of detecting secondary ions produced over a large mass range. The ions can originate from atoms, molecules or molecular fragments from the sample's surface. It has the potential of identifying unique chemical characteristics of the fingerprint. In TOF-SIMS a focused, short-pulsed scanned primary ion beam sputters the top surface layer of the sample. The secondary ions produced in this sputtering process are extracted from the sample surface and injected into a specially designed time-of-flight mass spectrometer. The ions are dispersed in time according to their velocity (which is proportional to their mass-to-charge ratio [m/z]). The TOF-SIMS technique is capable of detecting secondary ions produced over a large mass range (typically 0 to approximately 5000 atomic mass units) and performs this mass analysis at relatively high mass resolutions (> 6000 in terms of m/m), which allows specific identification of molecules and molecular fragments with the same nominal atomic mass. This technique is capable of generating a detailed image of the sample's surface using these molecules and molecular fragments, which can then be interpreted in terms of the chemistry of the fingerprint.

Figure 11.3 shows a fingerprint placed on a 5000 Å-thick gold strip evaporated on a stainless steel substrate. Experiments performed using this sample reveal much about ion-beam mixing and subsequent possible variations for analysis. Auger spectroscopy analysis was used to image the substrate material (gold) on which the fingerprint was placed. Since the

Figure 11.3   A fingerprint was placed on a metal strip of 5000 Å of gold over stainless steel. After implantation the fingerprint appeared very dark and could be rubbed without smearing.

Auger process is sensitive to only about 4–10 atomic layers, material from the fingerprint (even just a few atomic layers) would stop the escaping Auger electrons from leaving the substrate. The result was a negative image of the fingerprint, i.e., the fingerprint drawn in "not gold." This experiment revealed the sensitivity of the Auger process in mapping images of the fingerprint through use of the substrate Auger signal. Another, and more typical, Auger experiment was conducted using this same sample. In this instance, a secondary-electron image of the fingerprint was first produced. Choosing selected areas of the fingerprint image for small-spot Auger analysis produced different spectra from within the selected regions of the fingerprint. The chemical spectra varied for these different regions. An Auger peak (chlorine) unique to one of these regions was selected for further Auger mapping. The Auger mapping produces a surface image by assigning a light-intensity value that corresponds to the abundance of an atomic species. The result is an image drawn with an atomic element — in this case, the image is a fingerprint drawn in chlorine. Furthermore, this sample also illustrates the durability of a fingerprint following mixing when placed in a scanning electron microscope, or Auger analyzer.

This same sample (5000 Å of gold over stainless steel) was then re-introduced into the ion implanter for implantation with a very high dose of argon. The dose used was that required to completely sputter away and remove the 5000 Å of gold (on which the fingerprint resided) over the stainless steel. When the sample was visually inspected post-implantation, the gold had been completely removed (through sputter-etching by the argon beam). The surface had the appearance of shiny stainless steel. However, upon closer visual inspection, traces of the original fingerprint (in a gold color) could still be seen, even though the gold on which it was placed had been completely removed. The conclusion drawn from this work is that through the process of differential sputtering a representation of the fingerprint was ion-beam mixed through the gold layer into the stainless steel substrate. This startling discovery has many implications for forensic science and technology.

The same sample was then put into a scanning elctron microscope with energy-dispersive spectroscopy (EDS) for analysis. The scanning electron microscope revealed an image of a fingerprint shown in light and dark bands corresponding to the fingerprint ridges and grooves (Figure 11.4). EDS analysis of these light and dark areas produced spectra showing different chemical abundances (Figures 11.5–11.9). The light areas had a higher atomic concentration of gold compared with the dark areas. Both areas showed all of the elements expected from stainless steel. Laser profilometer analysis showed no difference in height between the light and dark areas, suggesting therefore that the compositional difference must be due to a combination of ion-beam mixing and ion-

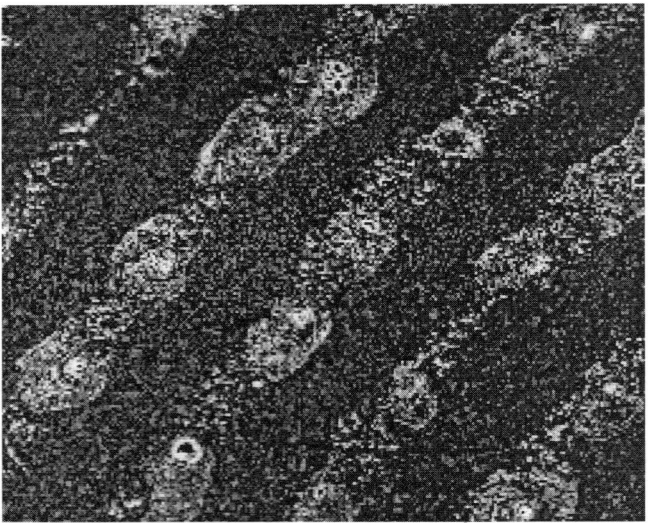

Figure 11.4 Scanning electron micrograph of a fingerprint placed on a 5000 Å gold layer deposited on a 5000 Å stainless steel substrate. The fingerprint was first ion-beam mixed into the gold layer, analyzed and imaged. The sample was then reintroduced into the implanter and ion implanted with a very high dose of argon ions, sufficient to remove the gold layer (see text). The light and dark areas are differential secondary electron emission due to a difference in composition.

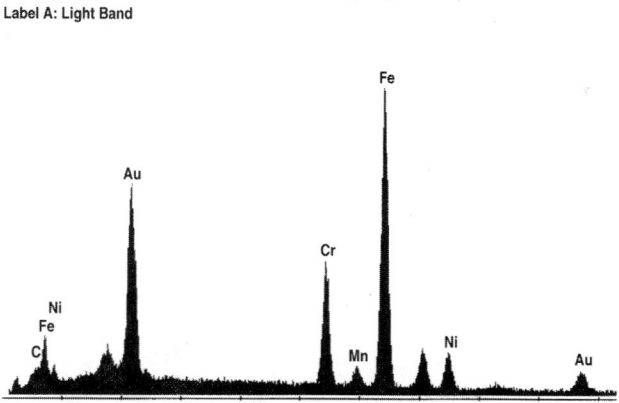

Figure 11.5 An EDS spectrum of the "light band" regions of the fingerprint. The expected components of stainless steel are present. Also present is a large gold peak. This can only be from the gold layer that has been sputtered away. It is interesting to note that the gold remains in the fingerprint pattern.

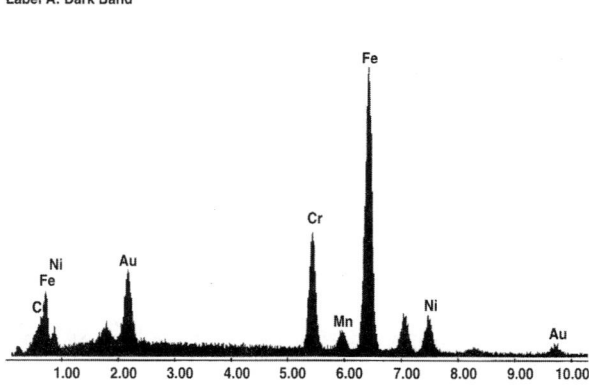

Figure 11.6 An EDS spectrum of the "dark band" regions of the fingerprint. Again, the expected components of stainless steel are present. The magnitude of the gold peak is now significantly smaller. This is most likely due to the differential sputtering effects from the different materials in the ridge versus groove areas of the fingerprint. Again also, even though diminished there is still gold present in the substrate, which suggests an ongoing differential ion-beam mixing process through the gold layer.

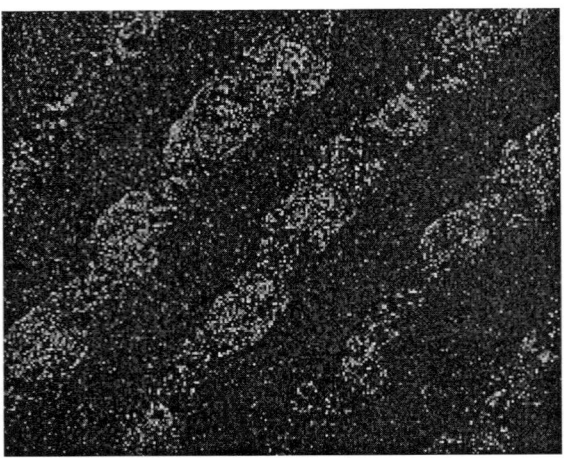

Figure 11.7   An Au M_α x-ray map.

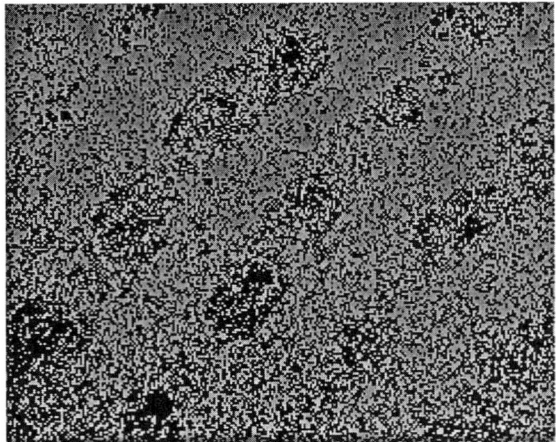

Figure 11.8   An EDS image produced by mapping the iron K-alpha emission from the same area as the previous map. The areas with a higher concentration of iron are lighter. These are the areas of stainless steel, which constitutes the substrate. The original 5000 Å gold layer has been sputtered away. The pattern of the fingerprint is still visible but as a negative of the previous image. Here the fingerprint ridges remain darker because they contain more gold and less iron.

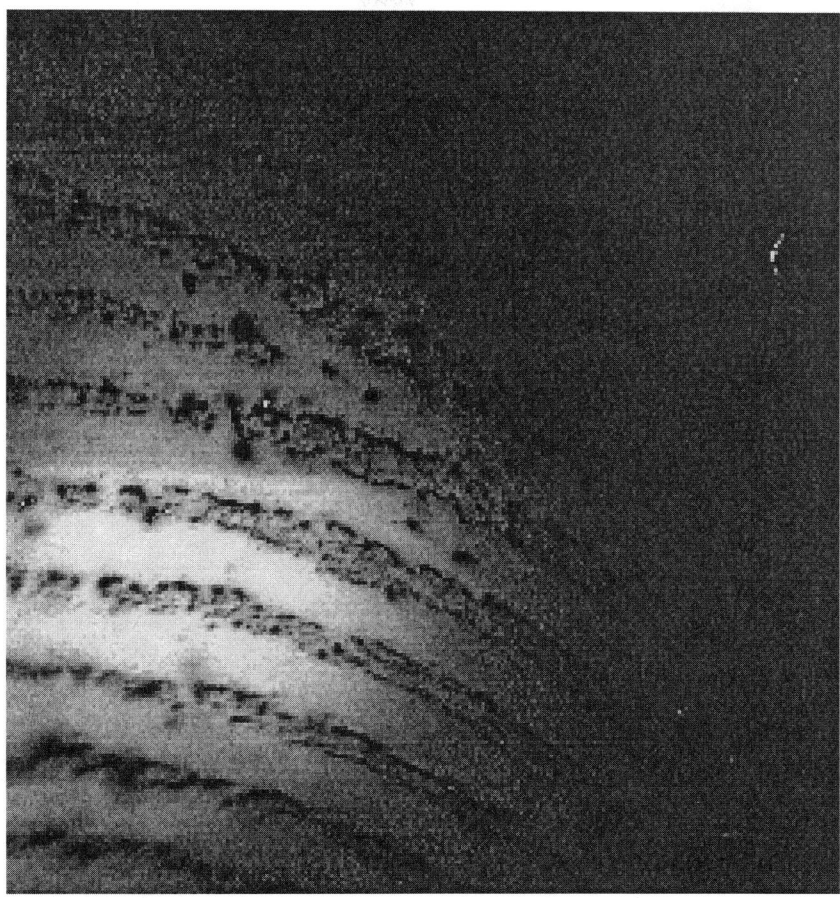

Figure 11.9 Auger computer mapping of the gold signal. The lighter areas represent the more abundant gold signal. This is effectively a map of a fingerprint drawn in "not-gold." This technique of acquiring a negative image is very sensitive. The Auger signal from the substrate can be blocked with just a few atomic layers of fingerprint material, allowing the imaging of a fingerprint whose concentrations may be very low, in fact, just a few atomic layers.

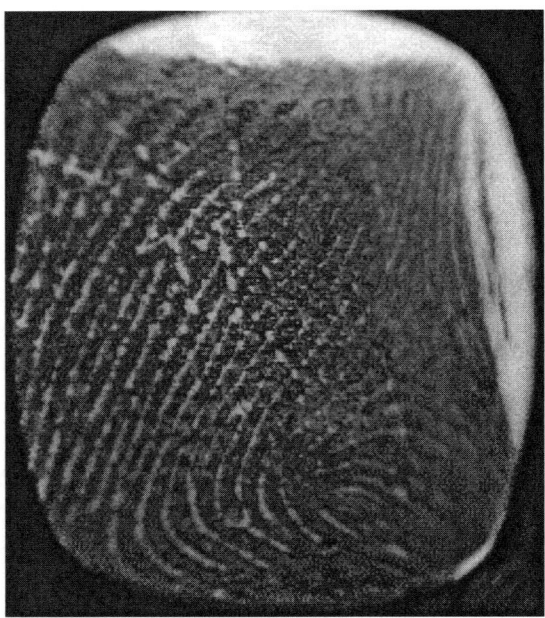

Figure 11.10   SED image of the fingerprint on gold from the Auger system has a three-dimensional effect. At low magnification the whole fingerprint was visible. The image remained unchanged with the electron beam on the same area for over one hour without any noticeable effects.

beam differential sputtering. It is interesting to note that the ion-beam mixing process continued through the 5000 Å layer of gold, and resulted in a recoverable fingerprint after all of the original substrate was removed.

The long-term exposure of the aforementioned fingerprint in a scanning electron microscope shows its enhanced ruggedness and durability (Figure 11.10). The ability to examine and capture micro-features of fingerprints and fingerprint fragments has a timely and significant advantage in forensic science. Recently, the human interpretation of the macro features of fingerprints called into question their validity as trial evidence   (http://www.news.cornell.edu/Chronicle/02/1.31.02/fingerprints.html). Scanning electron micros-

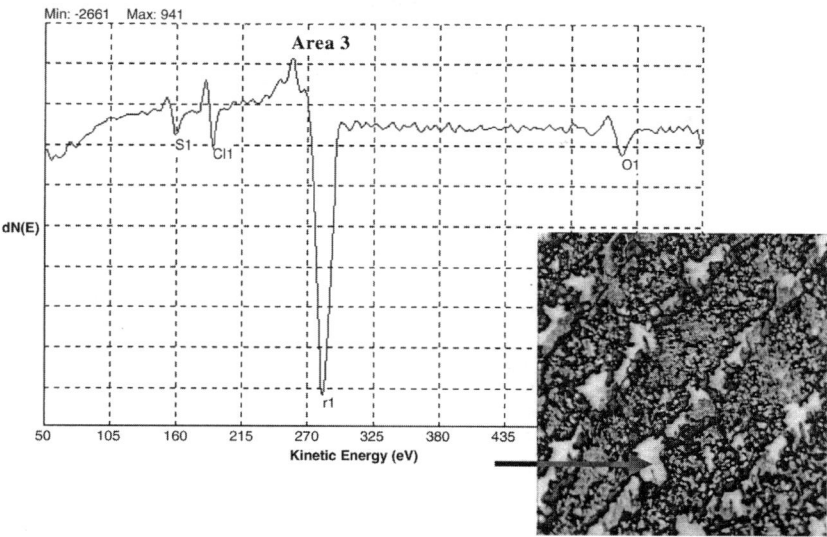

Figure 11.11 Closeup of a selected area from the previous SED (Figure 11.10). There are three distinct areas of grayscale: the arrow is pointing to a white area. The Auger beam spot was placed on this white area. The spectrum is the Auger spectrum from the white area. The spectrum has peaks of sulphur, chlorine, carbon, and oxygen.

copy is possible long after the ion-beam mixing takes place. This allows digital images of micro-features to be analyzed with sophisticated computer algorithms and produce a non-subjective matching of evidence. An added advantage of this technology is that it could be accomplished with partial fingerprints and even small fragments.

EDS analyses performed on this sample yielded the following results. Figure 11.9 shows an enlarged area of a fingerprint that is a computer-acquired digital secondary-electron map. There appear to be three major regions of contrast. These regions are the very light areas, the gray areas, and the very dark or black areas. These features are much larger than the 1000 Å spot size of the scanning Auger beam. Spectra were acquired from within each of the contrast regions (Figures 11.11and 11.12). Because these spectra were

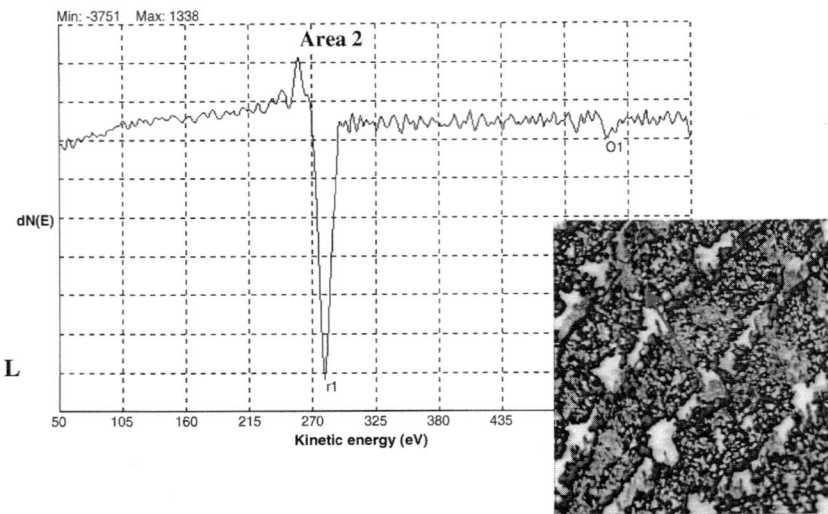

Figure 11.12　Closeup of a selected area from the initial SED. There are three distinct areas of grayscale. The arrow is pointing to a black area, on which the Augur beam spot was placed. The spot is very much smaller than the white area. The spectrum is the Auger spectrum from the black area. The spectrum has visible peaks of carbon and oxygen. More peaks would likely be visible if ion-beam sputtering were performed to reduce the adventitious carbon peak. It is important to note that different areas of the same fingerprint produce different spectra.

acquired without first sputter-cleaning the surfaces, a large carbon peak was present. The other peaks present were small by comparison. When the top atomic layers of carbon are removed with a sputtering ion beam, the elements below are revealed. A more detailed analysis of residual peaks would require this cleaning to be performed first. The resulting spectra were compared for peaks unique to specific regions. Area three exhibited two peaks that were not present in the other areas. These peaks were sulfur and chlorine. A chlorine Auger map was acquired assigning light intensity to the strength of the chlorine Auger peak (Figure 11.13). The resulting image clearly shows a fingerprint. This fingerprint is the result of the ridges of the finger depositing more chlorine than the grooves.

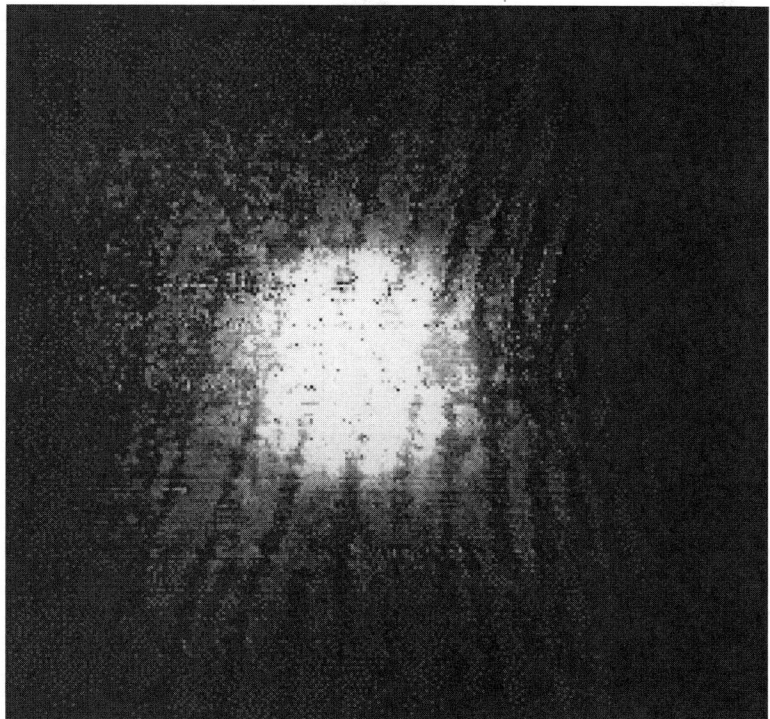

Figure 11.13 Computer mapping of the intensity of the chlorine Auger peak found in Figure 11.11 produced this image. Such imaging is possible because the fingerprint is non-uniform chemically, and one can produce a chemical map of it with Auger mapping. The lighter the area, the more abundant the chlorine signal. The bright spot in the middle is a result of the non-linearity of the Auger detector. A fingerprint is much larger than the anticipated area of analysis for most detectors. It would not be difficult to produce a detector with a linear response over a much larger area.

11.4 CONCLUSIONS

The ion implantation process takes volatile and fragile fingerprint material and imbeds those atoms, molecular fragments and molecules into a variety of substrate classes such as metals, paper, glass and plastics, making them more durable, visible and permanent. Following ion implantation the fingerprint is no longer only on the surface, but is now a permanent part of the substrate material. The most important aspect of this process is that the fingerprint material is now detectable by sophisticated material analysis and imaging techniques at very low concentrations. Using Auger spectroscopy we have shown that it is possible to produce an image of a fingerprint by mapping its chemical concentrations. Further, it is possible to produce a negative image of a fingerprint using Auger spectroscopy by mapping the concentrations of the substrate that are masked by a few atomic layers of fingerprint material above it. The ability to acquire scanning electron photomicrographs of the small-scale features of a fingerprint is made possible by the ion-beam mixing process. The ion-beam mixed fingerprint can withstand long-term examination and analysis with SEM/EDS. This is a promising new technique not previously available to the forensic community. In conjunction with SEM analysis, EDS imaging has produced positive and negative images of fingerprints.

These micro-features could be used in conjunction with computer software to analyze and catalog them, to provide a more positive identification and match than is now possible with interpretation of macro-features. At a time where interpretation of macro-features by human experts has been called into question, a treated fingerprint can now be greatly magnified to the point where unique micro-features can now be identified with sophisticated computer algorithms. Such micro-features may prove to be the definitive elements needed for a positive identification. These techniques are able to detect concentration levels as low as parts per billion.

It is also possible to obtain an image of a fingerprint with molecules and molecular fragments, thereby providing another link to the person leaving the fingerprint through the evidence of the materials they have handled.

11.5 DISCUSSION OF FUTURE WORK

Future work will include the application of SIMS for this analysis. The advantages of SIMS over Auger are critical for this application. Auger analysis can detect concentrations as low as parts per thousand. SIMS detection limits can be in parts per million. Auger detects atomic species only; SIMS is a mass spectrometer technique that can detect atomic species, molecules and molecular fragments. This detection of molecules and molecular fragments at very low concentrations opens up the possibility of computer mapping of fingerprints with molecules or molecular fragments. A SIMS mapping of a fingerprint containing trace material from items such as motor oil, gunpowder or TNT could further enhance the information obtainable from the fingerprint.

The described technology has been accepted for a pending patent application "Method and Apparatus for Fingerprint Detection and Analysis" notice of allowance on November 17, 2003. The patent has been licensed to a company in the United States. They intend to produce stand-alone units that will use these processes. These will be for sale to forensic laboratories.

REFERENCES

1. For further reading see F.A. Smidt, G.K. Hubler and B.D. Sartwell, Surface modification of metals by ion beams, *Surf. Coat. Technol.*, 51, 1991; J.W. Mayer, L. Eriksson and J.A. Davies, *Ion Implantation in Semiconductors*, Academic Press, 1970.

2. L.E. Davis, N.C. MacDonald, P.W. Palmberg, G.E. Riach and R.E. Weber, *Handbook of Auger Electron Spectroscopy, A Reference Book of Standard Data for Identification and Interpretation of Auger Electron Spectroscopy Data*, 2nd ed., Physical Electronics Industries, Inc., Eden Prairie, MN, 1976.

3. For further reading see A. Benninghoven, C.A. Evans, Jr., R.A. Powell, R. Shimizu and H.A. Storms (Eds.), *Secondary Ion Mass Spectrometry SIMS II*, *Springer Series in Chemical Physics*, Vol. 9, Springer-Verlag, 1979.

Glossary

Adatom, admolecule — An atom or molecule that is adsorbed onto and retained by a surface

AERE — Atomic Energy Research Establishment, Harwell, U.K., site of pioneering studies of ion implantation in metals in order to improve their surface properties

AES — Auger electron spectroscopy, a surface analytical technique that provides information about chemical composition typically to a depth of 5 nm with a lateral resolution of about 20 nm

AFM — Atomic force microscope, an instrument used to obtain the topography of a surface on atomic scale by monitoring the deflection of a sharp stylus as the latter is moved across the surface and in proximity to it (see also STM)

Allotropy — State of a substance wherein an element is present in two or more different forms

Ångström — Unit of length equal to 0.1 nm (10^{-10} m)

Anisotropy — That which exhibits properties with different values when measured in different directions

ARE — Activated reactive evaporation, a plasma deposition process

ATP — Advanced Technology Program, a research program of the U.S. Department of Commerce

BFTEM — Bright-field TEM (see TEM)

Brachytherapy — An advanced-cancer treatment wherein radioactive seeds or sources are placed in or near a tumor, giving a high radiation dose to the tumor while minimizing the radiation exposure in the surrounding healthy tissues

Bright-field TEM — see TEM

BSM — Beam-sweep magnet

CAD — Cluster-assisted deposition

Cathodo-luminescence — Emission of light by a substance that results from excitation by electrons

Channelling — In ion implantation of crystal structures, the tendency of ions that are injected parallel to and between atomic planes to travel much farther than they would in the corresponding amorphous solid

CMP — Chemical-mechanical polishing

Collision cascade — Succession of collisions between a projectile ion and host atoms, involving ionization and bond-breaking of the latter, or of nuclear collisions in which the incident ion is scattered elastically and the struck atoms recoil

Colloid — Substance that consists of particles too small to be seen with a basic optical microscope that are dispersed throughout another substance

COO — Cost of ownership

CTC — Concurrent Technologies Corporation (www.ctc.com)

CTF — Columnar thin film

Dark-field TEM — see TEM

Dislocation loop — A type of crystal defect in which a dislocation line comes around and closes in on itself; a combination of a screw and edge dislocation

DLC — Diamond-like carbon, a form of carbon synthetically produced on a surface with properties (e.g., hardness) that resemble those of diamond

EAM — Embedded atom method: a method of calculating interatomic potentials

ECR — Electron cyclotron resonance (refers to a certain motion of electrons subjected to high-frequency electromagnetic and static magnetic fields; refers also to a type of ion source, developed at Centre d'Études Nucléaires Grenoble in the 1980s, which is based on ECR)

EDS — Energy dispersive x-ray spectroscopy

EHC — Electroplated hard chromium

Ehrlich-Schwöbel barrier — In surface physics, a type of barrier-to-surface diffusion at step edges

Energy/Displacement Spike — Occurs during ion bombardment under certain conditions of high energy-deposition density, in which the mean kinetic energy of the bombarded atoms can reach up to several electron volts per atom

Epitaxial — Characteristic of the growth on a crystalline substrate of a crystalline substance that takes on the same orientation as the substrate

ESRF — European Synchrotron Radiation Facility

Etching — Removal of a certain portion of a solid by means of a preferential eating away of the surface by an acid or a laser beam

EXAFS — Extended x-ray absorption fine structure: a technique (associated with synchrotrons) that sweeps x-rays in wavelength past an absorption edge, which can be used to determine the atomic number, distance and co-ordination number of the atoms surrounding the element whose absorption edge is being examined

Extrinsic residual stress — Stress due to external factors than tend to change the stress-free length of the film subsequently to deposition

FEG-TEM — Field-emission gun transmission electron microscopy (see also TEM)

FEM — Final energy magnet

Ferrimagnetism — Characterized by a magnetization in which one group of atoms or ions tends to assume an ordered but nonparallel arrangement in zero applied field

FIM — Field-ion microscopy

Fission tracks — Ion tracks that are produced by the heavy ions that are released from nuclear fission

Forensic — That which relates to the application of scientific knowledge to support a legal (usually criminal) investigation

Frenkel pair/Frenkel defect — Vacancy interstitial defects formed by the distribution of vacancies, interstitial atoms and other types of lattice disorders sub-sequent to a cascade of atomic collisions

GCIB — Gas cluster ion beam (a surface-treatment technique discussed at length in Chapter 4)

Gibbs energy/function — Defined as the enthalpy minus the product of thermodynamic temperature and entropy; formerly called free energy or free enthalpy

GISAXS — Grazing-incidence small-angle x-ray scattering

GIXRD — Grazing-incidence x-ray diffraction

GIXS — Grazing-incidence x-ray scattering

GLAD — Glancing-angle deposition technique (see Chapter 9)

GMR — Giant magnetoresistance: results from electron-spin effects in ultra-thin multilayers of magnetic materials that cause large changes in their electrical resistance when a magnetic field is applied. GMR is many times stronger than ordinary magnetoresistance; enables sensing of significantly smaller magnetic fields, which in turn allows hard-disk storage capacity to increase significantly

HBM — Helicoidal bi-anisotropic medium; rotationally inhomogeneous and anisotropic dielectric material, prepared in thin-film form by inclining and rotating the substrate during evaporative deposition in order to obtain chiral symmetry; has many interesting optical and other properties, useful for example, as circular polarization filters

HREM — High-resolution electron microscope (see also TEM)

HRTEM — High-resolution transmission electron microscopy (see also TEM)

Hygroscopic — Characteristic of a substance that captures and retains moisture

IBAD — Ion-beam-assisted deposition, a technique in which an ion beam irradiates a substrate onto the material being vacuum-deposited

IECC — Ion Engineering Centre Corporation

II — Ion implantation (q.v.)

INFN — Istituto Nazionale di Fisica Nucleare (Italian research organisation)

Intrinsic residual stress — Stress due to microstructural relaxation processes that tends to change the stress-free length of the film during deposition

Ion-beam mixing — The use of an energetic ion beam to introduce disorder and therefore intermixing in surface or near-surface layers, allowing , for example, the creation of alloys outside of thermodynamically allowed equilibrium concentration ratios; differs from ion implantation (q.v.) in that the ion beam acts mainly as a catalyst in the process

Ion implantation — A process whereby atoms from an energetic beam penetrate into a material to improve the properties of the latter, be they tribological, optical, electrochemical or other.

Ion tracks — Needle-shaped regions of atomic disorder in solids that are produced by the ionization caused by rapidly moving charged atomic nuclei

LIGA — From the German acronym for "LIthographie, Galvanoformung und Abformung," a technology originally developed in Karlsruhe used in fabricating tall microstructures with minimal lateral dimensions combining x-ray lithography and electroplating

LSI — Large scale integrated circuit

Luminescence — Low-temperature emission of light by a chemical or physiological process

LWIR — Long-wave infrared

Magneto-resistance — Change in the electrical resistance of a substance in the presence of a magnetic field

Magneto-striction — Magnetic analogy to the piezoelectric effect, found in materials that expand in the direction of the applied magnetic field

MD — Molecular dynamics

MEMS — Micro-electromechanical systems: a generic term to describe micrometer-scale electrical/mechanical devices

MFM — Magnetic force microscope

Microhardness — Measure of the hardness of a material on a microscopic scale via the size of the impression produced by a sharp point forced into the material under a given load

MNCG — Metal nanocluster composite glass

Morse potential — One model for interatomic potentials in molecules

MOSFET — Metal-oxide-semiconductor field effect transistor

Mott-Hubbard type gap — A gap found in the energy bands of an electron gas that forms due to electron–electron interaction and is related to the formation of local magnetic moments

MWIR — Mid-wave infrared

Nanocluster — Clusters or aggregates of atoms having nanometer dimensions

NDCEE — National Defense Centre for Environmental Excellence (U.S.)

NEMS — Nano-electromechanical systems (see MEMS)

Nuclear tracks — Elongated, needle-shaped regions that are densely populated with atomic defects produced by accelerated ions that penetrate the solid

Nucleation length — Characteristic separation between islands on a surface

OEM — Original equipment manufacturer

Optoelectronics — Branch of electronics that studies electronic devices for the emission, modulation, transmission and detection of light

ORNL — Oak Ridge National Laboratory (U.S.)

Orowan mechanism — A hardening process in alloys in which a dislocation plane encounters an inclusion that it is unable to shear; it avoids it by circumventing it, leaving behind a dislocation loop

Over-focus — see TEM

Paramagnetism — Property of a substance that has a positive magnetic susceptibility that varies weakly as a function of an applied magnetic field

Passivation — Use of a certain procedure, coating or surface treatment, to protect (for example a solid-state device) against contamination

PIXE — Proton-Induced X-ray Spectroscopy

PKA — Primary knock-on atoms, the first set of atoms that are struck by impinging ions in a collision cascade

PL — Photoluminescence

Plasma resonance — Oscillation of an electron gas at the plasma frequency (which is proportional to the square root of the electron density)

Plasmon — A quantum of plasma oscillation of a free-electron gas (in metals), can be excited by electrons striking the surface of a thin metal film

PVD — Physical vapor deposition

QMS — Quadrupole mass spectrometer

RBS — Rutherford backscattering spectrometry, an analytical tool that uses elastic scattering of 0.1-3 MeV charged particles to depth-profile the first few micrometers of a solid surface

Recoil tracks — Regions of clumped, diffuse atomic disorder in solids that are produced by the heavy recoiling atoms that result from alpha decay of heavy radioactive atoms

Remanence — Magnetic induction that remains in a substance that is no longer subjected to an external magnetic field

RF — Radio frequency

RF-type ion source — An ion source that uses a plasma discharge created by applying a radio frequency signal to a gas at low pressure

SAED or SAD — Selected-area (electron) diffraction (see TEM)

SAM — Source analyzer magnet

SDS — Safety delivery system

SIMS — Secondary ion mass spectrometry: one of the most sensitive surface-analytical techniques in which a sample is bombarded with an energetic ion beam, resulting in the ejection of material in the form of positive and negative ions and neutral species. The

ejected ions are then analyzed with a mass spectrometer to yield information about the chemical composition of the sample surface. SIMS can be static, dynamic
or imaging, with lateral resolutions of 1 mm, 1μm and
250 Å respectively, and probing depths in the tens of Å.

Spallation — Nuclear reaction in which light particles are
ejected as the result of bombardment (as by high-
energy protons)

SPR — Surface plasmon resonance: peak in the energy-loss
spectra of electrons impinging and exciting a plasmon
(q.v.) on a metal surface. An excellent tutorial on this
and similar matters can be found at: http://
www.nims.go.jp/heavyion/English/e_tutorial/tutorial-
nano.htm

Sputtering — The dislodging of atoms from the surface of a
material by bombarding it with ions; also, to deposit
(e.g., a metallic film) by such a process

STEM — Scanning transmission electron microscope (see
TEM)

STM — Scanning tunneling microscopy (see AFM)

Super-paramagnetism — Phenomenon in which magnetic
materials may exhibit a behavior similar to paramagnetism at temperatures below the Curie or the Néel
temperature

TED — Transient-enhanced diffusion

TEM — Transmission electron microscopy; uses electrons (of
picometer wavelength) rather than light (of roughly
micrometer wavelength) to image materials prepared
in thin-film form. In bright-field imaging (the most
common), the image of a thin sample is formed by the
electrons that pass through the film without diffraction, the diffracted electrons being stopped by a dia-
phragm (see for example http://www.feic.com
/support/tem/bright.htm). In the dark-field imaging
mode, a diffracted beam is used for imaging, which can
provide higher contrast in some cases. In the selected-
area diffraction mode, an aperture is used to zero-in
on a very small area of the sample whose crystalline
structure is to be determined. Over-focus and under-

focus are techniques wherein the objective lens is made to go slightly out of focus in some cases to bring out details or features with greater contrast.

Thermal/Heat Spike — A phase in which the atoms within an ion-bombarded region have an effective temperature within the spike zone significantly above that required for melting

TM — Transition metal

TOF-SIMS — Time-of-flight secondary ion mass spectrometry (see SIMS)

Track etching — A method of revealing the damage sites of ion tracks or alpha-recoil damage

Tribology — Study that deals with the design, friction, wear and lubrication of interacting surfaces in relative motion (as in bearings or gears)

UHMWPE — Ultra-high molecular weight polyethylene

Under-focus — see TEM

XPS — X-ray photoelectron spectroscopy (also called ESCA, for "electron spectroscopy for chemical analysis"), a surface-analysis technique that measures the energy of photoelectrons emitted following x-ray excitation. Sensitive to chemical binding, it can measure up to about 5 nm depth with a lateral resolution that can vary from 5 μm to 5 mm, depending on the degree of refinement of the apparatus

XRD — X-ray diffraction

XTEM — Cross-sectional transmission electron microscopy (see TEM)

Index